10th International Scientific Conference on Aeronautics, Automotive and Railway Engineering and Technologies (BulTrans-2018)

MATEC Web of Conferences Volume 234 (2018)

Sozopol, Bulgaria
15 - 17 September 2018

Editors:

Boyko Gigov
Nikolay Nikolov
Svetla Stoilova

Ivan Kralov
Michael Todorov
Valeri Stoilov

ISBN: 978-1-5108-7488-6

Printed from e-media with permission by:

Curran Associates, Inc.
57 Morehouse Lane
Red Hook, NY 12571

proceedings.com

Some format issues inherent in the e-media version may also appear in this print version.

No changes have been made to the content of these proceedings. There may be changes to pagination and minor adjustments for aesthetics.

Printed by Curran Associates, Inc. (2019)

For additional information, please contact EDP Sciences – Web of Conferences
at the address below.

EDP Sciences – Web of Conferences
17, Avenue du Hoggar
Parc d'Activité de Courtabœuf
BP 112
F-91944 Les Ulis Cedex A
France

Phone: +33 (0) 1 69 18 75 75
Fax: +33 (0) 1 69 28 84 91

contact-edps@webofconferences.org

Additional copies of this publication are available from:

Curran Associates, Inc.
57 Morehouse Lane
Red Hook, NY 12571 USA
Phone: 845-758-0400
Fax: 845-758-2633
Email: curran@proceedings.com
Web: www.proceedings.com

TABLE OF CONTENTS

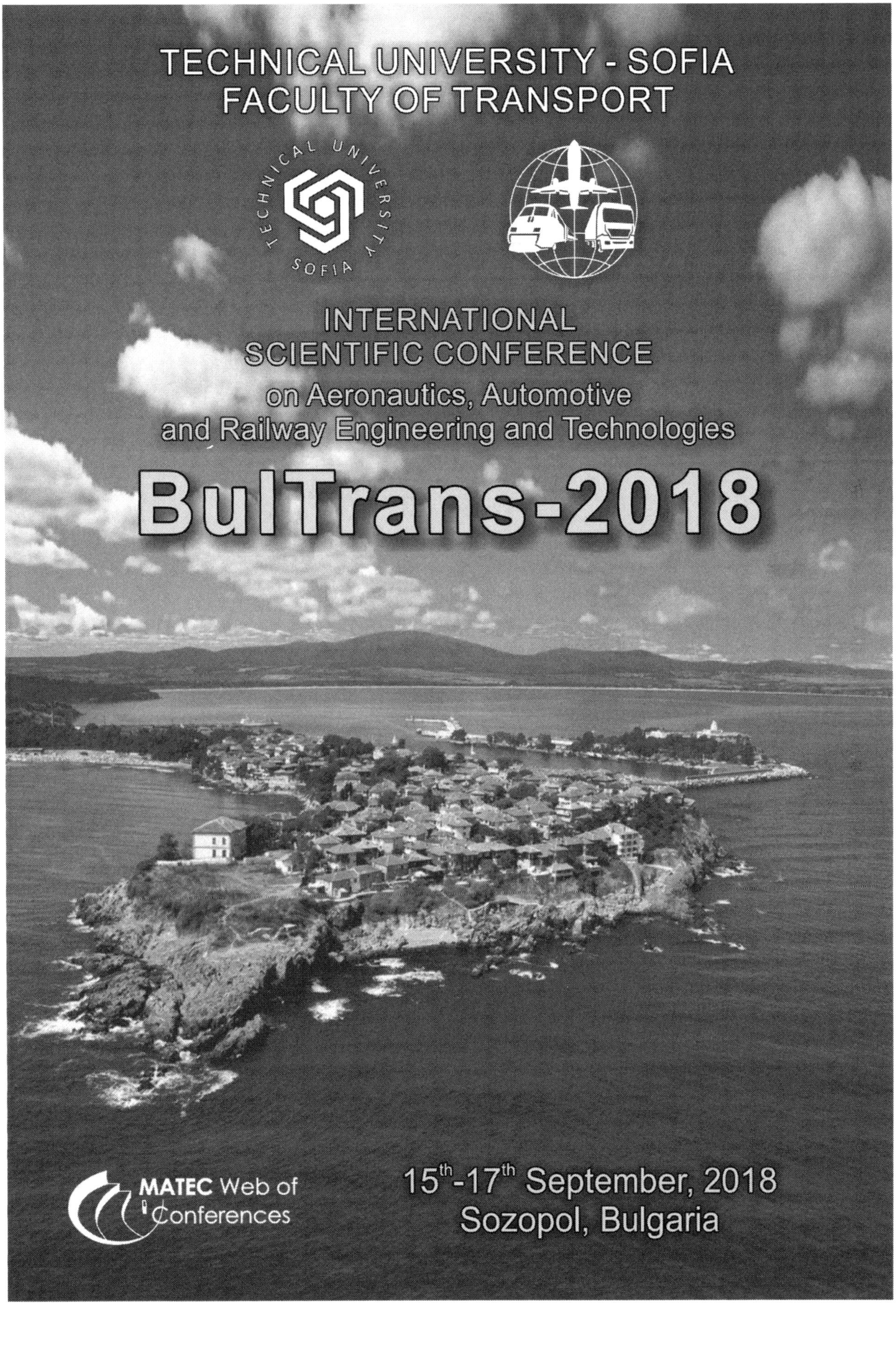
TECHNICAL UNIVERSITY - SOFIA
FACULTY OF TRANSPORT
TECHNICAL UNIVERSITY SOFIA
INTERNATIONAL
SCIENTIFIC CONFERENCE
on Aeronautics, Automotive
and Railway Engineering and Technologies
BulTrans-2018
MATEC Web of Conferences
15th-17th September, 2018
Sozopol, Bulgaria

PREFACE

BulTrans International Scientific Conference on Aeronautics, Automotive and Railway Engineering and Technologies is held for the first time in 2009, as part of *The Science Days* of Technical University of Sofia. Till now, ten editions of the Conference have been organized, nine in USOB "Lazur" and a jubilee forum in Sofia.

BulTrans-2018 was the 10th edition of the Conference. It was attended by over 140 participants from 12 countries.

Thorough scientific achievements in the field of transport are being presented at the conference. **BulTrans** is the most prestigious scientific forum for road, railway and air transport in Bulgaria.

Our sponsors have significant contribution for the realization of the event. Among them are: Siemens Mobility, Knorr-Bremse, Automotor Corporation and many others.

OPENING ADDRESS

***Assoc. Prof. B. Gigov, PhD**, Chairman of Organizing Committee*

Every year, the Faculty of Transport at the Technical University - Sofia is organizing the International Scientific Conference on Aeronautics, Automotive and Railway Engineering and Technologies BulTrans.

The event is aimed at presenting scientific research in the field of transport and transport technology and promoting collaboration between scientists from different countries. It also traditionally gives floor for young scientists, PhD students and students.

The proceedings include a selection of papers from the scientific forum based on their quality and relevance to the conference topics.

I would like to express my gratitude to the members of the scientific editorial board, the reviewers and the participants for increasing the scientific value of the conference every year.

Statement of Peer review

In submitting conference proceedings to *Web of Conferences*, the editors of the proceedings certify to the Publisher that

1. They adhere to its **Policy on Publishing Integrity** in order to safeguard good scientific practice in publishing.
2. All articles have been subjected to peer review administered by the proceedings editors.
3. Reviews have been conducted by expert referees, who have been requested to provide unbiased and constructive comments aimed, whenever possible, at improving the work.
4. Proceedings editors have taken all reasonable steps to ensure the quality of the materials they publish and their decision to accept or reject a paper for publication has been based only on the merits of the work and the relevance to the journal.

Title, date and place of the conference

BulTrans-2018 - 10th International Scientific Conference on Aeronautics, Automotive and Railway Engineering and Technologies

15th-17th September 2018, Sozopol, Bulgaria

Proceedings editor(s):

Assoc. Prof. **Boyko Gigov**, PhD
Assoc. Prof. **Nikolay Nikolov**, PhD
Assoc. Prof. **Svetla Stoilova**, DrSc
Prof. **Ivan Kralov**, PhD
Prof. **Michael Todorov**, PhD
Prof. **Valeri Stoilov**, PhD

Date and editor's signature

25 October 2018

MATEC Web of Conferences **234**, 01001 (2018) https://doi.org/10.1051/matecconf/201823401001
BulTrans-2018

Numerical simulation and performances evaluation of the pulse detonation engine

Vasile Prisacariu[1], *Constantin* Rotaru[2], *Ionică* Cîrciu[3], and *Mihai* Niculescu[4]

[1,2,3]Henri Coandă Air Force Academy of Braşov, Aviation Department, Romania
[4]INCAS Bucharest, Aerodynamic Department, Romania

Abstract. A pulse detonation engine (PDE) is a type of propulsion system that uses detonation waves to combust the fuel and oxidizer mixture. The engine is pulsed because the mixture must be renewed in the combustor between each detonation wave. Theoretically, a PDE can operate from subsonic up to hypersonic flight speed. Pulsed detonation engines offer many advantages over conventional propulsion systems and are regarded as potential replacements for air breathing and rocket propulsion systems, for platforms ranging from subsonic unmanned vehicles, long range transports, high-speed vehicles, space launchers to space vehicles. The article highlights elements of the current state of the art, but also theoretical and numerical aspects of these types of unconventional engines. This paper presents a numerical simulation of a PDE at *h*=10000 m with methane as working fluid for stoichiometric combustion, in order to find out the detonation conditions.

1 Introduction

According to scientific literature [1-4] there has been preoccupation, since 1940, with propulsion systems based on detonation, which presupposes the elimination of at least the rotating constructive elements placed after the combustion chamber (turbine). Implementation of the PDE concept has encountered difficulties due to the need for speed and frequency of fuel blending, especially at high propulsion system operating speeds, and support for a stable and controlled detonation process. Then serious research began with the 1950s at the University of Michigan when researchers published a series of scientific papers about PDE.

The concept of PDE operation is similar to the impulse jet engine, but the difference is the combustion rate of the fuel mixture and the oxidant (deflection versus detonation). The combustion mixture is ignited in an open chamber, the resulting combustion increases the mixture pressure to about 10 MPa, which then expands into a nozzle at thousands of meters per second, which can be numerically modelled as a constant volume combustion process, where more chemical energy is released as heat than the constant pressure process found in conventional turbine engines. All potential chemical energy stored in the fuel is transformed into the internal energy of the gas.

Experimental tests have encountered a number of difficulties regarding both the transition of subsonic deflagration into a supersonic detonation wave and the correct mixing of the fuel and the oxidant to result in uniform detonation. PDE is an intermittent combustion propulsion system in which detonations occur at high frequencies by burning and accelerating the fuel and oxidant mixture. Fired gases move either at subsonic speed or at supersonic speed based on the geometric parameters of the nozzle, the fuel injection rate and the frequency of detonation. PDE uses the detonation waves for combustion of the fuel and oxidant mixture; theoretically, this type of engine can operate from a subsonic regime to hypersonic regimes (Mach> 5), [1-5].

The operation of a detonation engine consists of four major stages: filling the tube with the desired mixture; initiating detonation; propagation of detonation along the tube and evacuating products to the atmosphere see Fig. 1, [6].

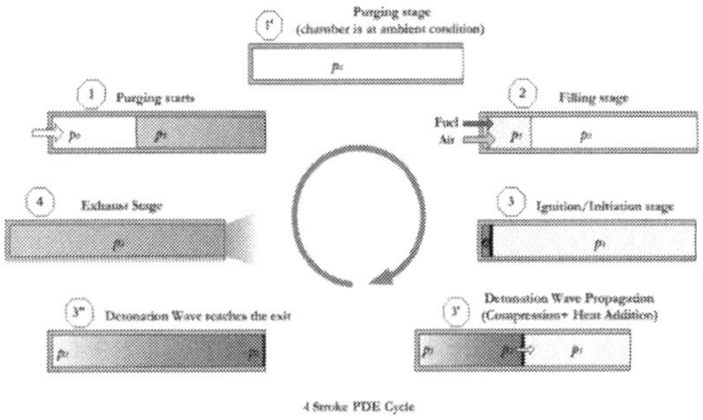

Fig. 1. PDE cycle, [6].

According to [7, 8] a positioning of the propulsion concept PDE according to Mach number versus traction can be as shown in Fig. 2, as the specific impulse of the PDE is greater than that of missiles and lower than that of turbojet engines.

*Corresponding author: aerosavelli73@yahoo.com

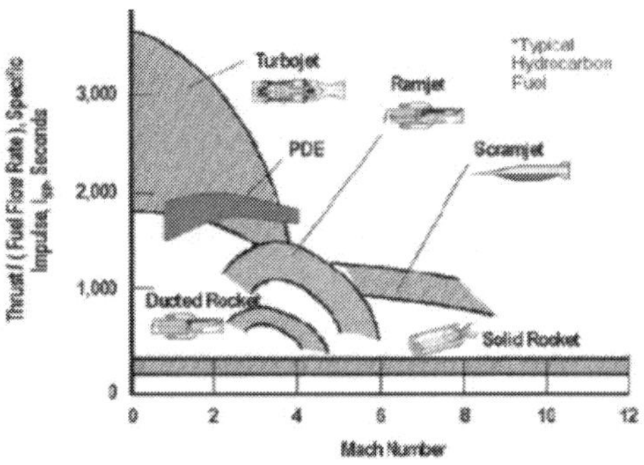

Fig. 2. PDE propulsion (Mach number vs. thrust/ specific impulse), [7, 8].

The need to optimize the mass and thrust values of jet engines used in the aerospace industry has led to the elimination of the rotating constructive elements that de-energized and drastically limited the maximum gas temperature (less mechanical complexity). Elimination of the turbine makes it possible to increase the temperature and speed of combustion gases through a more efficient supersonic combustion (detonation) and is capable of providing gas velocities and pressures far superior to conventional combustion (deflagration). Detonation bursting with shock waves almost doubles the thermodynamic efficiency of the engine cycle; high-frequency detonation has significant implications for noise, vibration and pollution levels.

For detonation, either the burning process must be strong enough, or downstream deflagration must be transformed into a supersonic wave in a process known as deflection-detonation transition (DDT). DTT can be induced by position internal ramps on the combustion wave path to increase the turbulence level, offering significant improvements (at least 27%) to the total pressure loss of a spiral with the same performance [9].

References to scientific literature reveal aspects of both the use of materials with optimal high temperature properties in view of the extremely high heat released on the fired combustion unit as well as the use of cooling elements and streams and the minimization of the DDT transition period [10].

Scientific references indicate the PDE classification by design type: pure (PDE with / without pre-dielectric, with / without ejector, single tube PDE, multi-tube PDE, PDE-turbojet PDE-rocket) of the fuel used [11, 12].

2 Theoretical considerations

2.1. Definition of detonation

Detonation takes place in a mixture of fuel and air moving at a subsonic speed that is compressed by a normal shock wave propagating to a high Mach, which is then followed by a rapid heat release and a sudden increase in pressure, see Figure 3. Detonation is a violent, rapid, exothermic reaction that produces a tough environment in which it is difficult to dynamically model and dynamically modify local pressures, speeds and temperatures. The detonation process results in pressure increases due to adiabatic compression (20-30 bar, 50 bar maximum pressure) and the combustion products exhibit a high temperature gradient due to inherent compression and lower molecular dissociation.

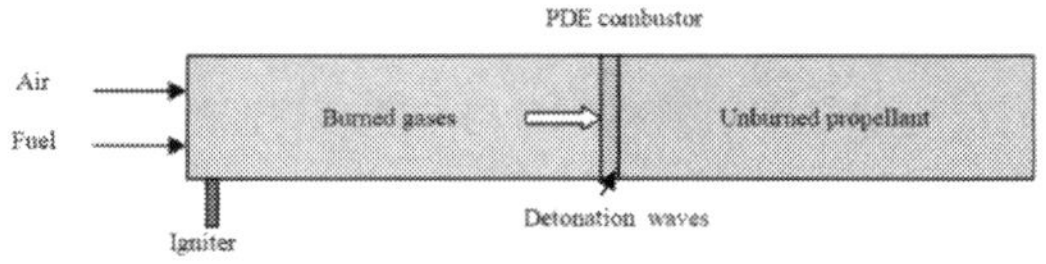

Fig. 3. Detonation process – PDE.

Specific thrust is one of the most important elements for designing the pulse detonator. According to the usual practice, the traction of a PDE can be characterized by the specific impulse (I_{sp}), defined as the impulse per unit of fuel mass [13]:

$$I_{sp} = \frac{I}{m_f g}, \quad (1)$$

where I_{sp} is specific impulse, I– impulse, m_f– fuel mass, g – gravitational acceleration.

2.2. Requirements and performance of PDE

Assuming PDE with a combustion chamber of an jet engine, we can define a series of general requirements that can characterize such a propulsion system, [14, 15]: ensure a stable combustion process; the efficiency of the combustion process is the highest possible with the least possible total and static pressure losses; have a low thermal load and a high operating resource; have a reduced dimensioning with positive implications for the mass and mechanical strength.

According to [6, 11] PDE performance is influenced by a series of parameters, the most relevant is: the diameter of the detonation tube d; the length of the detonation tube L; the frequency of the detonation cycle f; equivalence ratio r; cell size λ:

$$T = I_v V f = I_v L \frac{\pi d^2}{4} f, \quad (2)$$

where T is thrust, V – volume of the detonation tube, I_v–impulse per unit volume.

3 Numerical simulation of PDE

3.1. Geometry of PDE

According to [16, 17], for CFD analysis we propose a simple geometry (see Fig. 4) of pre-detonator PDE and Laval nozzle, having the characteristics of Table 1.

Table1.PDE characteristics.

Parameters	Values	Parameters	Values
Nozzle ramp	150	Diameter	100 mm
PDE length	1000 mm	Nozzle type	Laval

 https://doi.org/10.1051/matecconf/201823401001

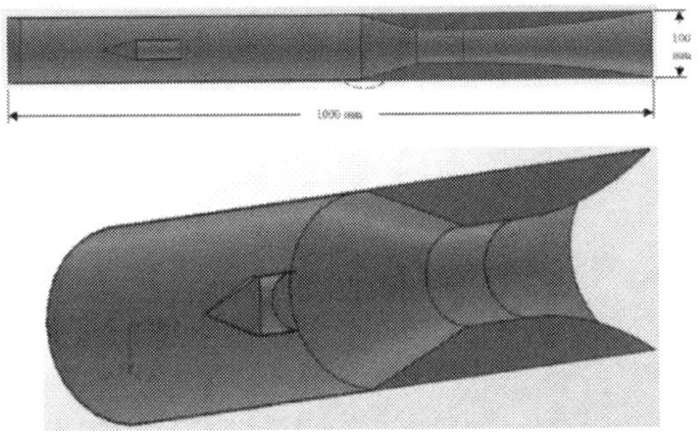

Fig. 4. PDE geometry.

3.2. CFD analysis conditions

For CFD 2D deeper flow analysis around the admission to the Laval nozzle was used Ansys 18 [18] and a simple geometry of the PDE propulsion system with the initial conditions in Table 2 for stoichiometric burning:

$$CH_4+2O_2=CO_2+2H_2O. \quad (3)$$

Table2. Initial condition for PDE analysis.

Parameters	Values	Parameters	Values
Altitude	10000 m	Fluid	CH_4
Admission pressure	26436 Pa	Admission speed	7 Mach
Admission temperature	223.16 K	Cells number/type	quadrilaterals / 33696
Wall type	adiabatic		

Nitrogen N_2 has been assumed to be inert gas and does not participate in chemical reactions. For stoichiometric burning ($\lambda = 1$) on input or imposed the following chemical compositions (molar): $CH_4 = 0.095$, $O_2 = 0.19$, $N_2 = 0.715$. The quadrilateral cell mesh grid is highlighted in Fig. 5.

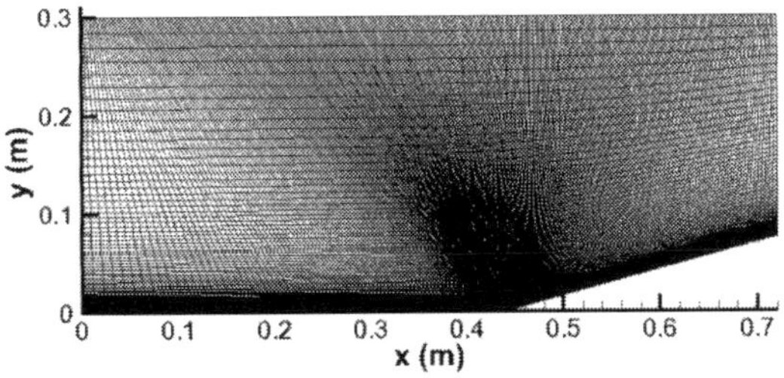

Fig. 5. The mesh grid CFD, Tecplot.

3.3. CFD 2D analysis

For the 2D analysis, we selected the Laval nozzle input region of Fig. 4, following the distribution of the molar fraction of the methane that is shownon Fig. 6, the pressure (Fig. 7) and the Mach number (Fig. 8).

The meshing scale required for 2D analyses was optimized on the Laval nozzle inlet wall and on the vertical zone between $x = 0.4 \div 0.5$. The distribution of the concentration level of CH_4 in the approach of the wall (boundary layer) depends on the wall distance (Fig. 6).

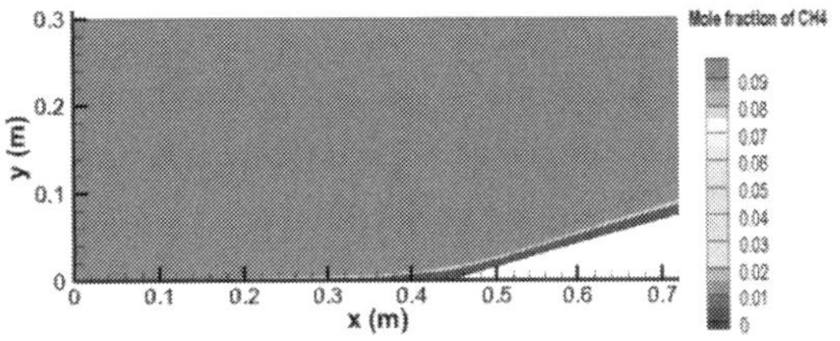

Fig. 6. Molar fraction, CH4Tecplot.

Fig. 7 shows the variation of the pressure gradient in the approach of the inlet nozzle along the x axis with a maximum of 2.4×10^5 Pa at $x = 0.6$ m. The Mach number drops to zero, which indicates the boundary layer in the vicinity of the nozzle ramp (Fig. 8).

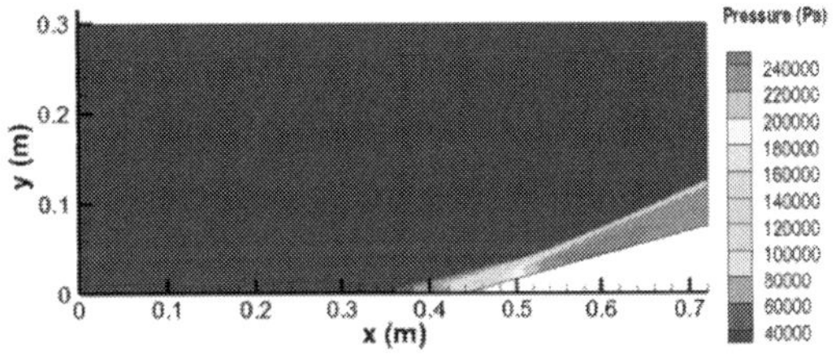

Fig. 7. Pressure, Tecplot.

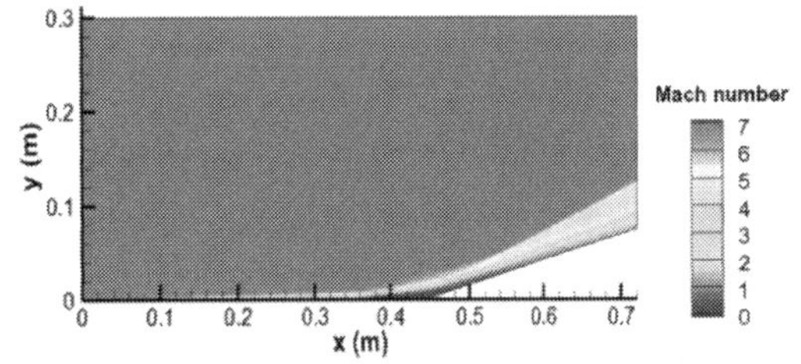

Fig. 8. Mach number, Tecplot.

Fig. 9 shows the variation of temperature in relation to the distance to the wall (nozzle ramp) with implications upon the thickness of the boundary layer.

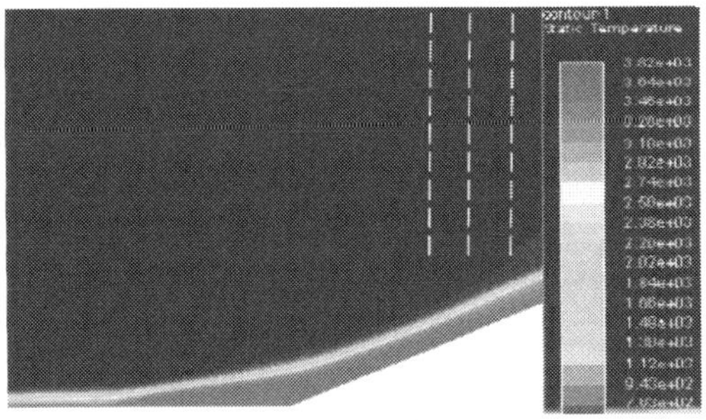

Fig. 9. Static temperature.

To better understand the 2D fluid behaviour in the approach of the nozzle ramp we extracted the values of the variation of the boundary layer and the variation of the total pressure to the corresponding vertical $x = 0.4$ m. The thickness of the limit layer corresponding to a Mach 99% of the flight speed (6.93 Mach) is 0.064 m (see Fig. 10) and the total pressure drops to 2.8×10^4 Pa (see Fig. 11).

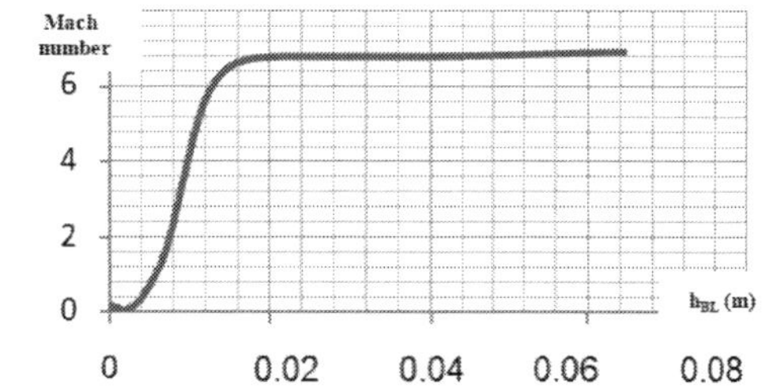

Fig. 10. Boundary layer thickness (h_{BL}) at x = 0.4 m.

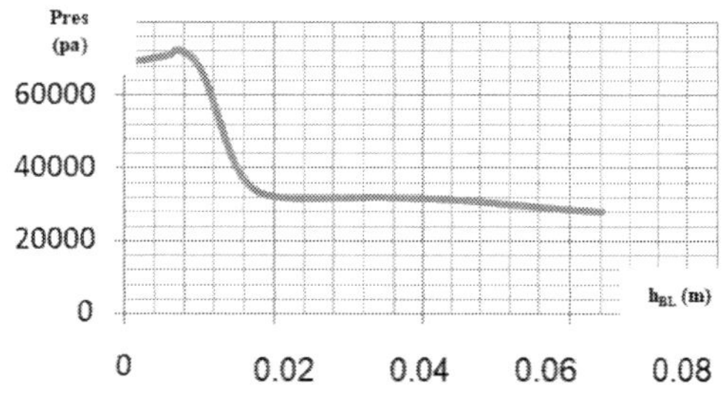

Fig. 11. Total pressure at x = 0.4 m.

Boundary layer thickness at x = 0.45 m (at the ramp application point of 15^0) is 0,069 m (Fig. 12), total pressure down at 2.79×10^4 Pa (see Fig. 13).

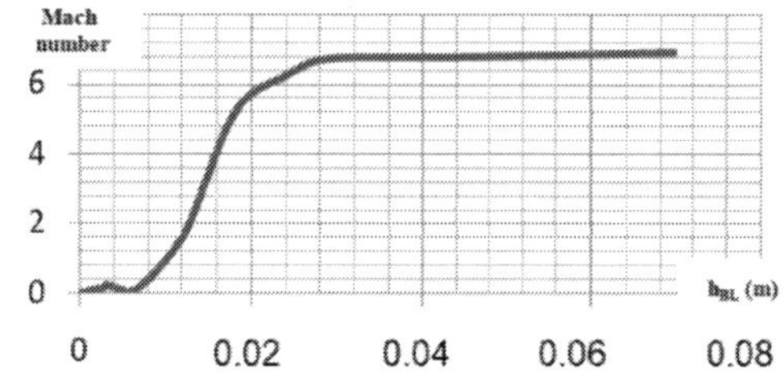

Fig 12. Boundary layer thickness (h_{BL}) at x = 0.45 m.

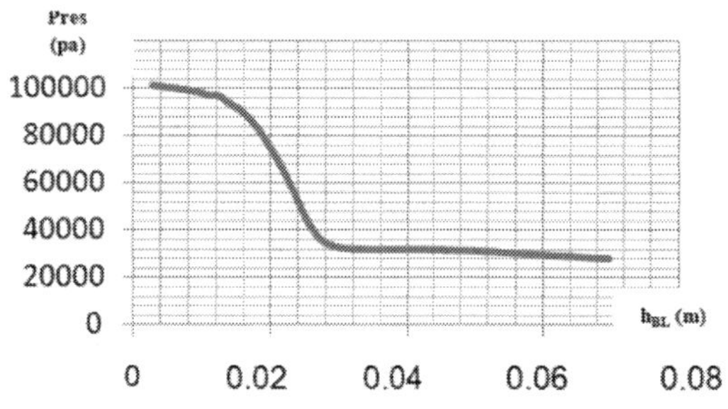

Fig. 13. Total pressure at x = 0.45 m.

The thickness of the boundary layer at x = 0.5 m is 0.076 m (Fig. 14) and the total pressure increases to 2.82×10^4 Pa (see Fig. 15). Thus, as is normal, the thickness of the boundary layer increases downstream with the increase of the Mach gradient, to a 0.95% molar methane fraction, Fig. 16.

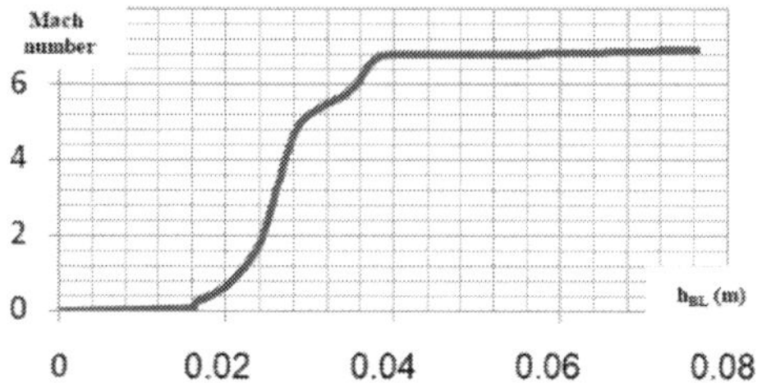

Fig. 14. Boundary layer thickness at x = 0.5 m.

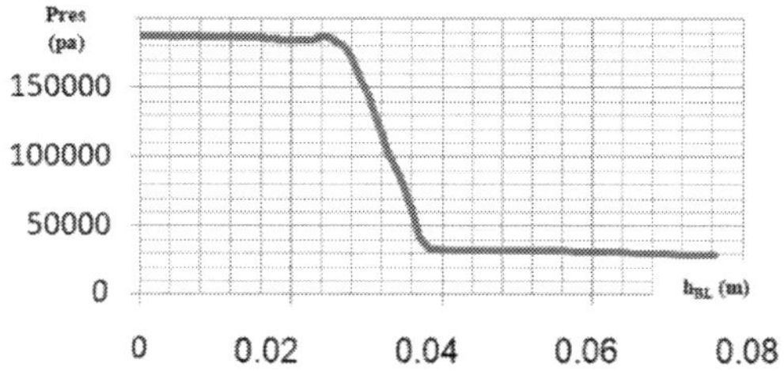

Fig. 15. Total pressure at x = 0.5 m.

Following the variation of the Mach number and the corresponding total pressures, a slight shock wave is observed at the extreme edge of the nozzle ramp.

The shock wave interacts with the boundary layer due to the upstream propagation of the pressure disturbances favouring the separation of the flow, depending on the Mach number and the angle nozzle ramp [16].

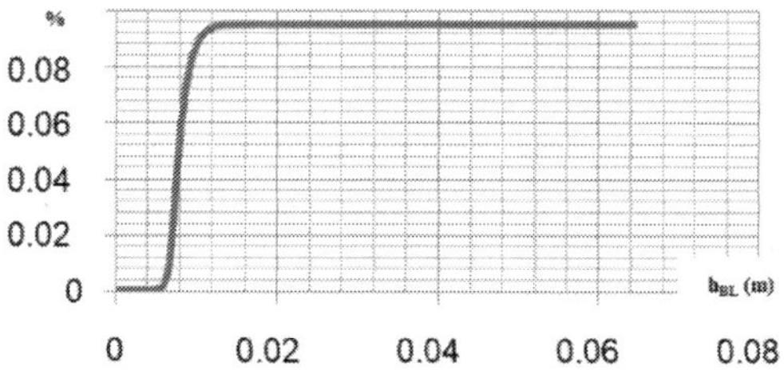

Fig. 16. Methane molar fraction at x = 0.4 m.

4 Conclusions

PDE studies require interdisciplinary concepts, methods and technologies, based on applied mathematics, combustion theory and computational fluid dynamics. Understanding the detonation process is essential for future assessments of heat transfer at constant volume and supersonic speeds of the combustion wave. Theoretically, PDEs offer many advantages over current turbo propulsion systems or missiles. Experimental development has faced many challenges, such as vibration, cooling or optimal detonation control.

The numerical simulation was focused on the angle of the nozzle ramp (15^0) at a Mach number high enough to ensure the detonation of the working agent (methane) without taking into account the three-dimensional effects.

Knowing that the fuel-oxidant mixture has narrower detonation limits than combustion limits, it is necessary to be considered in numerical simulations, a number of effects due to the ramp angle, wall temperature, initial mixing speed, and 3D space implications (finite effects).

This work was supported by a grant of the Ministry of National Education and Romanian Space Agency, RDI Program for Space Technology and Advanced Research – STAR-C3-2016-624, MILADEE project, contract 174/20.07.2017.

References

1. K. Kailasanath, Review of propulsion applications of detonation waves, AIAA Journal, **39**(9), 1698-1708 (2000).
2. T. Bussing, G. Pappas, Introduction to pulse detonation engines, *32nd Aerospace Sciences Meeting and Exhibit, Aerospace Sciences Meetings Reno*, NV, U.S.A., AIAA Journal 94-0263 (1994). https://doi.org/10.2514/6.1994-263.
3. http://large.stanford.edu/courses/2016/ph240/cassady1 , accessed at 20.06.2018
4. J. Nicholls, H. Wilkinson, R. Morrison, Intermittent detonation as a thrust-producing mechanism, Jet Propulsion, **27**, 534 (1957).
5. C. Goldenstein, I. Schultz, R. Spearrin, J. Jeffries, R. Hanson, Diode laser measurements of temperature and H2O for monitoring pulse detonation combustor performance, Stanford University, *24th ICDERS Taiwan*, (2013)
6. D. Valli, T. Jindal, Pulse detonation engine: parameters affecting performance, International Journal of Innovative Research in Science, Engineering and Technology, **3**(4), (2014)
7. C. Myers IV, Initiation mechanisms of low-loss swept-ramp obstacles for deflagration to detonation transition in pulse detonation combustors, N0001409WR20022, Office of Naval Research (ONR) Code 33 Ballstone Tower One 800 N. Quincy St Arlington, VA 22203-1995 (2009)
8. R. Friedman, American Rocket Society, **24**, 349, (November 1953).
9. W. Dvorak, Performance characterization of swept ramp obstacle fields in pulse detonation applications, M.S. Thesis, Naval Postgraduate School, Monterey, CA, (2010).
10. S. Lee, J. Watts, S. Saretto, S. Pal, C. Conrad, R Woodward, R. Santoro, Deflagration to detonation transition processes by turbulence-generating obstacles in pulse detonation engines, J. Propul. Power, **20**, 1026 (2004).
11. S. Bhavesh, S. Linsu, Design and development of pulse detonation rocket engine with predetonator, International Journal of Basic and Applied Science Research (IJBASR), **1**(1), 1-6 (2014)
12. P. Debnath, K. Pandey, Numerical investigation of detonation combustion wave in pulse detonation combustor with ejector, Journal of Applied Fluid Mechanics, **10**(2), 725-733, (2017) DOI: 10.18869/acadpub.jafm.73.239.27266.
13. E. Wintenberger, J. Austin, M. Cooper, S. Jackson, J. Shepherd, An analytical model for the impulse of a single cycle pulse detonation engine, AIAA 2001-3811, (2001).
14. V. Prisacariu, I. Cîrciu, Considerations regarding the performance of combustion chambers for turbo-jet engines, Review of the Air Force Academy, **2**(32), 53-60 (2016), DOI: 10.19062/1842-9238.2016.14.2.7.
15. S. Farokhi, *Aircraft propulsion*, second edition, (Wiley and sons, 2014).
16. M. Marini, Analysis of hypersonic compression ramp laminar flows under sharp leading edge conditions. Aerospace science and technology, **5**(4), 257-271 (2001)
17. F. Lu, D. Wilson, Some perspectives on pulse detonation propulsion systems. In: Z. Jiang (eds) *Shock Waves*. Springer, Berlin, Heidelberg, (2005), DOI https://doi.org/10.1007/978-3-540-27009-6_8.
18. ANSYS Fluent Tutorial Guide, Ansys 17, 2015, 724-746-3304, available at http://dl.racfd.com/ANSYS-Fluent-Tutorial-Guide_r170.pdf.

Experimental study of a helicopter rotor model in hover

Filip Panayotov[1], *Ivan* Dobrev[2], *Fawaz* Massouh[2], and *Michael* Todorov[1,*]

[1]Technical University of Sofia, Department of Air Transport, 8 Kliment Ohridski Blvd., 1000 Sofia, Bulgaria
[2]Arts et Métiers – ParisTech, DynFluid laboratory, blvd. L'Hôpital 151, 75013 Paris, France

Abstract. This paper presents the results from an experimental study of the aerodynamic performance and efficiency of a model helicopter rotor in steady hover at Reynolds numbers below 70000. Results are shown for a two-, three- and four-bladed rotor configuration for various pitch angles and speeds of rotation. The influence of the rotor solidity on the aerodynamic efficiency in terms of the Figure of Merit is highlighted. The profile drag component is evaluated as a function of the Reynolds number. The internal friction losses of the test stand are estimated and taken into account in all measurements. A brief description of the designed test stand is provided. The experimental setup allows for the measurement of thrust and torque of helicopter rotors with diameters of up to 1 meter, for pitch angles varying from –12° to +24° and for rotational speeds of up to 3000 RPM. Conclusions are drawn about the aerodynamic performance and efficiency of the studied rotor configurations. This experimental study provided a significant database, which will serve for validation purposes.

1 Introduction

In recent years, there has been a significant increase in the number of studies on the aerodynamic performance of small-scale helicopter rotors of unmanned aerial vehicles (UAVs). This is largely due to the recent increase of usage of UAVs for various commercial, leisure and military applications.

Although there is a large amount of data available from decades-long theoretical, numerical and experimental studies of full scale helicopter rotors, it is not applicable in cases of validation of newly developed models for low-Reynolds flows, due to the existing particularities regarding the aerodynamic performance of scaled rotors. Their overall efficiency, which in the case of hover is expressed with the Figure of Merit (FM), is reduced in low-Reynolds flow conditions.

According to [1-3] the following particularities apply for those flow conditions:

- the boundary layer is thicker and susceptible to flow separation at small angles of attack,
- the profile drag is significantly bigger as a share in percentage from the induced drag when compared with a full-scale rotor with the same geometry,
- the swirl effects on the flow downstream of the rotor are more pronounced than those of the full-scale rotors.

Different approaches are used to study the aerodynamic performance of UAV rotors. They can be classified into the following three main groups, namely:

- Classic aerodynamic theory,
- CFD simulations,
- Experimental studies.

The classic aerodynamic solutions are based on existing theories, initially developed for airscrews, which were later adapted for helicopter rotors. Those include: the Actuator Disk Theory (ADT); the Blade Element Theory (BET); the Blade Element Momentum Theory (BEMT); the prescribed-vortex wake method and the free-vortex wake method, adapted for helicopter rotors in hover [4].

Paper [5] is a good example of the capabilities of the free-vortex wake methods not only to precisely simulate the hover performance of helicopter rotors but also to allow for the study of unsteady flow phenomena.

Computational fluid dynamics (CFD) is based on the numerical resolution of inviscid Euler equations or Navier-Stokes equations with an appropriate turbulence model, in order to determine the parameters of the flow-field in the simulated region around the studied rotor. This approach provides a more precise estimation of the aerodynamic (pressure) loads distribution along the blades. Although computationally expensive, the CFD simulations allow for a more detailed study of the helicopter rotor aerodynamics. Those simulations are very useful when studying unsteady rotor operation. In such a way the interaction between the tip-vortices and the rotor blades can be examined, which in return provides a more precise estimation of the aerodynamic performance of the rotor.

The hybrid CFD models offer a good compromise between high degree of accuracy of the obtained solution and reduced amount of computational resources. The most common approach is to replace the studied rotor geometry in the CFD model with the equivalent surface or volume distributed forces. This results in a significant reduction in the complexity of the simulation. The authors in [6] have applied this approach in their study, which allowed them to evaluate the rotor performance, while

* Michael Todorov: Michael.Todorov@tu-sofia.bg

taking into account the complex dynamics of the lead-lag, torsion and flapping motions of the blades.

Each of the above mentioned approaches has its advantages and disadvantages in terms of the accuracy of the obtained results, the required computational resources and the necessary simulation time. In principle, all numerical models need to be validated against experimental data. Then only a validated solution can be used for aerodynamic performance studies or geometry optimization of UAV rotors.

Thus, experimental studies are performed in order to obtain large databases for validation and calibration purposes of newly developed numerical models. Due to the absence of experimental data for small-scale propellers, operating at low-Reynolds flow conditions, it is not a privilege but rather a necessity for universities and companies developing UAVs to possess their own test stands for rotors and propellers [7-9]. These facilities allow not only for the acquisition of experimental data but also provide the possibility to test commercially available model rotors and propellers, as well as to assess new designs and concepts. They provide the opportunity for research and development activities.

In [10] the authors evaluate the hover performance of a helicopter rotor. For this purpose a rotor test stand is built. The authors found out that while the scaled rotors operate in a low-Reynolds environment, there is between 15 to 20% lower aerodynamic efficiency in hover as compared to a full-scale rotor with the same rectangular untwisted blade with NACA 0015 aerodynamic profile.

The aim of the present work is to evaluate experimentally the aerodynamic performance of scaled helicopter rotors. This study will also reveal the capabilities of the specialized test stand, developed in the Department of Aeronautics of Technical University – Sofia. The other main objective of this work is to acquire a large experimental database, which will serve for validation and calibration of numerical solutions under development.

2 Experimental studies

With the use of a specific helicopter rotor test facility, different series of measurement for multiple speeds of rotation and pitch angles are performed in order to study the aerodynamic characteristics and hover efficiency of three rotor configurations with 2, 3 and 4 blades.

2.1 Experimental setup

Fig. 1 shows the helicopter rotor test stand, which provides the possibility to measure the produced thrust T, N, the torque Q, Nm and the angular velocity Ω, rad/s of scaled helicopter rotors with diameters of up to 1 meter and pitch angles varying from –12° to +24°. The following measuring devices are used:

- Torque transducer HBM T20WN with nominal (rated) torque of 1 Nm, accuracy class 0.2 and output voltage ±10 V,
- Force transducer HBM S2M with nominal (rated) force of 10 N, accuracy class: 0.02 and output voltage of 2 mV/V.

The scaled rotors are driven with a brushless electric motor Maxon ECi-40 with nominal (rated) power of 100W. It allows for rotational speeds of up to 4190 RPM, although the maximum permissible rotational speed is limited to 3000 RPM. The electric motor has built-in Hall sensors, which make it possible to measure the speed of rotation. The motor is driven by Maxon ESCON 50/5 servo controller, which allows for precise regulation of the rotational speed.

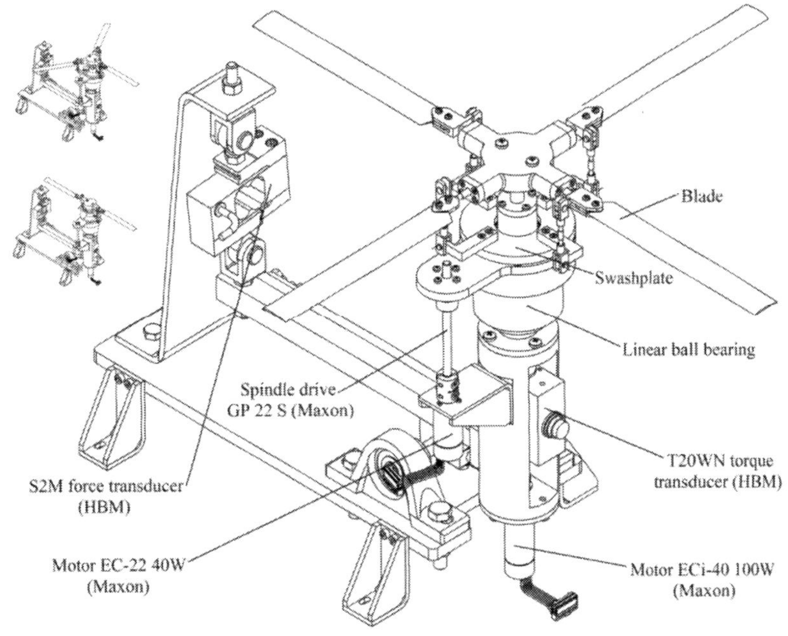

Fig. 1. Helicopter rotor test stand.

The test stand allows for the variation of the pitch angle of the blades. This is achieved by a spindle-driven mechanism, which uses the Maxon GP22S spindle with gear ratio of 1:370 and the EC-22 electric motor with nominal (rated) power of 40 W. This motor is also controlled by the ESCON 50/5 servo controller and can achieve a rate of change of ±3 °/s.

The pitch angle may be varied during rotation of the helicopter rotor. Thus, unsteady processes can be studied, in order to evaluate their impact on the aerodynamic characteristics of the rotor.

A specially developed LabVIEW driver permits both electric motors to be controlled automatically according to a pre-programmed test schedule. Thus, multiple operational regimes can be evaluated in one test session without the need to constantly stop, re-program and re-start the test bench. An automatic data acquisition system is organized, which stores the experimental data in a well-structured database.

The data acquisition from the force and torque transducers is ensured by the HBM PMX measuring amplifier system, shown in Fig.2. The PMX module is equipped with CODESYS V3 Soft-PLC, which allows for easy setup, configuration and calibration of each individual channel. It offers various analogue and digital inputs and outputs, thus providing large data acquisition capacity and control flexibility.

The entire test bench is powered by a DC power supply HY3005D2, which provides up to 300 W.

The construction of the test stand allows for easy and rapid change of the rotor configuration. The test stand can easily be modified for the attachment of hinged rotors.

2.2 Studied rotors

The present study concerns three-rotor configurations, differing in the number of blades, such as shown in Fig. 3. All three rotor configurations are hingeless with an inner diameter of 0.162 m and an outer diameter of 0.532m. The blades are rectangular and untwisted with a chord of 0.025 m. The blade profile is Goettingen 417 (GOE417).

2.3 Methodology

Three experimental studies were organized and carried out. The first one aimed to evaluate the torque Q_{nb} due to internal friction losses in the bearings of the main shaft and the air friction, created by the undetectable rotor hub. These measurements are important as they are later subtracted from all acquired torque data in order to deduce the real rotor torque without overestimation. The second one aims to estimate the coefficient of profile drag at zero thrust C_{D0}. This study is performed only with the two-bladed rotor configuration as it permits higher rotational speed and thus a broader range of Reynolds numbers can be reached. In this case the torque Q_0 is registered only if $T = 0$ N.

In order to obtain the C_{D0}, equations (1), (2) and (3) are used [4]:

$$C_{pd} = \frac{(Q_0 - Q_{nb})\Omega}{\rho A \Omega^3 R^3} = \frac{Q_{pd}\Omega}{\rho A \Omega^3 R^3} = \frac{P_{pd}}{\rho A V_{tip}^3}, \tag{1}$$

$$C_{D0} = \frac{8C_{pd}}{\sigma}, \tag{2}$$

$$\sigma = \frac{N_b c}{\pi R}, \tag{3}$$

where ρ is the air density of 1.225 kg/m^3; $A = \pi R^2$, m^2 is the area of the rotor; Ω, rad/s is the angular speed of rotational; R, m is the radius of the rotor; $V_{tip} = \Omega R$, m/s is the tip speed of the blades; N_b, - is the number of blades of the rotor configuration; c, m is the blade chord; σ, - is the rotor solidity; $Q_{pd} = Q_0 - Q_{nb}$, Nm is the rotor torque due to profile drag; $P_{pd} = Q_{pd}\Omega$, W is the power consumption due to profile drag; and C_{pd}, - is the coefficient of profile drag force.

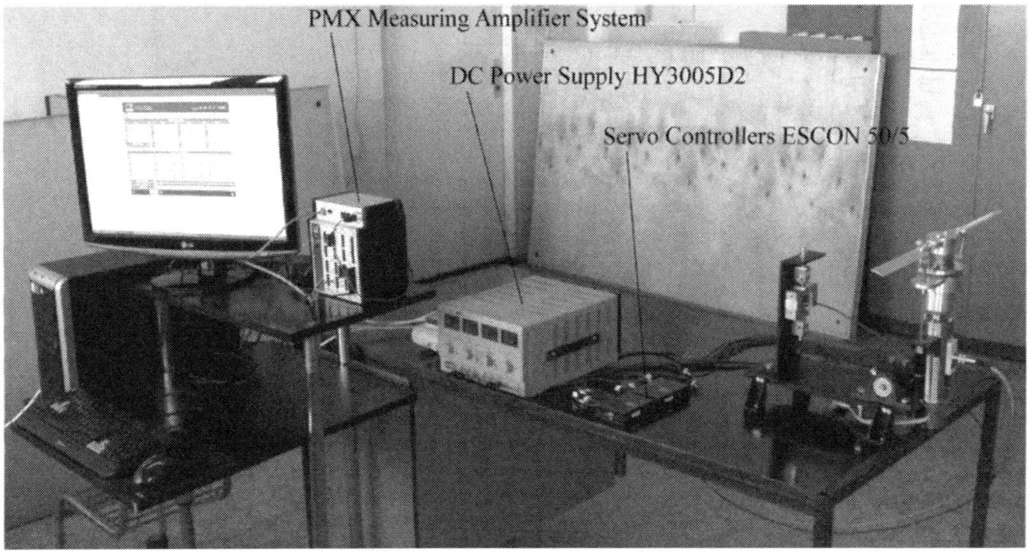

Fig. 2. Helicopter rotor test stand with the two-bladed rotor configuration.

 https://doi.org/10.1051/matecconf/201823401002

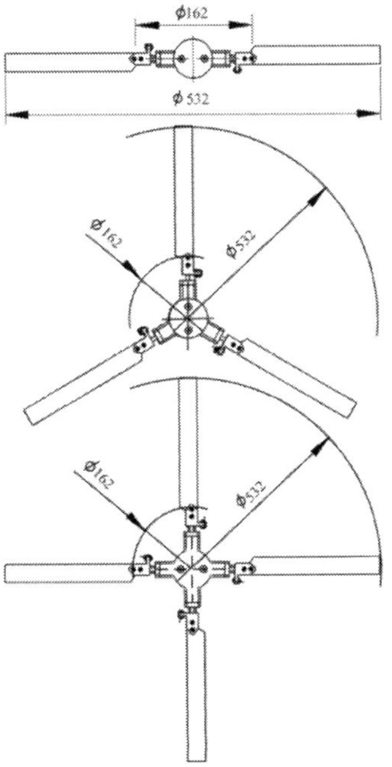

Fig. 3. Studied rotor configuration (dimensions in millimetres). Blade airfoil: GOE417; Blade chord: 25 mm.

The third study is such that for each rotor configuration a fixed pitch angle is set and then the thrust T, N and torque Q, Nm are measured for a multiple set of rotational speeds. This procedure is repeated for all preselected pitch angles $\theta = 0°, 3°, 6°, 9°, 12°$ and $15°$.

In [4] the coefficient of thrust C_T is given with (4); the coefficient of power C_P is given with (5); and the Figure of Merit (*FM*) is defined with (6):

$$C_T = \frac{T}{\rho A \Omega^2 R^2}, \tag{4}$$

$$C_P = \frac{Q\Omega}{\rho A \Omega^3 R^3} = \frac{P}{\rho A \Omega^3 R^3}, \tag{5}$$

$$FM = \frac{C_T^{3/2}}{\sqrt{2} C_P}. \tag{6}$$

In [4] the Figure of Merit computed with (6) is shown to have been introduced as a nondimensional measure of hovering thrust efficiency for helicopter rotors, which provides a common base for comparison between rotors with different geometries.

In order to ensure comparability between the results for helicopter rotors with different number of blades and thus with different rotor solidity, the following dimensionless parameters are used:

- C_T/σ, - – blade loading coefficient,
- C_P/σ, - – power coefficient.

2.4 Experimental results

Fig.4. shows the decomposition of the rotor torque for the two-bladed configuration at zero thrust. From Fig. 4 it can be seen that initially the profile drag torque is negligible right up to 400 RPM, after which it grows significantly. It can be seen that between 800 and 1300 RPM the profile torque Q_{pd} equalizes its value with the no blades torque Q_0, which is due to internal frictions in the main shaft bearings. After 1400 RPM the profile drag torque rises as a curve of second order, where the no blade torque is rather linear.

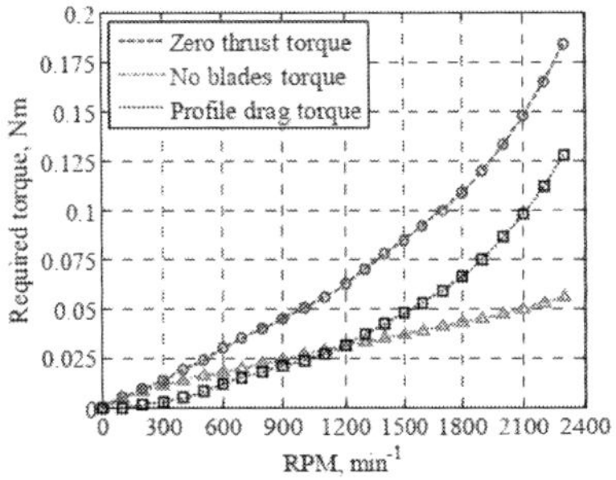

Fig. 4. Evaluation of the torque required to drive the test stand at zero thrust for the two-bladed rotor at various speeds.

The data in Fig.5 is derived from Fig.4. by using the simple relation between torque and power: $P = Q\Omega$. It can be seen that at 1000 RPM the power losses, due to bearing frictions, are around 3 W, where at 2000 RPM they rise to 10 W. At 1000 RPM the total power consumption at zero thrust is about 5 W.

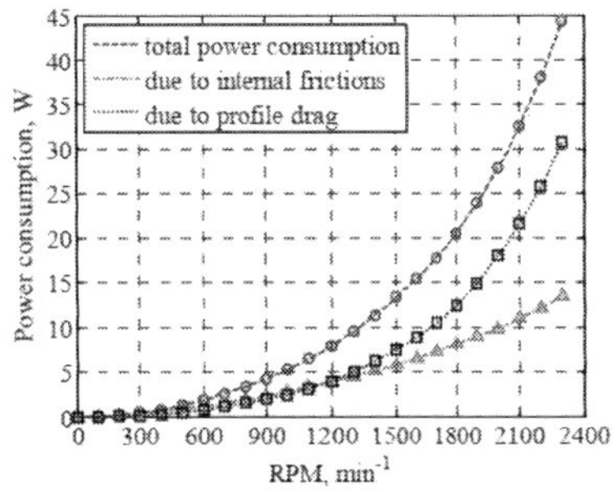

Fig. 5. Estimated power consumption required to drive the test stand at zero thrust for the two-bladed rotor at various speeds.

Fig. 6. shows the variation of the profile drag at zero thrust C_{D0} against the Reynolds number of the flow. Those results are closely comparable to those published in [11]. It can be seen that the profile drag reduces with 35% with the increase of Re from 20000 to 50000.

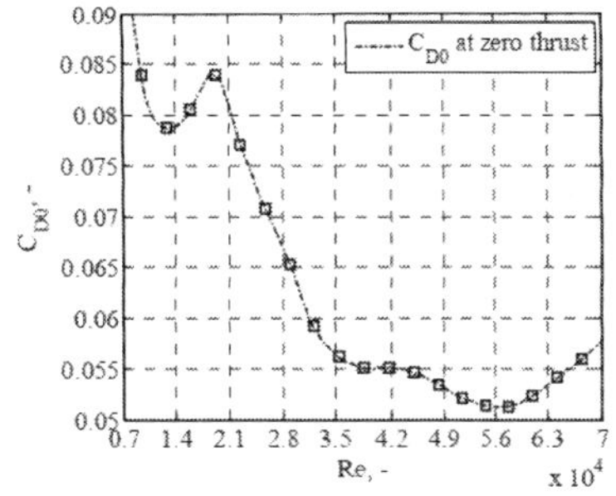

Fig. 6. Profile drag at zero thrust depending on Reynolds for profile Goettingen 417.

Fig.7. shows the hover polar for all rotor configurations. It can be seen that the three and four bladed-rotor configurations require lower hover power.

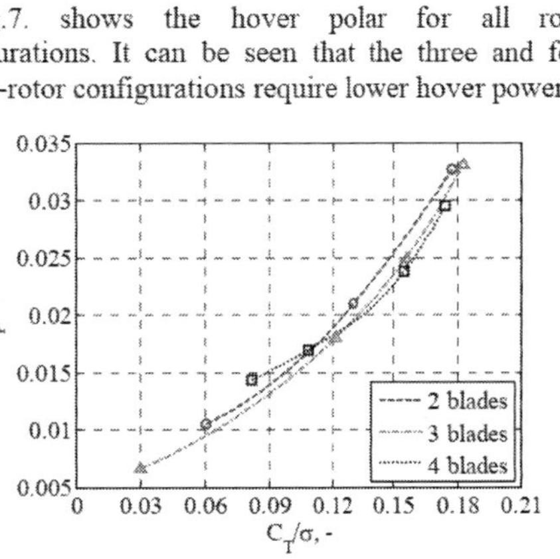

Fig. 7. Hover polar for all rotor configurations at 900 RPM and Re≈29000.

Fig.8. shows the blade loading coefficient variation from the pitch angle of the rotor blades θ. It can be seen that the lower the number of blades, the higher the blade load on each individual blade, which may lead to early stall conditions as it can be observed for the two-bladed configuration for $\theta = 12° \div 15°$.

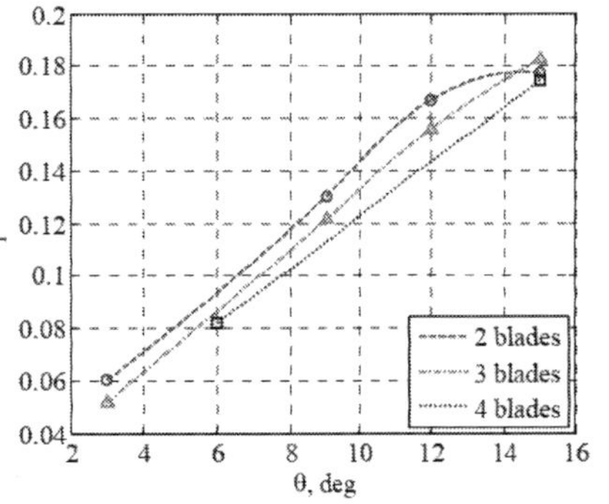

Fig. 8. Blade loading coefficient against the pitch angle of the blades for all rotor configurations at 900 RPM and Re≈29000.

Fig.9. shows the hover efficiency against the blade load coefficient. It is clear that the higher the number of blades, the higher the rotor solidity and thus the higher the hover efficiency of the rotor.

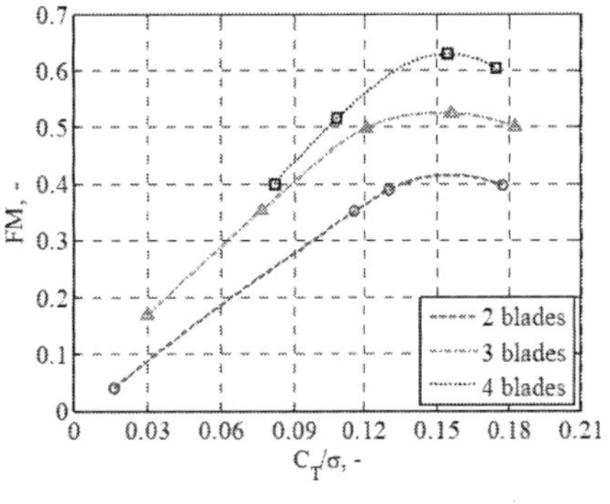

Fig. 9. Figure of Merit against the blade loading coefficient for all rotor configurations at 900 RPM and Re≈29000.

3 Summary and conclusions

This paper presents an experimental approach for the study of a helicopter rotor model at low-Reynolds flow conditions.

A fully operational test stand is designed and shown. It allows for the evaluation of the aerodynamic performance and efficiency of model helicopter rotors in steady hover. Thus, three rotor configurations with 2, 3 and 4 blades were considered.

Thanks to this study, a relatively large database was obtained. This experimental data is primarily for low-Reynolds flow conditions of hovering rotors. It will be quite useful for validation and calibration purposes in the case of newly developed numerical solutions.

The obtained results required the implementation of an important experimental and technical know-how.

References

1. M. Benedict, J. Winslow, Z. Hasnain, I. Chopra, Performance and flowfield measurements of a MAV-scale helicopter rotor in hover, Int. J. of Micro Air Vehicle, **7**(3), 231-255 (2015)
2. B. Hein, I. Chopra, Hover performance of a micro air vehicle: Rotors at low Reynolds number, J. of the American Helicopter Society, **52**(3), 254-262 (2007)
3. M. Ramasamy, B. Johnson, J. Leishman, Understanding the aerodynamic efficiency of a hovering micro□rotor, J. of the American Helicopter Society, **53**(4), 412-428 (2008)
4. J. Leishman, *Principles of Helicopter Aerodynamics*, 67-70 (Cambridge university press, 2006)
5. A. Karpatne, J. Sirohi, S. Mula, C. Tinney, Vortex ring model of tip vortex aperiodicity in a hovering helicopter rotor, J. of Fluids Engineering, **136** (2014)
6. I. Dobrev, F. Massouh, M. Todorov, A hybrid aerodynamic model for a study of pattern helicopter rotor dynamics, *NCAS-2013*, 11-14 (2013)
7. J. Brandt, M. Selig, Propeller performance data at low Reynolds numbers, *AIAA 2011-1255* (2011)
8. V. Serbezov, S. Dimitrov, K. Rangelov, Mobile test stand for evaluation of electric power plants for unmanned aircraft, *BulTrans-2017*, MATEC Web of Conferences, **133** (2017)
9. H. Panayotov, S. Pentchev, Determination of propeller characteristics for multirotor drone design, *TECHSYS-2017*, 264-268 (2017)
10. B. Lee, Y. Byun, J. Kim, B. Kang, Experimental hover performance evaluation on a small-scale rotor using a rotor test stand, J. of Mech. Science and Technology, **25**(6), 1449-1456 (2011)
11. C. Lyon, A. Broeren, P. Giguere, A. Gopalarathnam, M. Selig, Summary of low-speed airfoil data, **3**, 128-132 (1995)

Modeling of aircraft jet noise in airports

Hristina Georgieva[1*], *Krasin* Georgiev[2]

[1]Technical University – Sofia, Mechanical Department, 1756 Sofia, 8 Kliment Ohridski Blvd., Bulgaria
[2]Technical University – Sofia, Aeronautical Department, 1756 Sofia, 8 Kliment Ohridski Blvd., Bulgaria

Abstract. A mathematical model with 4 degree of freedom created in Matlab for aircraft final landing trajectory is described in this paper. A midsize commercial passenger aircraft similar to an Airbus A320 has been chosen as a reference aircraft. The parameters of model are obtained from Airbus, Eurocontrol and the approach procedure at the Munich airport is selected up from Jeppesen Airway manual. A semi-empirical model of Stone for predicting the jet noise has been used. The proposed model was validated against 10 real flights obtained from Aircraft noise monitoring at Munich airport. The computed error between the real data and modelling is reported on. Obtained results are presented numerical and graphically. The observed effects of aircraft speed, aircraft angle of descent and aircraft weight for reduction of aircraft jet noise in airports represent subjects of discussions in the paper.

1 Introduction

Contemporary society is difficult to imagine without the air transport and related improvements, facilitating everyday life. It causes noise pollution of the environment, which is considered today to be one of the most significant environmental problem affecting population and the environment. The aircraft noise is highest on the airports and communities around airports. It can lead to community annoyance, sleep disturbance and could increase the risk of high blood pressure, heart disease, heart attack for people living near airports. Significant research is currently being undertaken with the goal of reducing aircraft noise [1]. The Environmental Noise Directive (END) [2] is the legislative instrument at European Union (EU) under that environmental noise is monitored and necessary actions (standards and recommended) norms are taken. To reduce the noise levels in airports and place close to the airports the Directive applies noise criteria for noise mapping, developing and implanting action plants. The theoretical models allow predicting and managing the real object. There are two types of models. The theoretical models of the first group are based on empirical data and use specialized software products [3, 4]. This work is focused on the second type of models that use the basic principles of aircraft modelling. They represent the aircraft as a point-mass model including the aerodynamics of the aircraft itself. The differential equations describe the aircraft position in space and allow finding an appropriate relationship between the investigated parameters. They allow the optimization of aircraft trajectories and determination also of noise level from these trajectories. The models that are used in the literature are with 4 to 6 degrees of freedom [5, 6, 7]. Khardi and Abdallah propose a 4-degree model of aircraft [5] simulating the final trajectory of landing. In their work they use two approaches for minimization of aircraft jet noise. The first one suggests a constant throttle setting and variation of angle descent between 0° and –4.5°. The second one defines two stages of throttle section which variation is between 0 and 0.6 and the angle of descent also varies between 0° and –4.5°. Menendez describes an aircraft model [6] with 6 degrees of freedom for evaluation of aircraft noise abatement procedures. He is focused on the more flexible trajectories which will enable the definition of optimal flight procedures regarding the noise annoyance impact, especially in the arrival and departure phases of flights. In work [7] the noise reduction is obtained as coupling the aircraft nose. The two similar commercial aircraft are supposed to land successively on one runway without conflict. The aircraft movement of the two aircrafts is presented by mathematical model with 6 degrees of freedom. While maintaining an evolution of the flight approach procedures the same optimal trajectory has been achieved for the two-aircraft even though the procedures are different. Koenig et al. [8] describe the increase of the aircraft angle of descent to reduce the noise level without necessity of flight procedure change. In the above models the semi-empirical model of Stone [9] is used to predict the engine jet noise level for single and mixed dual jets up to bypass ratios of 15 as the main source. In general, the model can be applied to subsonic or supersonic flights up to Mach numbers of 2.5. J. E. Bridges et al. [10] analyse the method and prove its applicability to modern engines with a high bypass ratio.

The main sources of aircraft noise are the engine and its body. In the current work a mathematical model with 4 degree of freedom created in Matlab is described. It includes two sections. The first presents the modelling of aircraft final landing trajectory and the second one the modelling of engine jet noise level. The model permits to

* Corresponding author: hgeorgieva@tu-sofia.bg

investigate different variants of aircraft landing trajectories for reducing the aircraft jet noise level in airports through the variation of angle of descent, velocity of descent and aircraft weight. Hence it has been proposed a complete method for reducing the aircraft jet engine noise, including all the possible components on which it depends.

2 Methodology of data selection

2.1 For modelling of aircraft landing trajectory

The aircraft chosen for the simulation is similar to the Airbus A320 which is widely used for short to medium range flights [11]. According to performance training manual of Airbus [12] the aircraft velocity of descent varies from 82 to 118 m/s. Data for modelling of aircraft landing trajectory on Munich airport (Fig. 1) is taken from the Jeppesen Airway manual [13].

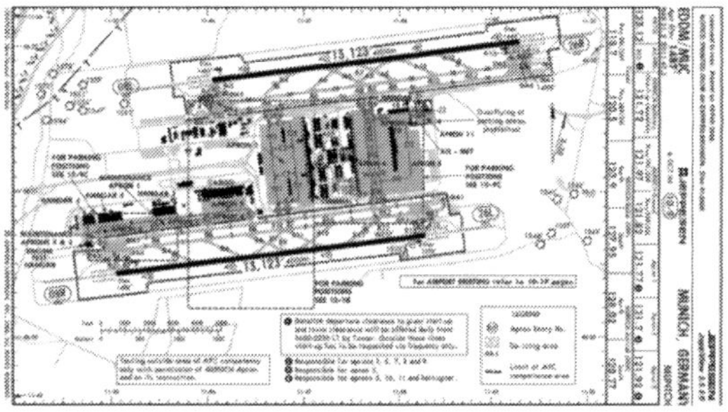

Fig. 1. Runway map of Munich airport.

In Table 1 is shown the selected parameters: the runway altitude, altitude of descent and the angle of descent. This aircraft is equipped by two types of engines: CFM565A and V2527A. According the data from Eurocontrol [14] widely used is CFM565A engine. This engine develops a maximum thrust force of 118 kN (Table 1).

Table 1. Aircraft landing trajectory modelling data.

Data from	Parameter	Value	Unit
Aircraft	Max landing weight, m_{MTLW}	64500	kg
	Wing span, b	35.8	m
	Wing are, S	122.6	m^2
	Wing aspect ratio, A	9.395	–
	Sweeper angle, χ	25	deg
	Velocity of descent, V	82-116	m/s
Runway	Runway altitude, h_{runway}	498	m
	Altitude of descent, h	900	m
	Angle of descent, γ	3	deg
Engine	Thrust, T	118	kN
	Thrust specific fuel consumption, $tsfc$	9.4444 $\times 10^{-5}$	N/sN

2.2 For modelling of jet engine jet noise

The engine jet noise model requires specific input data with respect to geometry and operational conditions of the selected engine (Table 2). From the Airbus technical data are taken the geometric parameters [11]. The aircraft altitude, aircraft speed and the engine thrust setting determine the engine operation. According to the engine operation, specific engine thermodynamic parameters have to be generated.

To calculate the parameters of ambient condition it has been used the International Standard Atmosphere (ISA) [15]. The required specific design parameter of the engine for primary and secondary flow a turbofan calculator is used [16, 17].

Table 2. Engine noise modelling data.

Data from	Parameter	Value	Units
Engine	Hydraulic diameter of inner contour, D_{19}	1.2	m
	Hydraulic diameter of outer contour, D_9	1.8	m
	Nozzle area of inner contour, A_{19}	Function of D_{19}	m^2
	Nozzle area of outer contour, A_9	Function of D_9	m^2
	Polar angle, α*	45	deg
	Jet velocity in inner contour, v_9	219-593	m/s
	Flow velocity in outer contour, v_{19}	230	m/s
	Jet temperature in inner contour, T_9	277-1132	K
	Flow temperature in outer contour, T_{19}	286	K
Ambien condition	Flow pressure, p_∞	90811.7	Pa
	Flow temperature, T_∞	282.206	K
	Flow density, ρ_∞	1.121	kg/m^3
	Sound speed, c_∞	336.766	m/s
	Mach number, M	0.219	–

3 Modelling of aircraft landing trajectory and jet engine noise

The proposed model includes 2 sections.

3.1 Modelling the aircraft landing trajectory

Fig. 2 shows the aircraft model used in this study and the aircraft landing trajectory.

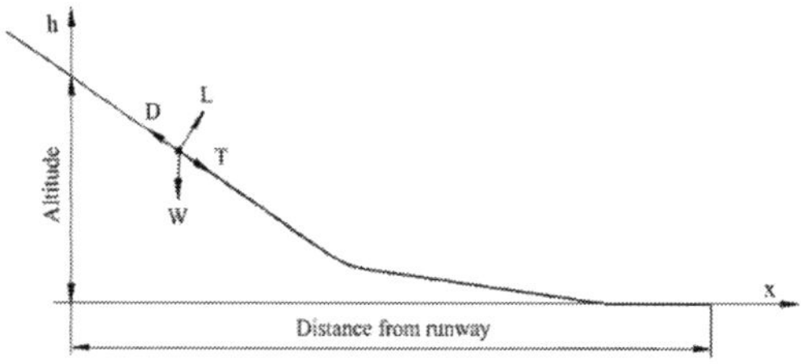

Fig. 2. Aircraft landing trajectory

For the differentials equations (1) describing the aircraft landing trajectory is valid the following four-dimension system derived at the aircraft centre of mass:

$$\dot{x} = V\cos\gamma;$$
$$\dot{h} = V\sin\gamma;$$
$$\dot{V} = \frac{T-D}{m} - g\sin\gamma; \tag{1}$$
$$\dot{m} = -tsfcT,$$

where (x, h) are the position of the aircraft, (V, γ) are respectively the velocity and the angle of the descent. The variables (T, $tsfc$, D, m and g) are respectively the engine thrust, thrust specific fuel consummation, the drag force, the aircraft weight and gravity acceleration. Those variables are expressed as:

$$T = T_0\delta_x \frac{\rho}{\rho_0}\left(1 - M + \frac{M^2}{2}\right);$$
$$D = \frac{1}{2}\rho S V^2\left(C_{x0} + k_i C_{Z\alpha}^2\alpha^2\right); \tag{2}$$
$$M = \frac{V}{c},$$

with T_0 the full thrust, δ_x the throttle setting, ρ the density of air at altitude, ρ_0 the density of air at ground, M the Mach number, S the wing span, $C_{Z\alpha}$ the gradient of the lift depends on the high lift device, C_{x0} the drag coefficient for $\alpha=0$, k_i the induced drag coefficient, c the speed of sound at the altitude h [18].

3.2 Modelling of aircraft engine jet noise

The jet noise consists of three principal components: turbulent mixing noise, broadband shock associated noise and screech tones. The jet noise is approximately omni-directional during the descent phase and the noise levels decrease as speed decreases when assuming that the thrust is constant. It is extremely complex to predict the jet noise.

In this work to model the aircraft engine jet noise is used a semi-empire model of Stone [9].The model assumes a symmetric noise emission with respect to the engine reference axis.

$$L_{jet} = L_{norm} + \Delta L_{dir/spec}\left(Str_{c-jet}, \alpha_{cor}^*\right) + \Delta L_{c-jet} \tag{3}$$

L_{norm} is a normalized sound pressure level and it is defined according to the nozzle geometry and the operating condition.

$$L_{norm} = 141.0 + 10\log 10\left[\left(\frac{\rho_\infty}{\rho_{\infty,ISA}}\right)^2\left(\frac{c_\infty}{c_{\infty,ISA}}\right)^2\right] +$$
$$+10\log 10\left(\frac{A_9}{d}\right) + 10\omega\log 10\left(\frac{\rho_9}{\rho_\infty}\right) + 75\log 10\left(\frac{\bar{v}_9}{c_\infty}\right) -$$
$$-15\log 10\left[\left(1 + Ma_{con}\cos(\alpha^*)\right)^2 + 0.04Ma_{con}^2\right] - \tag{4}$$
$$-10\log 10\left[1 - Ma\cos(\alpha^*)\right] + 3\log 10\left(\frac{2A_9}{\pi(D_9)^2} + 0.5\right),$$

where $\bar{v}_9$ is the effective jet speed defined as:

$$\bar{v}_9 = v_9\left[1 - \frac{V}{v_9}\cos(\alpha^*)\right]^{\frac{2}{3}} \tag{5}$$

ω is the exponent applies to hot jets:

$$\omega = \frac{3.0(\bar{v}_9/c_\infty)^{3.5}}{0.6 + (\bar{v}_9/c_\infty)^{3.5}} - 1 \tag{6}$$

Ma_{con} is a convective Mach number:

$$Ma_{con} = 0.62\frac{v_9 - V\cos(\alpha^*)}{c_\infty} \tag{7}$$

S_{tr9} is an effective Strouhal number.

$$Str_9 = f\frac{\sqrt{4A_9/\pi}}{\bar{v}_9}\left(\frac{D_9}{\sqrt{4A_9/\pi}}\right)^{0.4}$$
$$\left(\frac{T_{t,9}}{T_{t,\infty}}\right)^{0.4\left[1+\cos(\alpha^*)\right]}\left(1 - Ma\cos(\alpha^*)\right) \tag{8}$$
$$\left[\frac{\left(1 + 0.62\left(\frac{v_9 - v_\infty}{c_\infty}\right)\cos(\alpha^*)\right)^2 + 0.01538\left(\frac{v_9 - v_\infty}{c_\infty}\right)^2}{\left(1 + 0.62\left(\frac{v_9}{c_\infty}\right)\cos(\alpha^*)\right)^2 + 0.01538\left(\frac{v_9}{c_\infty}\right)^2}\right]^{0.5}$$

with α^*_{cor} is a corrected directivity angle defined as:

$$\alpha_{cor}^* = \alpha^*(v_9/c_\infty)^{0.1} \tag{9}$$

ΔL_{c-jet} is defined as:

$$\Delta L_{c-jet} = 5\log 10\left(\frac{T_{t,9}}{T_{t,19}}\right) +$$
$$+10\log 10\left[\left(1 - \frac{v_{19}}{v_9}\right)^m + 1.2\frac{\left(1 + \left(\frac{A_{19}}{A_9}\frac{v_{19}^2}{v_9^2}\right)\right)}{\left(1 + \frac{A_{19}}{A_9}\right)^3}\right] \tag{10}$$

where the exponent m is defined as a function of the area ratio:

$$m = \begin{cases} 1.1\sqrt{A_{19}/A_9} & for \quad A_{19}/A_9 < 29.7 \\ 6.0 & for \quad A_{19}/A_9 > 29.7 \end{cases} \tag{11}$$

4 Results and discussion

An option which has to be evaluated in order to reduce the engine jet noise level in airports is the aircraft angle of descent.

In this work is considered an aircraft landing which model uses the following reference values [16] for:

• angle of descent $\gamma = 3°$;
• velocity of descent $V = 67$ m/s;
• nominal weight $m = 56000$ kg.

The landing duration is 140 s. The throttle setting δ_x is constant during the landing. Graphics and numerical results are obtained from the proposed model. Fig. 3 shows the simulated aircraft landing trajectory for different angle of descent γ which varies between 2.8° and 3.4° and different velocity of descent V which varies between 53 m/s and 67 m/s.

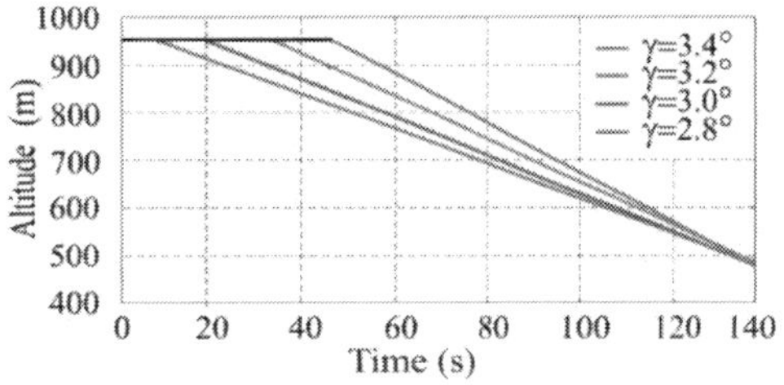

Fig. 3. Results from the simulated aircraft landing trajectory.

Table 3 shows the numerical values for calculated aircraft engine jet noise levels from the model for:
• different value of angle of descent γ;
• different velocity of descent V.

Table 3. Obtained numerical value from the jet engine model.

Param.	Value					Unit
Angle	3.0	2.8	3.2	3.4	3.0	deg
Weight	56000	56000	56000	56000	64000	kg
Speed	67	74	60	53	67	m/s
Noise	72.63	73.54	70.98	69.88	75.08	dB

As can be seen in table, according to the model:
• at a reference angle of descent $\gamma = 3°$ and a reference velocity of descent $V = 67$ m/s, the obtained level of aircraft jet engine noise is 72.630 dB;
• at an angle of descent $\gamma = 2.8°$ and a velocity of descent $V = 74$ m/s, the noise level from the aircraft jet engine increases as compared to the reference angle by 0.92 dB;
• at an angle of descent $\gamma = 3.2°$ and a velocity of descent $V = 60$ m/s, the noise level from the aircraft jet engine decreases as compared to the reference angle by 1.65dB;
• at an angle of descent $\gamma = 3.4°$ and a velocity of descent $V = 53$ m/s, the noise level from aircraft jet engine aircraft decreases as compared to the reference angle by 2.75 dB.

Obviously, the results show that there is a trend to reduce the aircraft jet noise after changing the angle of descent γ and velocity of descent V. Larger variations in descent angle γ and velocity of descent V are limited by the difficulties of change in the landing procedure. When the aircraft landed, it spent much of its fuel and the weight of the passengers and baggage in the cargo compartment stays constant. Thus study with changing the aircraft weight m is not necessary. However sometimes immediately after the aircraft takes-off, there is an emergency situation (incident) that causes an unplanned landing. In this cause the aircraft has a maximum landing weight. The model permits to calculate the noise level in this case also (Table 3).

Mathematical modelling is a modern research tool, but only when the proposed model is validated. To validate the calculated noise level of the aircraft jet engine it has been used experimental data from ten real flights (Table 4, Fig. 5) from Aircraft noise monitoring system at Munich airport (Fig. 4). The system operates currently with 16 stationary measuring stations which are positioned at a radius of some 20 km around the airport. Additionally three mobile measurement stations are also used at places where no stationary measuring station provides information about aircraft noise pollution [19].

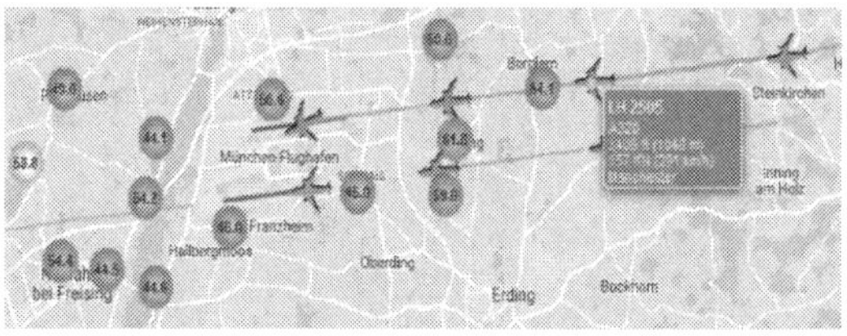

Fig. 4. Noise monitoring system at Munich airport.

Table 4: Data from the measurement station NMT7 at Munich airport runway 08L/26R on 30.04.2018 and angle of descent $\gamma = 3°$.

№	Flight		Speed	Unit	Noise	Unit
1	Dublin-Munich	EI 352	66	m/s	73	dB
2	Hanover-Munich	LH 2093	69	m/s	72	dB
3	Berlin-Munich	LH 2033	72	m/s	74	dB
4	Frankfurt-Munich	LH 100	78	m/s	72	dB
5	Palma de Majorca-Munich	DE 1509	80	m/s	73	dB
6	Moscow-Munich	SU 2322	70	m/s	71	dB
7	Berlin-Munich	EZY 5665	75	m/s	73	dB
8	London-Munich	LH 2473	82	m/s	74	dB
9	Kiev-Munich	LH 2545	77	m/s	73	dB
10	Berlin-Munich	LH 2047	95	m/s	72	dB

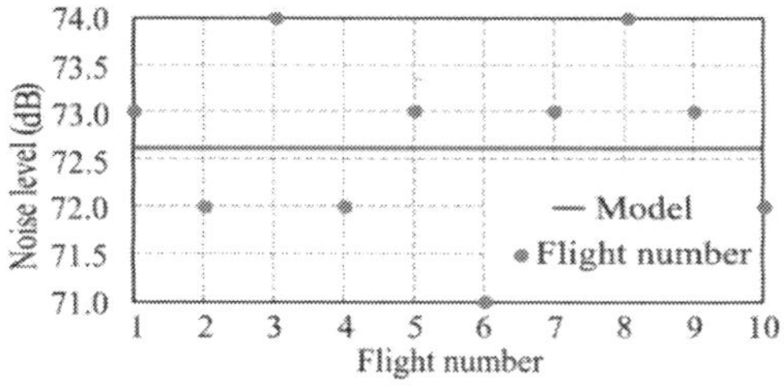

Fig. 5. Obtained results from the engine noise from the model and the real flight from the airport monitoring system.

The average deviation in the results from the jet engine model with the data from noise monitoring system is 1.1 %.

5 Conclusion

A main source of noise from a low fly-over aircraft in the airports is its engine. In this work for simulating the aircraft landing trajectory is used: aircraft Airbus A320 widely used for short to medium range flights equipped with two CFM565A engines and runway of Munich airport which is one of the busiest airport in Europe.

The suggested model:

- is validated;
- shows graphical and numerical results;
- allows to study different possibilities of reducing the aircraft jet the noise level during landing in airports;
- allows to calculate the aircraft jet noise level in emergency situations;
- no complicated mathematical methods which reduces the calculation time;
- is general and gives the possibility to investigate the reduction of jet engine noise on aircraft equipped with engine without mixing flows.

In the future, the research will focus on creating a model to investigate the noise level of airframe and the engine model will be developed by adding the fan noise.

This work has been supported by Research and Scientific Centre of Technical University of Sofia (Grand № 181ПР0012-04).

References

1. http://www.futuresky.eu/projects/noise (30.07.2018)
2. EC, 2002, *Directive 2002/49/EC of the European Parliament and of the Council of 25 June 2002 relating to the assessment and management of environmental noise,* OJ L 129, 12-25 (18.7.2002)
3. E. Boeker, E. Dinges, B. He, G, Fleming, C. Roof, P. Gerbi, A. Rapoza, J. Hermann, *Integrated noise model (INM) version 7.0 technical manual,* FAA-AEE-08-01 (2008)
4. W. Krebs, *Sound source data for aircraft noise simulation*, Acta Acoustica united with Acustica, **90**(1), 91-100 (2004)
5. S. Khardi, L. Abdallah, *Optimization approaches of aircraft flight path reducing noise: Comparison of aircraft modelling methods*, Applied Acoustics, **73**, 291-301 (2012)
6. X. Menedez, *Contributions to the optimisation of aircraft noise abatement procedures*, Doctoral Thesis, Càtedra abertis de Gestión de Infraestructuras del Transporte Universitat Politècnica de Catalunya (2011)
7. F. Nahayo at al., Optimal control of two-commercial aircraft dynamic system during approach, The Noise Levels Minimization. Gen. Math. Notes, **3**(2), 27-49, (2011)
8. R. Koenig, E. Schubert, On the influences of an increased ILS glide slope on noise impact, fuel consumption and landing approach operation, *AIAC14 Fourteenth Australian International Aerospace Congress*, 28 February – 3 March, Melburn (2014)
9. J. Stone, D. Groesbeck, C. Zola, Conventional profile coaxial jet noise prediction, AIAA Journal, **21**(1), 336-342 (1983)
10. J. Bridges, A. Khavaran, C. Hunter, Assessment of current jet noise prediction capabilities, *14th Aeroacoustics Conference*, Vancouver, Canada, May 5-7 (2008)
11. Airbus-AC-A320, *Aircraft Characteristics - Airport And Maintenance Planning. AIRBUS S.A.S. Customer Services*, Technical Data Support and Services, 31707 Blagnac Cedex, France (2016)
12. Airbus, Training&Flight Operation support and services, Flight crew performance course, *A318/A319/A320/A321, Performance Training Manual,* 31707 Blagnac Cedex, France (2005)
13. http://ww1.jeppesen.com/personal-solutions/aviation/vfr-charts.jsp (30.07.2018)
14. https://www.aircraftnoisemodel.org/ (30.07.2018)
15. https://www.digitaldutch.com/atmoscalc/ (30.07.2018)
16. D. Adolfo, D. Bertini, A. Gamannossi, et C. Carcasci, Thermodynamic analysis of an aircraft engine to estimate performance and emissions at LTO cycle, *72nd Conference of the Italian Thermal Machines Engineering Association, ATI2017*, 6-8 September, Lecce, Italy (2017)
17. https://www.particleincell.com/2014/turbofan-calculator/ (30.04.2018)
18. A. Bos, *Aircraft performance summary tables for the base of aircraft data* (BADA) revision 3.0. EEC Technical / Scientific Reports
19. https://travis-web01.munich-airport.de/data/travis.php?lang=en&_ga=2.41081378.185368683.1524577872-2134088401.1513440516 (30.04.2018)

Challenging low Reynolds - SWT blade aerodynamics

Damian Kądrowski[1,*], *Michał* Kulak[1], *Michał* Lipian[1], *Małgorzata* Stępień[1], *Piotr* Baszczyński[1], *Karol* Zawadzki[1] and *Maciej* Karczewski[1]

[1] Institute of Turbomachinery, Lodz University of Technology, 90 - 924 Lodz, 219/223 Wolczanska Street, Poland

Abstract. One of the main issues related to the design and development of small wind turbines (SWTs) is the low Reynolds number. Operation in the transitory regime makes the rotor aerodynamic analysis a challenging task. Project GUST (Generative Urban Small Turbine) realized currently at the Institute of Turbomachinery (Lodz University of Technology, Poland) is devoted to the development of SWT ($D = 1.6$ m) for low-Reynolds number (low wind speed) flow conditions. The emphasis is on the blade design, aiming at improving the rotor aerodynamic efficiency. The paper will highlight the rotor design process, based on contemporary methods of experiment-simulation integration approach and use of rapid manufacturing techniques. In-house wind tunnel measurements of a scaled model performance were executed. A numerical analysis using dedicated software (QBlade) was conducted in parallel. A comparison between the obtained results indicated that the chosen numerical tools are capable of providing a reliable output, even in complex, transitional flow conditions. Bearing in mind the above observations, QBlade was incorporated into the development process of a completely new blade geometry which would increase rotor performance. The selected design has indeed prove to show better power outcome in an additional experimental campaign.

1 Introduction

Nowadays, the objective of fulfilling the world energy requirements and at the same time of taking into account global climate changes is a challenging task. The aim is for traditional, non-renewable power sources to be aided and ultimately replaced by sustainable ones. Although wind energy, as an omnipresent and renewable source, is a promising solution, its utilization in large scale wind turbines, frequently organized into big wind farms, can negatively affect both climate and atmosphere [1]. Herein, the dispersed and decentralized small scale wind turbines show a significant potential for power production. It can be sufficient for urban applications or domestic needs without influencing the weather and environment conditions [2].

It is essential to understand small wind turbine characteristics, while having in mind their drawbacks. One should take into consideration such difficulties as: relatively high initial price, the need for proper site assessment resulting in optimal wind turbine placement [3], recognition of wind conditions on site (i.e. wind velocity, wind direction, temporal conditions) [4], national policy and aero-acoustic noise. On the contrary, when placed and sized correctly and operating at design conditions, small wind turbines can be considered as a convenient, effective and environment-friendly power supply. Taking economics into account, two factors are of concern: initial cost per kW power and the unit cost per kWh it produces [2].

To address the above issues, students' project GUST (Generative Urban Small wind Turbine), focuses on building horizontal axis wind turbine of 2 m^2 swept area and nominal power around 750 W. The team has presented the prototype during the International Small Wind Turbine Contest, being its winner for three times in a row (2016-2018). Besides the self-constructed mechanical components (such as hydraulic brake), alterations of power generator (system of coils switching), as well as self-designed safety and control systems, the most important innovation proposed by GUST is related to the aerodynamic design of wind turbine blades followed by their production through rapid prototyping (3D printing) methods.

The aim of the paper is to present a rotor design process. It is based on experiment-simulation integration (utilizing the in-house wind tunnel and simulation in dedicated software – Qblade). A correspondence between the obtained results indicates that the selected numerical tools are capable of granting a reliable output, even though the flow is complex (conditions of laminar-turbulent transitional flow). As a next step, QBlade was incorporated into the process of developing completely new wind turbine blade geometry, aimed at increasing rotor performance. The selected design was subjected to an additional experimental process, in order to prove the better power outcome.

* Corresponding author: damian.kadrowski@gmail.com

1.1 Reynolds number definition – flow similarity criterion

In order to consider the air movement around the aerofoil it was necessary to estimate the Reynolds numbers in particular cross-sections (25 uniformly distributed stations along blade span):

$$Re = u\, c_i \,/\, \nu_{air}\,, \qquad (1)$$

where c_i denotes assumed chord length of a specific cross section, ν_{air} is kinematic viscosity of the air and u represents a vector sum of axial wind speed and rotational speed of the blade element on the analysed radial position. Note that, in order to simplify the analysis at this early stage, the induced velocity is not taken into account. The estimation resulted in Reynolds number values in the range of 50000 ÷ 150000.

It is also possible to define the global Reynolds number basing on the rotor overall size, whence c in numerator becomes *D*, the rotor diameter. This criterion is especially useful when comparing rotors of different sizes or subjected to different wind velocity.

In the current study the default Reynolds number is the local one, based on chord length *c*, unless explicitly specified.

2 Numerical analysis – Qblade

QBlade is open source wind turbine calculation software, integrating various codes and models for a thorough simulation of wind turbine performance [5]. It integrates tools for e.g. aerofoil design and evaluation, BEM rotor simulation, simplified structural analysis of the blade and results post-processing.

The performed aerodynamic study included complex blade and rotor geometry design and analysis. The in-house designed blade, originally inspired by the model from NTNU Norway [6] is based on NREL S826 and S834 aerofoils. As a result of the differences in the rotor diameters between the two rotors, the original chords were scaled by a factor defined as a ratio between modified and original rotor radius.

The main criterion considered during the design process was to maximize the Annual Energy Production (AEP). Having the wind conditions imposed [6], the whole construction needed to be adjusted to the specific set of parameters. The geometry was also optimised to lower the cut-in wind speed and widen the operational range of the Tip Speed Ratios (*TSRs*, formula (4)). It is important to point out that the rotor geometry is characterized by relatively small scales (notably chord length). Thus the blade design criteria should also take into account the mechanical limitations of materials load concentration areas.

An initial part of the procedure was the aerofoil selection. The NREL S820 and S830 aerofoil families are dedicated for small wind turbines [7], having in mind the low Reynolds number observed for the flow. The assessment of the profiles' performance was done with the QBlade-integrated XFoil software, a panel-method-based code for inviscid and viscous flow analysis around aerofoils [8]. Further analysis and the validation process depended on data acquired from XFoil polars, visible in Fig. 1 (S826) and Fig. 2 (S834). The plots show *Cl/Cd* (lift-to-drag ratio) as a function of angle of attack (*AoA*) and *Cl* versus *Cd*. The simulations were performed for different Reynolds numbers, corresponding to local values observed at different blade stations. The flow is considered incompressible (i.e. $Ma \approx 0$).

In every figure all considered curves share roughly the same trace. However, it is noticeable how important the influence of the applied *Re* is. With decreasing Reynolds number the aerofoil performance falls drastically. In case of S826 aerofoil, for $Re = 62$k maximum $Cl/Cd \approx 41$, while for $Re = 31$k maximum $Cl/Cd \approx 20$; this means that by doubling the Reynolds number it is possible to double the aerofoil performance. A closer analysis reveals that for both aerofoils the maximum *Cl/Cd*(*AoA*) is roughly proportional to *Re*.

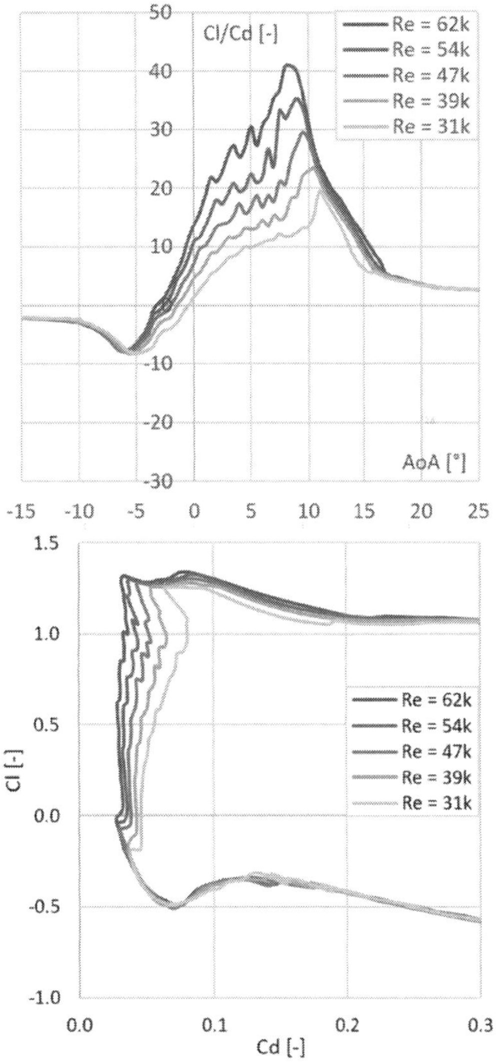

Fig. 1. S826 aerofoil polar plots: *Cl/Cd* (*AoA*) (top) and *Cl*(*Cd*) (bottom).

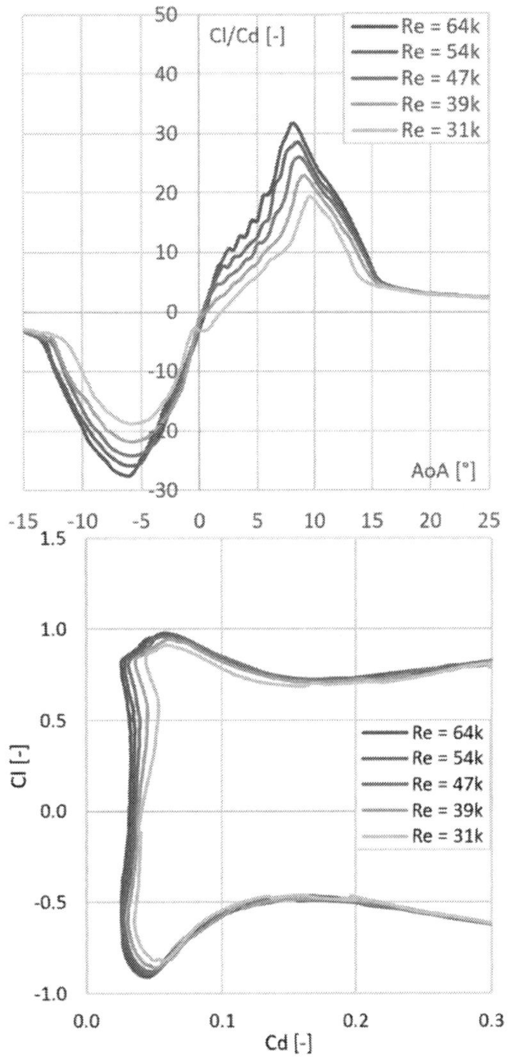

Fig. 2. S834 aerofoil polar plots: *Cl/Cd* (*AoA*) (top) and *Cl*(*Cd*) (bottom).

The same may be stated for the *AoA* corresponding to maximum lift to drag ratio, which value is shifted towards higher values as Reynolds number decreases.

It is also visible that the actual numerical analysis is difficult to conduct properly, judging by the unstable traces of the presented characteristics. This issue may be addressed by interpolating and smoothing the obtained results, however it is important to know of its presence.

Taking into account the rotor aerodynamic performance, it would be best to use S826 along the entire blade span (as it displays the highest lift-to-drag ratio). Thus this aerofoil was applied in the middle and tip sections of the blade. However, due to its relatively low thickness, S826 may be unfit for blade regions exposed to high loads, such as root. To ensure sufficient stiffness of the blade in that region, a higher *b*/*c* ratio aerofoil was chosen, that is S834. Selected aerofoil properties are given in Table 1, while the two selected profiles can be seen in Fig. 3.

The blade was constructed on the basis of the existing geometry mentioned above, modified using numerical tools: CFD solver (ANSYS package) and simplified codes (FAST, XFoil). A parallel examination in Qblade included examination of rotor performance over a range of wind speeds and *TSR* (2-7). The obtained blade design was validated in an experimental process, as presented in the following section.

Table 1. Features of selected NREL low-Re aerofoils.

	S822	S823	S826	S834
Thickness-to-chord ratio *b/c*	0.16	0.212	0.14	0.15
Max *Cl/Cd*	42.84	42.52	60.72	47.42
Corresponding *AoA* [°]	8.4	9.1	8.3	7.1
Corresponding *Cl*	0.921	1.157	1.409	0.909
Corresponding *Cd*	0.022	0.027	0.023	0.019

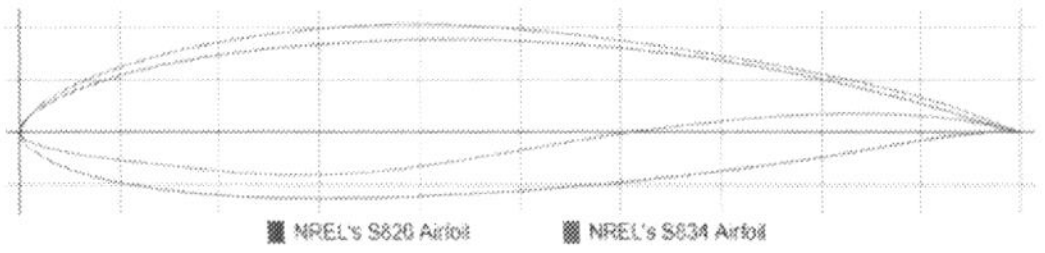

Fig. 3. NREL S826 (green) and S834 (red) aerofoils.

3 Wind tunnel test

The experimental examination of SWT scaled model was conducted at the subsonic, open-test section wind tunnel of the Institute of Turbomachinery, Lodz University of Technology (IMP TUL). The reference velocity value was provided by pneumatic measurements. Prandtl-type Pitot tube was used to measure dynamic pressure at the outlet of the tunnel. The ambient temperature, atmospheric pressure and humidity were used to compute the air density. The whole experimental method aims to evaluate the wind turbine performance c_p as a function of the tip-speed ratio (rotor blade tip velocity divided by wind freestream velocity). Wind turbine speed was actively regulated by a motor-generator connected to a microcontroller in order to control precisely the device rotational velocity. The latter, along with the shaft torque (measured with a torquemeter) provided information concerning the rotor aerodynamic performance. The stages of the experimental procedure are presented in Fig. 4.

The first stage was to complete and prepare the test bench, before commencing the research. Afterwards, the behaviour of the free stream was investigated at selected distances from the outlet, to determine the optimum location of the turbine model in the aerodynamic tunnel. In the next steps, the measurement technique and approach were verified, and then followed the actual study of selected wind turbine models. The results were further recomputed and interpreted, which included determining the uncertainty, the characteristics and conclusions made from the research. A view of the measurement site after mounting the turbine and connecting the measuring apparatus is shown in Fig. 5. Referring to the annotations, the measurement setup consisted of: 1 – rotational speed meter, 2 – digital-analogue converter and data acquisition module, 3 – differential pressure sensor, 4 – temperature-relative humidity sensor, 5 – reference pressure sensor, 6 – rotor, 7 – torque meter, 8 – motor, 9 – servo motor controller, 10 – Prandtl pressure probe.

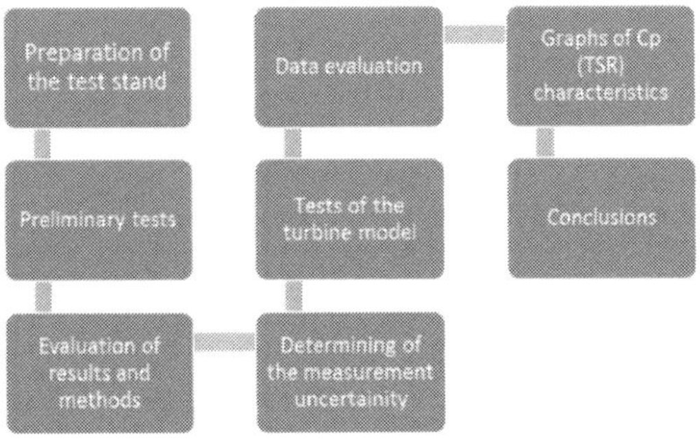

Fig. 4. Sample procedure of wind turbine characteristic experimental assessment.

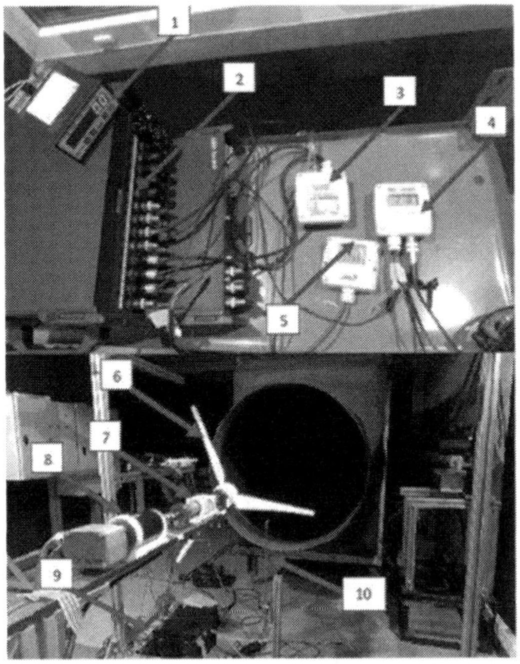

Fig. 5. Wind turbine test bench in IMP TUL wind tunnel.

The measurement procedure commences with setting the flow velocity v and turbine rotational speed ω at the desired values. Data is collected from all transducers (temperature T, humidity RH, dynamic pressure p_d, atmospheric pressure p_a, turbine shaft torque M, rotational speed ω). Flow velocity is computed as:

$$v = \sqrt{2p_d / \rho}, \quad (2)$$

with flow density ρ corrected for the thermodynamic properties of flow. Rotor power is computed as:

$$P = M\,\omega. \quad (3)$$

Dimensionless parameters – tip speed ratio TSR and power coefficient c_p – are computed as:

$$TSR = \omega\, R / v, \quad (4)$$

$$c_p = P / (p_d\, v\, \pi\, R^2), \quad (5)$$

where R is the rotor diameter.

The procedure was repeated over a desired range of v and ω to trace the rotor characteristic. The tests of the scaled wind turbine model were performed in order to verify the results obtained in the numerical simulation process. The models used in the measurements had a diameter of 450 mm and 400 mm. They were manufactured through 3D printing, an emerging method offering excellent repeatability at relatively low cost and time expenses. The datasets were determined separately for each wind speed; the data was approximated by a 4th degree polynomial to accurately reflect the outline of the curve.

At a specific wind speed, the power obtained by the rotor is mainly dependent on the relative speed between the end of the blade and the wind speed (the tip-speed ratio). Both too low and too high TSR results in adverse interaction between rotor and flow. Consequently, part of the wind kinetic energy remains unexploited and/or the rotor performance decreases. Thus each rotor has an optimal tip-speed ratio for which the wind turbine performance attains its maximal value.

4 Results and discussion

These aforementioned characteristics can be seen in Fig. 6. This graph shows the wind turbine performance c_p as a function of wind turbine rotational velocity and wind speed (TSR) for different wind velocities (different Reynolds number calculated based on the rotor diameter). The tested rotor has a diameter of 450 mm. Both experimental and numerical results are presented.

As it was expected, higher Re allows the turbine to reach higher c_p values. One can see that both experimental and numerical results are coincident for sub-optimal rotational velocities ($TSR <$ 2.5). The differences appearing for TSR between 2.5 and 4 may be caused by uncertainty in calculating TSR in the experiment. Along the whole range of rotational velocities, there is a good correlation between simulation and experiment for Re = 360k and Re = 420k. The discrepancies noticeable for Re = 300k may result from the fact, that it corresponds to the aerofoil transitional region, where its performance is especially difficult to assess or predict.

One should bear in mind that the local Re choice for aerofoil characteristics is based on arithmetic average of its predicted values along the entire blade span for optimal TSR; therefore, the discrepancies between the values of optimal TSR may result from the necessity of local Re estimation as a fixed value for the whole blade span. This approach was selected by the authors on the basis of previous experience.

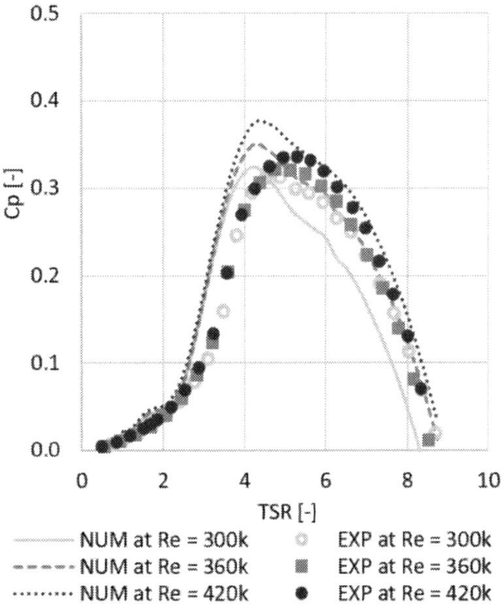

Fig. 6. C_p (*TSR*) characteristics for different *Re*; simulation vs. experiment for NREL-based blade.

The value of rotor idle speed (at which $c_p = 0$) is in agreement between the experimental and numerical results (especially for higher *Re* – 360k and 420k). Quantitatively, the differences between simulated and experimental optimal *TSR* are in the range of 10% to 15%, maximum c_p differences lie in range of 5% to 10%.

It can be concluded that the proper choice of Reynolds number for profile characteristics determination is the most crucial aspect; if chosen too low it may result in significant underprediction of wind turbine performance; if chosen too high – the performance may be overpredicted. Nonetheless, the curve shape could be preserved.

Summing up the above, it may be stated that three main areas of interest have been addressed:

- QBlade applicability in wind turbine (existing) blade analysis at low *Re* was confirmed;
- Verification of the correctness of numerical model assumptions and simplifications was performed; the model limitations were determined;
- Quantification of the level of correspondence of the results in relation to the experiment was carried out and proved to give satisfactory results as far as the design process is concerned.

On the basis of the above observations, we undertook to establish a more aerodynamically efficient blade. In order to do that a new set of aerofoils was chosen, the Wortmann FX series for low-Reynolds applications [9]. The chosen aerofoil, FX 63-137, has a significantly higher camber, as compared to S826. The newly designed rotor was tested in the IMP TUL wind tunnel and compared with the original design (Fig. 7). However, this time the models have a diameter of 400 mm.

It can be seen that the experimental optimal *TSR* is very well predicted by the simulation, as is the maximum c_p value. However, there are significant dissimilarities in *TSR* region above optimal. We should note that all geometrical features were scaled down with respect to earlier considerations; while the printing resolution remained the same. Thus, the quality of model representation deteriorated, especially in regions where small geometrical features are present (such as the blade tip). Also, due to scaling down the chord length, rigidity of the blade, influenced by the thickness, decreased. At increased tip speed ratio (above optimal value) it resulted in sudden drop of performance, due to increasing mechanical efforts. Note that the slenderness and susceptibility to deformation of the newly designed blade was relatively lower than in the previous model, thus the EXP-NUM discrepancies are less significant in this case.

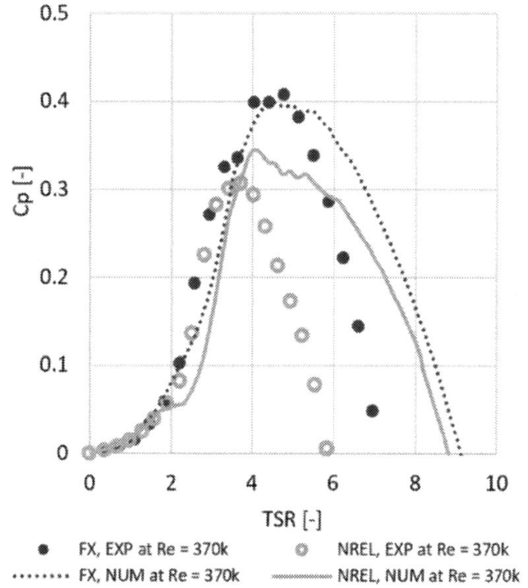

Fig. 7. C_p (*TSR*) characteristics for the same *Re*; simulation vs. experiment for different blade geometries.

Comparing the experimental data for the NREL-based blade at *Re* = 360k, 370k it is visible that the optimal *TSR* differs between the two cases, even though *Re* stays roughly the same. This was attributed to the possible issues with geometrical similarity preservation. Once again, the 3D printing process resolution is the same in both cases. Thus, the obtained relative blade surface roughness is higher for the smaller rotor; also, the quality of blade geometry is lower, as noted above.

It is interesting to point out the significant gain in the performance of the newly-designed blade. In the previous considerations QBlade was used in the analysis of a design developed using different techniques. In this case it remained the principal tool for design and investigation. Thanks to that the rotor performance was increased by as much as 15%, and blade stiffness was strengthened. Although it is doubtful that for a full-scale wind turbine that much increase would be observed, the case proves the principle of applicability of using QBlade as both an analysis and design tool.

5 Conclusions

This paper discusses observations concerning the design and assessment of wind turbine rotors in low-Reynolds number conditions.

The presented data firstly concerns basic blade geometry analysed experimentally (IMP TUL wind tunnel) and numerically (QBlade). It showed the importance of Reynolds number value for SWTs. It also proved that QBlade is a useful tool for assessment of existing rotor blade results. In order to perform a more thorough analysis it is advisable to supplement it with an experimental, independent study method.

In the second part of the paper a new blade design was discussed, based on a new set of aerofoils and designed entirely using the QBlade software. The results seem very promising, as the model tested experimentally showed c_p increase by as much as 15%. This proves the interest in incorporating the simplified model in the design process.

An important conclusion also comes from comparing two rotors of (supposedly) the same geometry, but manufactured in different scales. One should have in mind that especially in the case of small wind turbines, low Reynolds is connected with low-size geometry scales; therefore the Reynolds number similarity criterion may be insufficient. The described technology of manufacturing (3D printing) offers the perfect repeatability of produced objects (blades). However, the scaling of the models may lead to deterioration in the quality of small features representation, what in turn leads to unanticipated geometrical similarity discrepancies. This will project on discrepancies between simulation and the actual, empirical results.

References

1. C. Wang, R. Prinn, Potential climatic impacts and reliability of very large-scale wind farms, Atmos. Chem. Phys., **10**, 2053-2061 (2010)
2. A. Tummala, R. Velamati, D. Sinha, V. Indraja, V. Krishna, A review on small scale wind turbines, *Renew. Sustain. Energy Rev.*, **56**, 1351-1371 (2016)
3. B. Grieser, Y. Sunak, R. Madlener, Economics of small wind turbines in urban settings: an empirical investigation for Germany. Renew. Energy, **78**, 334-350, (2015)
4. K. Sunderland, T. Woolmington, J. Blackledge, M. Conlon, Small wind turbines in turbulent (urban) environments: A consideration of normal and weibull distributions for power prediction, J. Wind Eng. Ind. Aerodyn., **121**, 70-81, (2013)
5. http://www.q-blade.org/ (20.07.2018)
6. L. Sætran, J. Bartl, Invitation to the 2015 "Blind test 4" Workshop, Technical document NTNU, www.sintef.no/globalassets/project/nowitech/events/bt4_invitation2015.pdf (15.07.2018)
7. https://wind.nrel.gov/airfoils/AirfoilFamilies.html (18.07.2018)
8. http://web.mit.edu/drela/Public/web/xfoil/ (18.07.2018)
9. H. Riblett, What Are Wortman Airfoils??, J. Sport Aviation, Nov. 1988

Small wind, big potential: HAWT design case study

Filip Grapow[1,*], *Dominika* Raszewska[1], *Rafał* Skalski[1], *Jeremiasz* Czarnecki[1], *Katarzyna* Telega[1], *Marcin* Miller[1], *Paweł* Rogowski[1] and *Małgorzata* Prociów[1]

[1] Institute of Turbomachinery, Lodz University of Technology, 90 - 924 Lodz, 219/223 Wolczanska Street, Poland

Abstract. In this paper the project of a small wind turbine for urban application has been described. The project of a horizontal small wind turbine presented in this paper was made for the purposes of taking part in ISWTC, but the main target was to develop a turbine for urbanized regions starting in conditions of low wind speed. In this paper a description of aerodynamic and mechanical design are presented; also, control and safety systems are discussed. This turbine has the potential to become a cheap energy generator for household use and can be a competitor of vertical small wind turbines.

1 Introduction

World energy demand is growing drastically every year. Conventional energy sources are no longer sufficient for electricity consumption and renewable energy sources, like wind power or photovoltaics, are becoming an important part of the energy mix [1]. Moreover, in the energy systems all over the world there are more and more "micro installations" which are power sources below 40 kWp [2]. A lot of people want to have their own power plant in their yard or on the roof of their houses so the interest in small wind turbines is now rocketing [3]. The International Small Wind Turbine Contest can be considered as proof of this statement.

The International Small Wind Turbine Contest is held by NHL University of Applied Sciences, Leeuwarden, Netherlands [4]. The goal of this contest is to design and create a SWT (Small Wind Turbine), which has optimal power production in a pre-specified wind regime. In assessment not only generated power is important but also safety during operation, cost efficiency, emergency braking systems and even the visual aspect of the design.

This paper presents the project of GUST (Generative Urban Small wind Turbine) Team from Lodz University of Technology and consists of technical design, aerodynamic design and safety and control system description.

2 Technical specification

The members of the GUST team faced the real-life problem, where, on the one hand, the design of a small wind turbine should be innovative and efficient in order to cope successfully with low wind speeds frequently encountered onshore at hub height just a few meters above ground level, but, on the other hand, because of economic aspects and universality the prototype should not be complex. The team decided to construct and optimize a traditional Horizontal-Axis Wind Turbine (HAWT) [5].

After a sequence of numerical simulations of aerodynamic performance and in-depth analysis of the proposed designs from mechanical point of view, the final three-bladed upwind configuration was chosen (see Fig. 1). A short summary of its key operational parameters is presented in Table 1.

At this point it was impossible to predict the exact rated wind speed of the designed wind turbine so, as explained in the following sections, a value of 8.4 m/s was determined on the basis of calculations done within Simple Load Model analysis proposed in IEC 61400-2 standard [1]. The resulting design parameters like power and torque correspond to this estimated value of wind speed and determined numerically optimal tip speed ratio (TSR).

Fig. 1. 3D CAD model of the designed small wind turbine.

* Corresponding author: filip.grapow@p.lodz.pl

Table 1. Summary of the characteristic features of the designed small wind turbine.

Parameter	Value	Unit
Rotor diameter	1.6	m
Rated wind speed	8.4	m/s
Cut-in wind speed	3	m/s
Cut-out wind speed	11	m/s
Survival wind speed	30	m/s
Rated rotor power	350	W
Rated TSR	5	-
Rated rotational speed	500	rpm
Rated rotor torque	6.2	Nm

3 Aerodynamic design

The aerodynamic part of the design process involved complex blade and rotor geometry design and analysis, based on the design originally introduced by the team of Professor Lars Sætran (NTNU) [6]. As a result of the differences in rotor diameters, the geometry was scaled by a factor defined as a ratio between modified and original rotor radius. The main criterion considered during the design process was to maximize the Annual Energy Production (AEP). Having the wind conditions imposed, the whole construction needed to be adjusted to the specific parameters. The geometry was oriented at lowering cut-in wind speed, increasing the power coefficient and widening the operational range of the Tip Speed Ratios.

An initial part of the procedure was the airfoil selection. Due to a scale factor influence the existing geometry needed to be examined in terms of aerodynamics and material durability. Two substitutional numerical methods (ANSYS CFX and FAST) were used to calculate the turbine's performance, both of them implementing Blade Element Momentum theory.

In order to properly consider the air movement around the airfoil it was necessary to estimate the Reynolds numbers in particular cross-sections:

$$Re = u\, c_i / v_{\text{air}}, \tag{1}$$

where c_i denotes the assumed chord length of a specific cross section, v_{air} is the kinematic viscosity of the air and U_{rated} represents the vector sum of axial wind speed and rotational speed of the blade element on the analysed radial position. In order to simplify the analysis the induced velocity is not taken into account. The evaluation resulted in Reynolds number values within the range of 50000÷150000. Further analysis and the validation process depended on data acquired from XFoil polars (see Figures 2 and 3) [7].

The blade geometry was originally entirely built on NREL S826 airfoil [8], which after taking the size factor into account invariably displayed the highest lift-to-drag ratio. Therefore, midspan and tip cross-sections remained unchanged regarding the airfoil shape. Significant modifications needed to be made in the sections located closest to the hub. Decreasing the chord length resulted in reduced stiffness of the blade root so that a thicker airfoil needed to be implemented there. The NREL S822, S823, S834 was examined to reinforce above mentioned location of the blade. The S834 airfoil was chosen because of the maximum value of C_L to C_D ratio among the above mentioned airfoils (see Table 2.).

The obtained blade design was validated in a series of numerical simulations. The analysis was carried out using FAST software for the whole operational range of wind speeds and the Tip Speed Ratio values between 2 and 7.

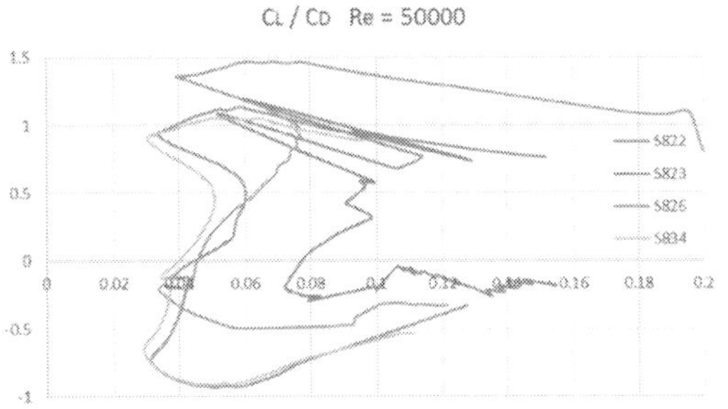

Fig. 2. Polar diagrams of four studied airfoil profiles with $Re = 50000$.

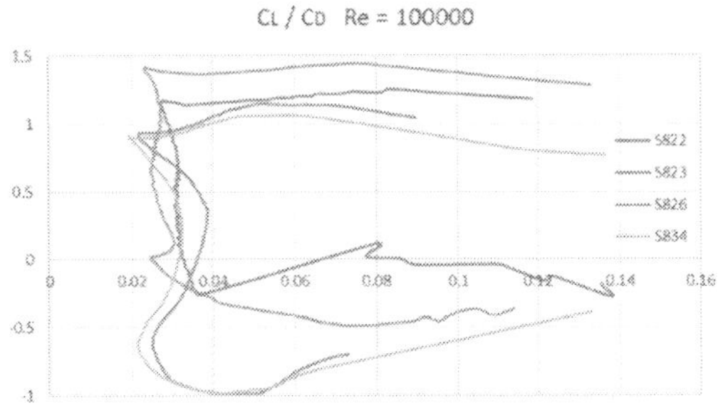

Fig. 3. Polar diagrams of four studied airfoil profiles with $Re = 100000$.

Table 2. Summary of performance parameters of studied airfoils.

	S822	S823	S826	S834
max C_L/C_D	42.84	42.52	60.72	47.42
AoA, °	8.4	9.1	8.3	7.1
C_L	0.921	1.157	1.409	0.909
C_D	0.022	0.027	0.023	0.019

4 Mechanical design

The main tasks of the mechanical design were to:

- Choose appropriate materials for elements,
- Connect generator shaft with the blades (via hub),
- Mount other components of the small wind turbine (like encoder and brake discs),
- Design brake and yaw systems,
- Evaluate stresses in the most vulnerable elements.

The element subjected to the stresses was the hub, which also had to have possibly low moment of inertia so it was chosen to use PA6 aluminum alloy. The blades have to be lightweight but it must also be possible to produce a complicated shape of the blades with a possibly low cost. Technologies like CNC milling and SLS- or SLA 3D printing were considered. Finally, the last method was chosen because SLA 3D printing was as accurate as SLS technology; moreover, it was the most economical

https://doi.org/10.1051/matecconf/201823401005

method. As a material, Protogen White (ABS-like, liquid photopolymer) was chosen. Also parts where stresses are negligible (nacelle, spinner) were 3D printed (less expensive FDM technology) [9].

The shaft of the generator on the rotor side had a keyset. At the end of it a screw with washer were screwed to prevent the hub from axial displacement. The hub was fabricated using CNC machine, which made it possible to get complicated shapes and reduced mass. The blades are mounted in the asymmetric cuts (which prevent improper assembly) using four screws put in the holes in the blades and in the hub. The brake and encoder discs are mounted on the sleeve with a collar placed on the shaft at the second end of the generator. (see Figures 4-6).

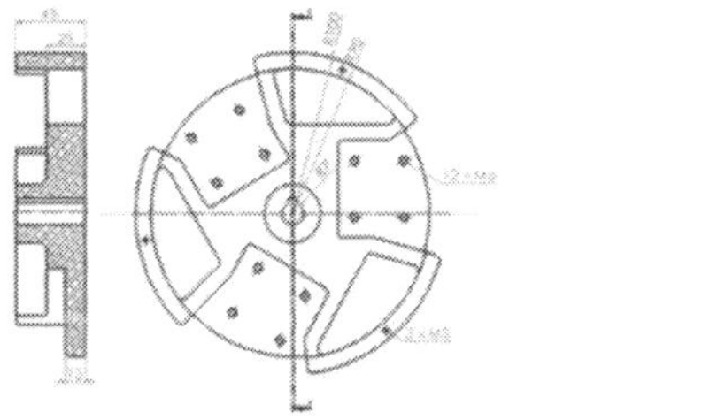

Fig. 4. Rotor hub: cross-section view (left), front view (right).

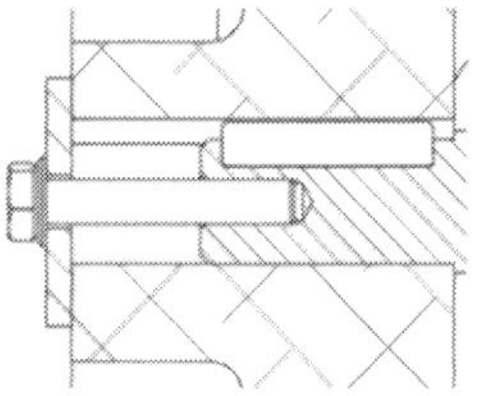

Fig. 5. Hub mounting.

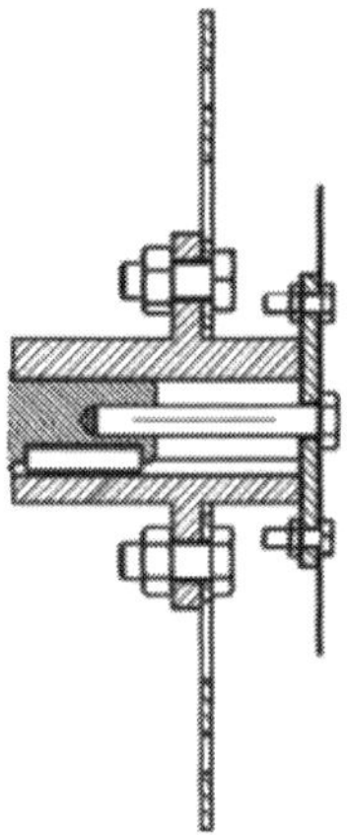

Fig. 6. Brake and encoder disc mounting.

As a brake was chosen a disc brake, which has low moment of inertia. The brake is controlled by a linear actuator through a hydraulic system. The dimensions of the hydraulic systems were calculated assuming that the volume of the hydraulic fluid is constant and its pressure is the same in all places of the hydraulic system [10]. The brake consists of a few simple-shaped elements. Drawings of the brake are presented in Fig. 7. and Fig. 8.

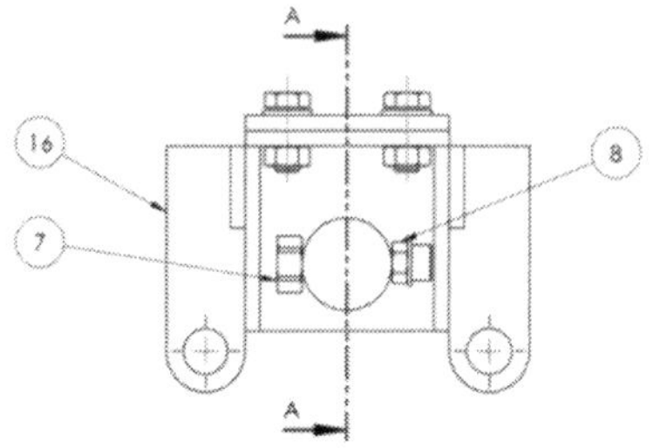

Fig. 7. Designed brake – side.

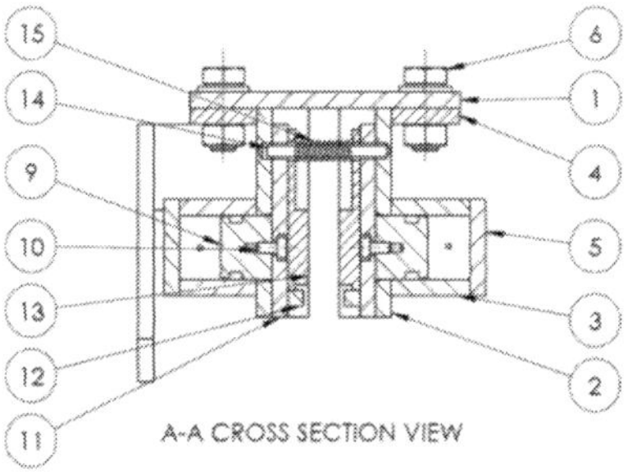

Fig. 8. Designed brake – section.

The yaw system, which adjusts the turbine position in the wind direction, includes the turntable and the tail. The tail was set perpendicularly to the rotor. When exposed to the wind stream, the tail spins into a position parallel to the wind direction. The turntable enables the turbine to rotate thanks to the bearings placed in it. The slip rings were also placed in the turntable to transmit the power from a rotating nacelle outside the turbine (Fig 9.) [11].

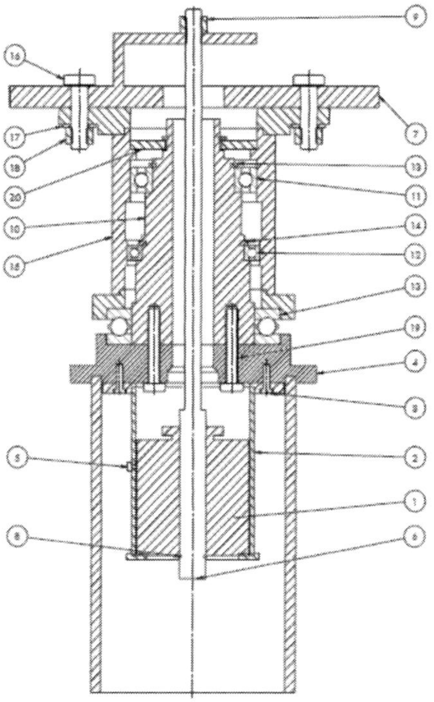

Fig. 9. Designed brake – section.

Blades stresses were assessed using the Simple Load Model. It is presented in the International Electrical Committee (IEC) 61400-2 standard and is applicable for horizontal axis machines with rated power below 50 kW. In all analysed loading conditions described in IEC standard designed turbine has not exceeded permitted values.

5 Generator

The electric generator is a device that makes the conversion from mechanical energy to electrical energy. Handling the intermittency of wind and maintaining the grid reliability are the most important points of small wind turbine generator development [12]. Because of insufficient resources for buying new generator it was decided to use one that was in stock of Institute of Turbomachinery, Lodz University of Technology. The main problem of that solution was that there was no information about its parameters. The test of Power and Torque versus RPM was made using the setup presented in Fig 10.

The generator was driven by a motor with controlled rotational speed measured by the RPM-counter. Also current and voltage were measured. The obtained results were used to create the power (Fig. 11) and torque (Fig. 12) functions.

The power of the tested generator reached around 400 W and the highest torque was around 6 Nm, which means that the generator is acceptable for usage in turbine design conditions.

6 Safety systems

Safety of the proposed SWT design is a crucial part of the construction and a lot of effort was exerted in order to propose a solution that would meet the requirements stated by the organizers of the ISWT Contest 2017. The safety system should:

- be independent and serve as a superior automatic protection for the turbine from single failures or faults in any component within the control system and from faults in the load connection;
- prevent the rotational speed and produced power from exceeding design limits for all load cases;
- allow the turbine to be stopped manually (remotely from at least a distance of 10m) in case of emergency and keep the rotor blocked;
- be integrated within the interface boundaries.

The main requirement for safety system was to design it without overcomplicating it. Simplicity is an advantage because it helps to reduce the number of the potentially weakest links in the system.

The following components make up the prototype of the safety system:

- Microcontroller with an independent battery and charging system;
- RPM and voltage / current monitoring sensor;
- Servomotor for brake actuation;
- Main battery serving as energy buffer for braking system;
- Simple data logger with micro SD card;
- Manual emergency button;
- Set of LEDs.

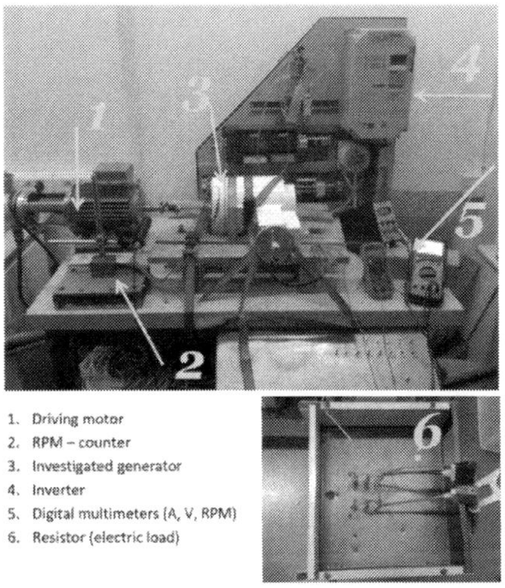

Fig. 10. Test setup for collecting generator power and torque characteristics.

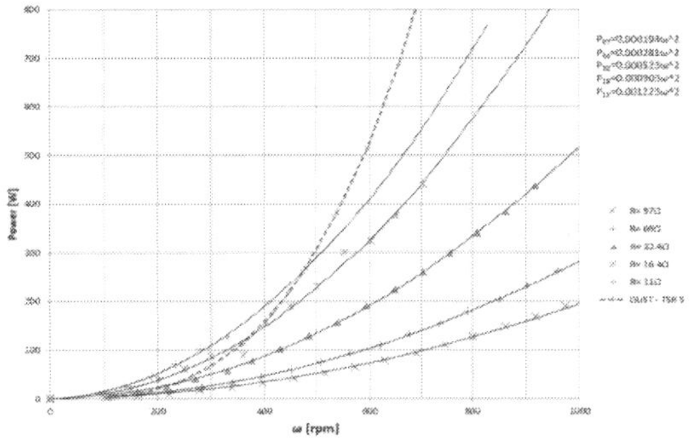

Fig. 11. Results of power vs RPM measurements of the generator supplemented with designed SWT power curve for optimal design TSR = 5.

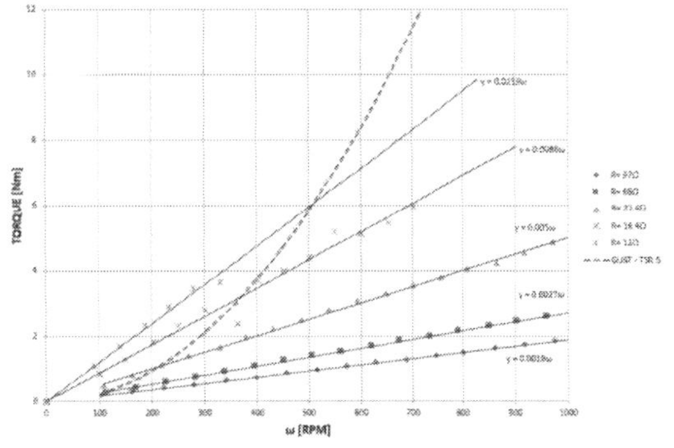

Fig. 12. Results of torque vs RPM measurements of the generator supplemented with designed SWT torque curve for optimal design TSR = 5.

The flow chart illustrating the logic of the whole control system and how it is managed via the microcontroller is presented in Fig. 13.

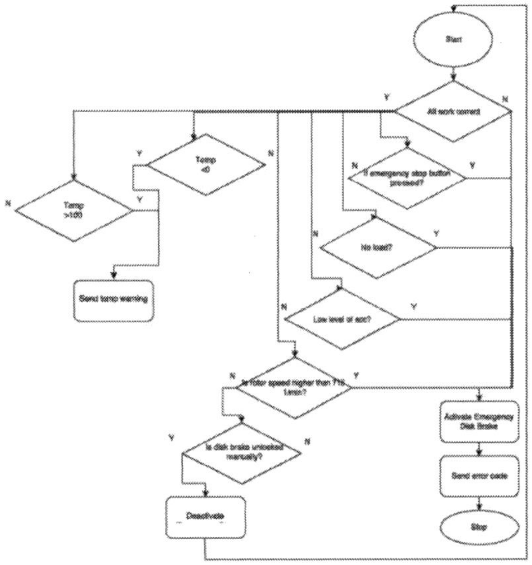

Fig. 13. Flow chart showing the proposed operational procedure for microcontroller.

As a microcontroller, which serves as a superior controller for the entire system, the Arduino pro mini 328 3.3V 8MHz was chosen, one of the most common solutions in these applications [13]. It may be easily programmed via USB, serves both digital (RPM sensor) as well as analog inputs (voltage measurement with a use of simple amplifier). Arduino would be equipped with an independent battery (Li-Po 3.7V) and charging system that would replenish its energy level from the main battery if it drops below a certain fixed level. The safety level was set at 50 percent of the total controller battery capacity.

At first the microcontroller checks whether all the sensors are active and then if any of measured parameter is not off the limit. For example checks if high temperature of the electronic system is detected that may lead to its malfunction. This might happen if the machine is exposed to sunlight on a scorching day for too long. Simultaneously, the controller checks if the manual emergency button is pressed or if there is enough energy stored in order to keep the electronics running. If any of the above conditions is not met the emergency brake will be activated preventing the turbine from starting to spin. This action is signalled by turning on an adequate LED in order to inform the users about the fault.

The same will happen if the maximum rotational speed of 800 RPM is exceeded or the voltage monitoring sensor indicates lack of load. The servomotor will start to clamp the actuators on the brake disc and bring it to stop. In order to unblock the rotor, the operator has to do it manually (using a remote control system – one featuring an emergency button).

Additionally, a measured data acquisition system was installed. For this task, a data logger with micro SD slot was used in the control system. For example, a record of the rotational speed measurement by a photo interpreter located just behind the brake disc is saved in the mentioned memory card.

7 Conclusions

The prototype of a small wind turbine, whose design was presented in this paper, was built and tested in the aerodynamic tunnel in Delft University of Technology, the Netherlands. The results of the test showed that all systems of the prototype worked properly, but there were some aspects that still needed improvement.

The first one was to design a system which would permit to adjust the aerodynamic power curve to the electrical power curve to increase the efficiency of producing electrical energy.

The second one was to decrease the inertia of the rotor and friction in the bearing system to decrease the cut-in wind speed.

The last one was to improve the control and safety system to make it possible to collect information from the sensors and save it in the cloud to have the possibility to access that information online.

References

1. WindEurope, Wind in power 2017, Annual combined onshore and offshore wind energy statistics, windeurope.org, Brussels, Belgium, (2018)
2. I.61400-2, Wind turbines – Part 2: Design requirements for small wind turbine (2006)
3. A. Tummala, R. Velamati, D. Sinha, V. Indraja, V. Krishna, A review on small scale wind turbines, Renew. Sustain. Energy Rev., **56**, 1351-1371 (2016)
4. http://www.iswtc.com/ (28.07.2018)
5. P. Gipe, *Wind energy for the rest of us: a comprehensive guide to wind power and how to use it*, (Bakersfield, CA: wind-works.org, 2016)
6. L. Sætran, J. Bartl, *Invitation to the 2015 "Blind test 4" Workshop*, Technical document NTNU, www.sintef.no/globalassets/project/nowitech/events/bt4_invitation2015.pdf (15.07.2018)
7. M. Drela, *XFOIL: An analysis and design system for low Reynolds number airfoils. Low Reynolds number aerodynamics*. Lecture Notes in Engineering, **54** (Springer, Berlin, Heidelberg 1989)
8. https://wind.nrel.gov/airfoils/AirfoilFamilies.html (18.07.2018)
9. K. Olasek, P. Wiklak, Application of 3D printing technology in aerodynamic study, J. Physics: Conference Series, **530** (2014)
10. D. Elger, B. Williams, C. Crowe, J. Roberson, *Engineering Fluid Mechanics*, 1-692 (2013)
11. L. Kurmaz, *Projektowanie węzłów i części maszyn*, Kielce, Poland, 1-449 (2007)
12. M. Hosain, M. Ali, *Future research directions for the wind turbine generator system*, Renewable and Sustainable Energy Rev., **49**, 481-489 (Elsevier, 2015)
13. F. Mahmuddin, A. Yusran, S. Klara, On the use of an Arduino-based controller to control the charging process of a wind turbine, *AIP Conference Proceedings*, **1814**(1) (2017)

Identification of BDS Registers

Tereza Topková[1,*], and *Stanislav* Pleninger[1]

[1]Czech Technical University in Prague, Department of Air Transport, Horská 3, 128 03 Prague, Czech Republic

Abstract. The SSR Mode S technology is currently one of the major means of surveillance in ATM. Addressed surveillance enables the elicitation of replies from Mode S transponder on board of an aircraft containing different flight information in the so-called BDS registers. The type of BDS register as well as its content is characterised by a number of the BDS registers which is determined according to SSR interrogation by default. The study deals with the development of a heuristic algorithm for identification of a particular BDS register with the assumption that the radar query is unknown. A BDS number is assigned to the appropriate message on the basis of the unique bit sequences and other criteria. The algorithm provides the opportunity to obtain various flight data from the aircraft by only receiving addressed surveillance replies on 1090 MHz frequency.

1 Introduction

SSR Mode S addressed surveillance is being used across Europe and other regions with high traffic density to ensure supplementary information which exceeds ATM data and is suitable for ATS. There is an assumption for the system functionality: both subjects shall be equipped with corresponding Mode S equipment, which consists of the SSR Mode S interrogator and the aircraft SSR Mode S transponder. The carrier frequency of the interrogation is established at 1030 MHz and the carrier frequency of the reply transmission shall be 1090 MHz [1, 2].

In response to Mode S interrogation there are four Mode S reply categories transmitted:

- Mode S all-call replies (DF 11);
- Surveillance and standard-length communication replies (DF 4, 5, 20 and 21);
- Extended length communication replies (DF 24);
- Air-air surveillance replies (DF 0 and 16) [3].

The category with downlink format 20 and 21 is represented by long interrogation and replies which contain the BDS (Comm-B Data Selector) register, providing Comm-B is considered a 112-bit reply containing the 56-bit message field. This field is used by the downlink standard length message, ground-initiated and broadcast protocols [2]. The BDS register represents the 56-bit MB part of the received message, which may contain different information from real operation depending on a BDS register number, for example, velocity, heading, meteorological information, etc. [4].

Required flight data can be provided to the ground station through the GIBS protocol which elicits a Comm-B message from a Mode S transponder on board the aircraft. This transponder may obtain a 56-bit buffer with real-time information from aircraft avionics, such as aircraft state data within 255 transponder registers. These 56-bit sequences are known as BDS registers such as GIBS[20]. If the contained information has not been updated within the allocated time, the register shall be deleted [2, 5].

The replies may be obtained by any receiver getting data on 1090 MHz frequency. However, the number of the received BDS register is not possible to decode from specific part of the message, because there is no identification which may determine the interrogated register number in the 112 bits of the reply. The required BDS register is recognised by knowing the initial interrogation.

The ATM laboratory at the Department of Air Transport operates ADS-B receivers which constantly collect any messages transmitted by Mode S transponders on frequency 1090 MHz. Only the replies induced by the SSR Mode S interrogation are received. Thus, the query with the initially required BDS register number is unknown. In order to acquire the information from the replies on the SSR Mode S addressed interrogation, a heuristic algorithm is created to distinguish passively received BDS codes. The algorithm is evaluated through a comparison with data from real operation provided by ANS CR.

2 Examined BDS registers

All aircraft operating IFR/GAT in Europe are required to carry and operate Mode S Level transponder(s) with Mode S Elementary Surveillance (ELS) capability. All State aircraft operating IFR/GAT in Europe are required to have aircraft with the Mode S ELS equipment by the 7th December 2017. All aircraft operating IFR/GAT in Europe, MTOW exceeding 5700 kg or with a maximum cruising TAS capability greater than 250 knots are required to be equipped with and operate Mode S transponder with Elementary Surveillance (ELS) and

* Corresponding author: topkoter@fd.cvut.cz

Enhanced Surveillance (EHS), (as well as ADS-B 1090 MHz extended squitter). This regulation is applicable for aircraft with a certificate of airworthiness first issued on or after 8th June 2016, and the rest of the aircraft shall be retrofitted by 7th June 2020 [6-8]. Besides BDS registers involved in ELS and EHS, there is a concentration on MRAR (Meteorological routine air report) for obtaining more accurate meteorological data. There are currently no European regulations requiring MRAR equipment. Nevertheless, there is a small number of SSRs using MRAR technology and a small proportion of aircraft (around 2-3%) equipped with Mode S transponder providing MRAR registers [1].

Almost the entire European region is covered by the SSR system, which enables Mode S ELS to interrogate. Mode S EHS is based on the same principle as ELS, and the number of SSR EHS interrogation has enormous potential to be upgraded from ELS in the future. The MRAR BDS register $4{,}4_{16}$ is currently interrogated by every second radar rotation over the Czech Republic [1]. Unfortunately, there is no interrogation of BDS code $4{,}5_{16}$.

2.1. ELS

Mode S Elementary Surveillance shall provide transponder parameters such as ICAO 24-bit aircraft address, SSR Mode 3/A, altitude reporting in 25ft increments and flight status (airborne/on the ground). In addition, BDS registers shown in Table 1 are required to be transmitted within ELS.

Table 1. ELS BDS registers [1, 9].

No.	BDS register	BDS register Name
1	$1{,}0_{16}$	Data link capability report
2	$1{,}7_{16}$	Common usage GIBC capability report
3	$2{,}0_{16}$	Aircraft identification
4	$3{,}0_{16}$	ACAS active resolution advisory

2.2. EHS

The main objective of implementing EHS BDS register is to improve ATC systems of current and real-time information in order to predict the future trajectory of the flight. Every Mode S Enhanced Surveillance transponder shall provide all services of ELS automatically. EHS is extended by parameters known as DAPs (Downlink Airborne Parameters), it ensures additional data for the air traffic controller (CAPs – Controller Access Parameters), and for the upgrade of ATM systems functionality (SAPs – System Access Parameters) [2]. GIBS protocol shall enable transmission of BDS registers, which are specified in Table 2.

Table 2. EHS BDS registers [1, 9].

No.	BDS register	BDS register Name
1	$4{,}0_{16}$	Selected vertical intention
2	$5{,}0_{16}$	Track and turn report
3	$6{,}0_{16}$	Heading and speed report

2.3. MRAR

Meteorological Routine Air Report registers are neither part of ELS, nor of EHS. They may however provide meteorological data obtained directly from sensors, which an aircraft may be equipped with. There is no need for further data processing in a ground station. Two BDS registers may be transmitted in Mode S replies shown in Table 3. Based on the contained parameters in BDS registers $5{,}0_{16}$ and $6{,}0_{16}$, some meteorological information may be obtained by indirect calculation. The resulting data does not ensure high level of accuracy [10].

Table 3. MRAR BDS registers [1, 9].

No.	BDS register	BDS register Name
1	$4{,}4_{16}$	Meteorological routine air report
2	$4{,}5_{16}$	Meteorological hazard report

3 Heuristic algorithm

The heuristic algorithm was created to determine the relevant BDS register number and is specified by a MATLAB script and its supplementary function. Before a received BDS register is tested by the algorithm, the initial processing is necessary to obtain data samples in a required form. A suitable input for the algorithm is considered a 56-bit array of a binary sequence from a received message with DF 20 or DF 21, which specifies the reply on the Mode S addressed interrogation. The result of the algorithm assigns a BDS register number to the examined 56-bit register when all appropriate conditions are met.

3.1. Assigning a BDS number according to unique parts

Initially, the bit analysis was made to find unique bits and combinations of bits which determine a specific BDS register to enable distinguishing particular BDS registers. According to the data structure in every register, different conditions were found and implemented into the algorithm. They may be represented, for example, by the group of the first eight bits of the message, so-called BDS1 code (1.-4. bit of the 56-bit sequence) and BDS2 code (5.-8. bit). These codes characterise the number of BDS register. Therefore, the first four bits of BDS register $1{,}0_{16}$ must be decoded as hexadecimal value "1" and the BDS2 code as "0". Unfortunately, this condition is not applicable for all registers. Other unique parts create bits stated as reserved which have to be set at zeros, or some groups of bits which can take only a limited value. In addition, availability of flight information in part of the register is specified by status bit and its following data field containing flight information [11]. This implies a status bit condition, where the status bit is set at "0", the following bits representing appropriate data field must be padded with zeros as well, otherwise the status condition is not met. Afterwards, some of the data fields are decoded and compared with a limitation of their values which were determined according to decoded BDS registers from real operation provided by ANS CR. The status bit test and structure formed by status bits and data field is typical for EHS and MRAR register.

A 56-bit binary sequence is tested by every possible condition to assign or exclude an assignment of the all examined nine BDS registers in one for-loop. Gradually, a message is examined whether it meets all conditions for one BDS register number after another. While all tested bits are set at required values, the number of tested BDS register is saved into a variable corresponding with a row of the bit sequence. It follows that there can be more BDS numbers assigned to one message. The basic structure of the created algorithm is shown in the Fig 1.

3.2. Final BDS number assignment

In order to assign only one final BDS number to every message, part of the script is created for resolution of conflicts where more than one BDS number is estimated to the tested binary sequence.

The direct BDS number assignment is processed by messages where no BDS number was estimated and the final BDS number is regarded as "Unknown", on the assumption that this BDS register is not included in any group of tested BDS codes or the message is incorrectly received. If there is more than one estimated BDS number by BDS code $1{,}0_{16}$, as well as by BDS $2{,}0_{16}$, it is always considered the final number. The last direct assignment is applied when there is only one guessed BDS number, with the exception of BDS register $4{,}5_{16}$. While this register is not transmitted, the messages with its assigned number are also marked as "Unknown". When a conflict of two or more estimated BDS numbers occurred, there was another data analysis processed to find out, whether there are any other conditions to be explored to make a final decision. Based on conflict analysis of provided data from real operation, an additional condition was not needed in most of the cases and the double assignments had been solved as the final allocation of one of the registers.

Variable "BDS final" represents the output of the algorithm where one appropriate BDS number is assigned to the corresponding row of the tested message. All original messages of distinguished BDS registers may be saved into a MAT-file which enables easy accessibility of the data for possible decoding or further processing.

4 Evaluation of the algorithm results

A sample of data from real operation provided by ATN CR was chosen for evaluation of the heuristic algorithm results. Every assigned BDS number by the algorithm is compared to the known number of an appropriate message from the ATN data sample. It makes it possible to evaluate the algorithm for six of nine examined registers.

On the condition that the BDS number identified by the heuristic algorithm corresponds to the known BDS number from the ASTERIX data, this BDS number is regarded as a correctly assigned one. Conversely, a wrong assignment is considered to be a different BDS number assigned to the known BDS register by the program. In other words, the BDS register is incorrectly identified or marked as unknown.

The number of an erroneously allocated BDS is calculated as the difference between a number of the BDS register assigned by the algorithm and a number of a known BDS number within the ATN data enlarged by the difference between a number of the known BDS number from ATN data and a number of the correct assigned BDS number. For this reason, the number of wrong assignments may exceed the original number of the BDS number allocations. Thus, the sum of both rates may exceed 100%, as well as the error rate itself.

The success rate is simply calculated according to (1). Similarly, an error rate can be evaluated as in (2).

$$Success\ Rate = \left(\frac{Correctly\ Assigned\ BDS\ Number}{ASTERIX\ BDS\ Number}\right) \cdot 100\% \quad (1)$$

$$Error\ Rate = \left(\frac{Wrongly\ Assigned\ BDS\ Number}{ASTERIX\ BDS\ Number}\right) \cdot 100\% \quad (2)$$

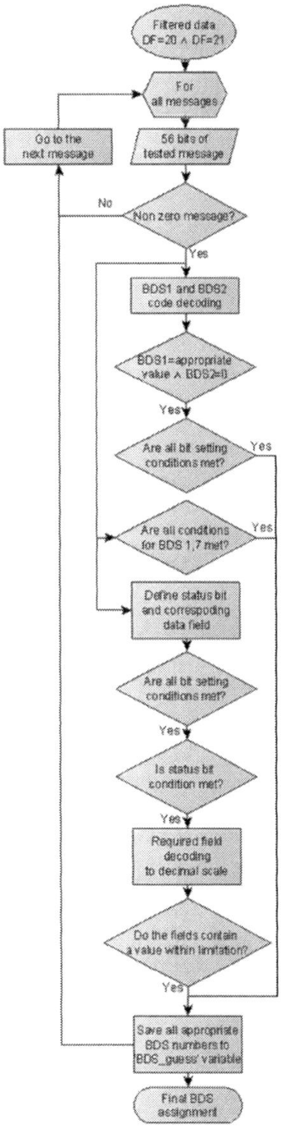

Fig 1. The Algorithm basic structure.
Note: No solutions are not processed.

Table 4. Algorithm evaluation.

BDS code number	Algorithm Assigned	ASTERIX BDS	Correctly Assigned	Wrongly Assigned	Success rate	Error rate
BDS code $1{,}0_{16}$	122	50	48	74	96,00%	148,00%
BDS code $2{,}0_{16}$	0	0	0	0	-	-
BDS code $3{,}0_{16}$	0	0	0	0	-	-
BDS code $4{,}0_{16}$	14033	14046	14033	13	99,91%	0,09%
BDS code $5{,}0_{16}$	7806	7843	7805	75	99,52%	0,96%
BDS code $6{,}0_{16}$	15163	15124	15124	39	100,00%	0,26%
BDS code $4{,}4_{16}$	239	239	239	0	100,00%	0,00%
BDS code $4{,}5_{16}$	0	0	0	0	-	-
BDS code $1{,}7_{16}$	404	308	308	96	100,00%	31,17%
Unknown codes	564	721	551	327	76,42%	45,35%
Total	38331	38331	38108	624	99,42%	1,63%

Both rates were evaluated for the six BDS numbers which were possible to compare, as well as for unknown BDS registers. With the exception of the unknown BDS numbers, the success rate exceeds 95% on a case by case basis. The total success rate of the BDS number assignment by the heuristic algorithm is equal to 99,42%. In less than two percent of the cases, the BDS number may be assigned incorrectly. The overview of the algorithm results is given in Table 4.

5 Discussion

According to the expectations, most of the identified registers were identified as BDS registers of the Enhanced Surveillance, followed by the messages of the Elementary Surveillance and the least occurring registers were the meteorological BDS codes. The probability of the correct BDS number assignment may be increased with further analysis of the remaining registers which is of no interest for the present research, although they are certainly transmitted within the SSR Mode Addressed surveillance.

The algorithm output makes it possible to obtain additional flight information from the aircraft by further data decoding. Distinguishing BDS registers are used for the next research activities for the ATM laboratory at the CTU in Prague. The decoded data is considered crucial for further analysis, statistics or databases.

6 Conclusion

The algorithm provides the opportunity to determine a response to Mode S addressed interrogation without knowledge of an uplink query, based only on the data analysis of ELS, EHS and MRAR BDS registers. It is not necessary to have any other information about the aircraft with the identical ICAO 24-bit address. The algorithm is able to assign the BDS number to an appropriate message for eight BDS registers transmitted by default, as well as BDS number $4{,}5_{16}$ for messages containing meteorological hazardous data. One of the biggest advantages of the algorithm is the evaluation of correct assignment of BDS number with high quality of achieved results.

The success rate of the BDS number correct assignment was calculated at more than 99%. Conversely, the wrong BDS number may be assigned in less than 2% of the cases. The algorithm is suitable for analysis of large amount of received messages and it can provide essential sources for other research of detailed flight characteristics.

References

1. J. Frei, Moznosti ziskani a vyuziti meteo dat z BDS registru pro potreby uzivatelu vzdusneho prostoru, *Doctoral thesis - CTU in Prague*, (2016)
2. ICAO, Doc. 9684: Manual of the Secondary Surveillance Radar (SSR) Systems, *AN/951*, ISBN 92-9194-333-9, (2004)
3. ICAO, Annex 10 Aeronautical Telecommunication, Volume IV (2014)
4. ICAO, Doc 9871: Technical Provisions for Mode S Services and Extended Squitter, *AN/464,* ISBN 978-92-9231-117-9, (2008)
5. ICAO, Doc 9688: Manual on Mode S Specific Services, AN/952, ISBN 92-9194-407-6 (2004)
6. Commission Implementing Regulation (EU) No 1207/2011 of 22 November 2011 laying down requirements for the performance and the interoperability of surveillance for the single European sky (22.03.2017)
7. Commission Implementing Regulation (EU) No 1028/2014 of 26 September 2014 amending Implementing Regulation (EU) No 1207/2011 laying down requirements for the performance and the interoperability of surveillance for the single European sky (22.03.2017)
8. Aircraft Equipage Requirements in the European Commision IRs 1207/2011 and 1028/2014 (22.03.2017)
9. ICAO, Annex 10 Aeronautical Telecommunication, Volume III (2016)
10. S. de Haan, J. Sondij, H. Philippi, B. Strajnar, A. Hoff, F. Besson, H. Vedel, J. a. Garcia-Moya., E. Stone, E. Riskila, Final report of the EUMETNET ADDD FS ET on the feasibility to initiate a new EUMETNET activity for aircraft derived observations under the EUMTENET observation program (2015)
11. R. Grappel, G. Harris, M. Kozar, R. Wiken, Elementary Surveillance (ELS) and Enhanced Surveillance (EHS) Validation via Mode S Secondary Radar Surveillance, *Project report ATC-337, ESC-TR- 3007-054, Lincoln Laboratory, Massachusetts Institute of Technology* (2008)

Checking the correct operation of main measuring instruments on the flight model and prototype of AIST small spacecraft

Andry Sedelnikov[1], *Ekaterina* Khnyryova[2,*], and *Tatiana* Ivashova[3]

[1]Samara National Research University, Department of Space Engineering, 34 Moskovskoyeshosse, Samara, Russia
[2]Samara National Research University, Department of Further Mathematics, 34 Moskovskoyeshosse, Samara, Russia
[3]Samara National Research University, Faculty of Electronics and Instrument Engineering, 34Moskovskoyeshosse, Samara, Russia

Abstract. Flight model and prototype of AIST small spacecraft were launched in 2013 and 2014 respectively. They were used for carrying out various scientific researches, as well as testing the operation of AIST series small spacecrafts. They were operated in an uncontrolled flight and did not have a fully-featured orbital motion control system. However, it has been possible to reduce the angular velocity by magnetorquers. To obtain the control laws, the angular velocity of small spacecraft was estimated by means of magnetometer sensors. On the flight model of AIST small spacecraft, the angular velocity reduction mode was used twice. In the first case, the angular velocity was reduced, in the second case, it did not change significantly. On the prototype of AIST small spacecraft, the angular velocity reduction mode was used three times. In all three cases, a significant increase in the angular velocity of small spacecraft rotation was observed. The paper describes a test for checking the magnetometer sensors correct operation. Based on this test, conclusions about the measuring equipment correct operation are made. Ineffective control of the angular velocity of small spacecraft rotation is most likely due to failure to take account of impact of scientific and supporting equipment on the measuring instruments.

1 Introduction

Flight model and prototype of AIST small spacecraft were identical in design. However, during their operation, the magnetorquers showed different efficiency in solving the problem of reducing the angular velocity of small spacecraft rotation. In [1] it is noted that the prototype had a significant initial angular velocity when it fell-away. In all other aspects, the operating conditions of both small spacecrafts were the same. It is assumed that one of the reasons for the magnetorquers ineffective work is insufficient quality of raw data. The parameters estimation errors of the small spacecraft rotational motion led to the obtaining a non-optimal control law. One of the possible explanation for this situation is the incorrect operation of the measuring instruments. Such incorrect operation can be caused by the failure of the measuring instruments and by the impact of scientific and supporting equipment on the measuring instruments. The proposed test for checking the correct operation will allow to determine possible failures of measuring instruments. At testing, it is necessary to study in detail the impact of scientific and supporting equipment in all normal modes of its operation on the measuring instruments in order to answer the question about measuring data accuracy.

2 Methods

A series of investigations is devoted to the checking the correct operation of magnetic measuring instruments on the flight model and prototype of AIST small spacecraft [2-5]. In the paper we propose a test for checking the correct operation based on the following assumptions.

The average orbital value of the induction vector's modulus of Earth's magnetic field is constant.

The average value of the measuring error is approximately the same for all measurement sessions. Then the following limitations must be satisfied with an unlimited increase in the measurements number:

$$\lim_{n\to\infty}\left(|\bar{\vec{B}}|-|\hat{\bar{\vec{B}}}|\right)=0,$$

$$\lim_{n\to\infty} D|\hat{\bar{\vec{B}}}|=0$$

where $|\bar{\vec{B}}|$ is the average orbital value of the induction vector's modulus of Earth's magnetic field; $|\hat{\bar{\vec{B}}}|$ is sample average value of the induction modulus; n is the measurements number; D is variance.

*Corresponding author: khnyryova@gmail.com

2.1. Test for measuring instruments of the AIST small spacecraft flight model

For the flight model of AIST small spacecraft it is typical a stable decrease in the average orbital value variance of the induction vector's modulus of Earth's magnetic field. However, the number of measurement sessions was insufficient to prove the correspondence between the average orbital value variance and the measuring error. There was much more measuring data for the prototype of the AIST small spacecraft, which allowed more detailed investigation the correct operation of measuring instruments.

In Table 1 the main data parameters are shown which were used to check the correct operation of measuring instruments on the flight model of AIST small spacecraft.

Table 1. The samples parameters of Earth's magnetic field measurement sessions on the AIST small spacecraft flight model.

Date	Sample size	Number of turns per measure-ment session	Total sample size	
			1	2
27.04.13	1150	1.2	1149	1149
29.04.13	400	0.6	1549	1548
10.05.13	1350	1.7	2899	2898
14.05.13	2350	2.6	5248	5246
16.05.13	1699	2.1	6726	6717
20.05.13	1750	2.3	8434	8426

The dynamic pattern of the average orbital value of the induction vector's modulus of Earth's magnetic field and its variance are shown in Figures 1 and 2.

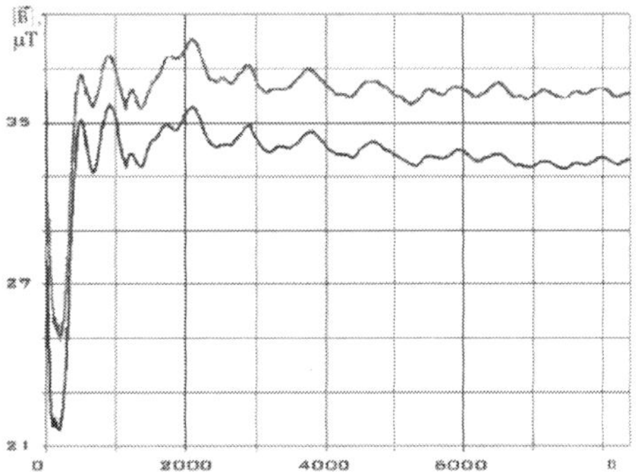

Fig. 1. The dynamic pattern of the average value of the total sample for sensor No. 1 (lower curve) and sensor No. 2 (upper curve).

Analysis of these dependences shows that the sample variance decreases with increasing sample size. The error of indirect measurements of the induction vector's modulus of Earth's magnetic field is $0.7\mu T$ approximately. The dependencies presented in Fig. 2 do not allow to conclude that the variance of the average orbital value of the induction vector's modulus of Earth's magnetic field for all used measurements is explained by the declared measuring error of separate measuring instruments. Measuring error is $\pm 0.5\mu T$ on the passport. If we assume that the indirect measurements error is approximately equal to the mean-square deviation, then in Fig. 2 the variance of $0.5\mu T^2$ can be explained by the measurement error. The minimum variance for sensor No. 2 by the total sample (Table 1) is $2.4\mu T^2$ approximately. The variance estimate of the average orbital value is tenable, but it can be shifted on small samples. Therefore, the sample size of 8400 points does not allow us to state unambiguously that the variance in Fig. 2 is the measurement error.

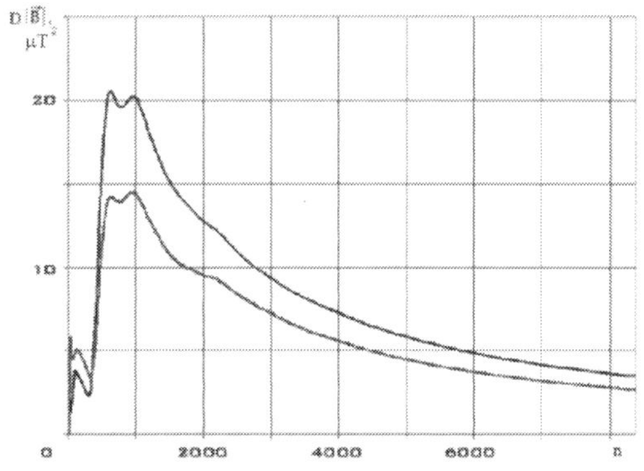

Fig. 2. The dynamic pattern of sampling variance of the total sample for sensor No. 1 (upper curve) and sensor No. 2 (lower curve).

2.2. Test for measuring instruments of the AIST small spacecraft prototype

For the AIST small spacecraft prototype, all the measurement data were divided into two types. The first type contained the orbital eclipse part of the small spacecraft, the second type included data obtained only in the orbital sunny part of the small spacecraft. The measurements from sunny part were used to check the correct operation of main measuring instruments. The electrical battery was always charged and it can be concluded that the supporting equipment had a uniform effect on the measuring instruments. In the case of both sunny and eclipse parts of the orbit, that effect would be non-uniform due to the lack of charging current in the eclipse part of the orbit.

It is also necessary to take into account the feature of the measurement channel X1 of the sensor No. 1, noted in [3]. The investigations presented in this paper do not allow us to recognize measurements as abnormal. However, in almost all normal modes of operation, the average value for the corresponding component of the magnetic induction vector was the last in the ranked series of average values of six measurement channels. In addition, the variance was the minimum for all channels. None of the three criteria for checking abnormal values used in [4] made it impossible to recognize the average value and variance of the channel X1 as the outlier. Such feature on a small part of the measurements could be explained by the low angular velocity of small spacecraft rotation [5]. As a result, the components of the induction vector's modulus of Earth's magnetic field in the instrumental bound coordinate system would change insignificantly. However, in [1] was noted that the angular velocity of the AIST small spacecraft prototype rotational motion was higher than the angular velocity of the AIST small spacecraft flight model rotational motion due to giving it the initial angular velocity. Measurements were made over a long period.

In Table 2 the main measuring data parameters are shown for the AIST small spacecraft prototype.

Table 2. The samples parameters of Earth's magnetic field measurement sessions on the AIST small spacecraft prototype.

Date	Sample size	Number of turns per measurement session	Total sample size	
			1	2
11.01.14	1850	2.0	1848	1849
14.01.14	1704	1.9	3550	3552
17.01.14	1459	1.8	5004	5007
24.01.14	1500	1.8	6500	6506
28.01.14	3328	3.7	9825	9830
31.01.14	2000	2.4	11820	11828
07.02.14	3449	3.7	15265	15274
11.02.14	1850	2.0	17111	17122
15.02.14	392	0.3	17499	17509
19.03.14	1863	2.7	19360	19372
25.03.14	2187	2.8	21543	21563
28.03.14	1331	2.7	22843	22860
07.04.14	1561	2.3	24401	24422
27.06.14	3115	4.0	27513	27538
18.07.14	2789	3.9	30297	30326
23.07.14	2241	4.2	32519	32546
25.07.14	3005	4.0	35520	35553
01.08.14	2725	4.0	38240	38274
05.08.14	2643	4.0	40880	40915
12.08.14	2617	3.9	43490	43522
15.08.14	3141	4.0	46626	46666
21.08.14	2321	2.9	48943	48985
14.10.14	3425	4.0	52358	52400

The dynamic pattern of the average orbital value of the induction vector's modulus of Earth's magnetic field and its variance are shown in Figures 3 and 4.

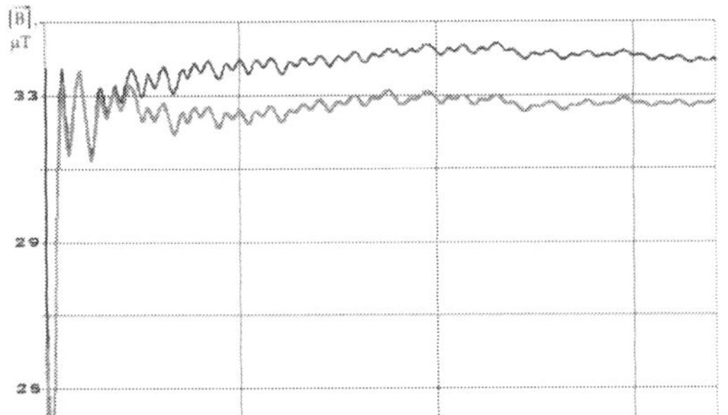

Fig. 3. The dynamic pattern of the average value of the total sample for sensor No. 1 (upper curve) and sensor No. 2 (lower curve).

Measuring instruments on the flight model and prototype of AIST small spacecraft were identical. And the error of indirect measurements of the Earth's magnetic field induction vector's modulus is also $0.7\mu T$. Unlike Fig. 2, in Fig. 4, the variance minimum value for sensor No. 2 is about $0.4\ \mu T^2$. As measurement error we can still explain the variance of $0.5\mu T^2$. Consequently, a significant increase in the sample size from 8,400 to 52,400 measurements changed the situation. The presented results allow us to make a conclusion about fault-free performance of the sensor No. 2. To investigate the effect on the sensor of scientific and supporting equipment it is necessary to test the separate vector components, in addition to the integral test for the induction vector's modulus of Earth's magnetic field.

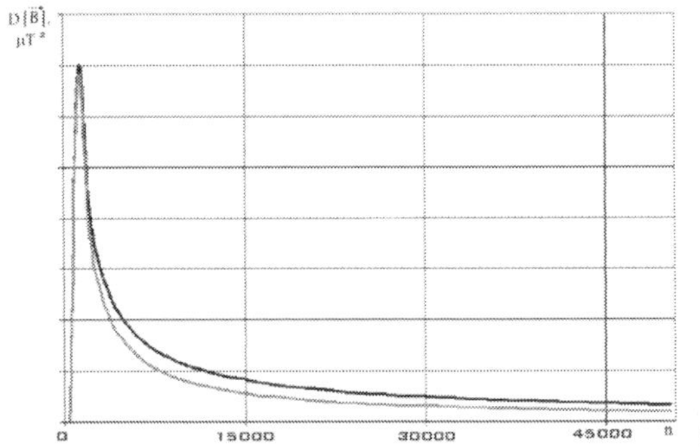

Fig. 4. The dynamic pattern of sampling variance of the total sample for sensor No. 1 (upper curve) and sensor No. 2 (lower curve).

3 Results and discussion

The variance of the sensor No. 1 is about $0.7\mu T^2$ and cannot be explained completely by the measurement error. The reason for the difference in the two sensors variances can be the features of the X1 measurement channel operation. We already said about it and now we can make several assumptions:

1. Failure in calibration of the channel X1. Perhaps after installing the sensors in the small spacecraft internal environment, an abnormal situation arose, and as a result the channel calibration changed. The consistency of the related channels X1 of sensor No. 1 and Y2 of sensor No. 2 is the worst;

2. The scientific and supporting equipment design was made in such a way that the maximum effect was on the channel X1 of the sensor No. 1. As a result, measurements on this channel were most significantly distorted by the magnetic fields of the elements and devices of the small spacecraft.

The magnetorquers operated in the microacceleration compensation mode to reduce the angular velocity module of the small spacecraft rotation. The second assumption is supported by the fact that in the microacceleration compensation mode, the average values and variances of the corresponding component of the induction vector's modulus of Earth's magnetic field were not the last ones in any of the three cases of using this mode. It can be explained by the fact that the magnetic remanence of the magnetorquers reduced the influence on the channel X1 of the sensor No. 1 other scientific and supporting equipment.

Thus, based on the results of the investigation, we can state that:

1. Failures of the standard measuring instruments of the Earth's magnetic field induction vector's components on the flight model and prototype of AIST small spacecraft were not detected during the testing of correct operation.

2. For the flight model of AIST small spacecraft the number of measurements were insufficient to explain the variance of the average orbital value of the Earth's magnetic field induction vector by the measurement error.

3. In both cases, the variance of sensor No. 1 was greater than the variance of sensor No. 2. It can be explained by the similarity in design of two small spacecrafts, which determine the same impact degree of the scientific and supporting equipment on the measuring instruments.

4. In the analysis of variances, the hypothesis [4] about the features of the measuring channel X1 of the sensor No. 1 was confirmed.

5. The hypotheses [1, 6] were confirmed. The data from the sensor No. 2 on both small spacecrafts allows to reconstruct more accurately the rotational motion of these small spacecrafts.

6. To investigate the impact of scientific and supporting equipment on the correct operation of measuring instruments, it is necessary to test the channels. At the same time, it is necessary to give a physical explanation to the value to which the average sample values of the separate measurement channels will tend with an unlimited increase in the sample.

This work is supported by the Ministry of Education and Science of the Russian Federation in the framework of the State Assignments to higher education institutions and research organizations in the field of scientific activity (the project # 9.1616.2017/ПЧ).

References

1. V. Abrashkin, Yu. Puzin, A. Filippov, K. Voronov, A. Piyakov, N. Semkin, V. Sazonov, S. Chebukov, Uncontrolled rotational motion of the AIST small spacecraft prototype, Cosmic Research, **55**(2), 128-141 (2017)
2. M. Belyaev, O. Volkov, M. Monakhov, V. Sazonov, Estimating the accuracy of the technique of reconstructing the rotational motion of a satellite based on the measurements of its angular velocity and the magnetic field of the Earth, Cosmic Research, **55** (5), 345-360 (2017)
3. A. Sedelnikov, Fast analysis of onboard measurements of the Earth magnetic field for the purpose of microaccelerations decrement on board of the "AIST" small spacecraft during its uncontrolled orbital flight, International Review of Aerospace Engineering, **11**(2), 76-83 (2018)
4. A. Sedelnikov, Mean of microaccelerations estimate in the small spacecraft internal environment with the use of fuzzy sets, Microgravity Science and Technology, Articleinpress (2018)
5. A. Sedelnikov, Modeling of microaccelerations caused by running of attitude-control engines of spacecraft with elastic structural elements, *Microgravity Science and Technology*, **28** (5), 491-498 (2016)
6. V. Abrashkin, Yu. Puzin, A. Filippov, K. Voronov, A. Piyakov, N. Semkin, V. Sazonov, S. Chebukov, Uncontrolled attitude motion of the small satellite AIST. Cosmic Research, **53** (5), 360-373 (2015)

General approach to optimize settings of ship radars repair procedures

Evgeni Guglev[1,*], and *Chavdar* Alexandrov[2]

[1]Nikola Vaptsarov Naval Academy, Faculty of Navigation, 9026 Varna, Bulgaria
[2]Nikola Vaptsarov Naval Academy, Department of Electronics, 9026 Varna, Bulgaria

Abstract. A number of possible formal methods for constructing an approach to solve the problem of global optimization of the setting process of ship radars repair procedure are introduced in this paper. The basis of the formalized task assignment is the selection of a sequence of operations related to determining the corresponding values of the electronic (electrical) components of radars, providing the global optimum of the created imitation model of the whole task with the corresponding subtasks. The algorithm of the global optimization task is built in the form of imitation experimental procedures with a multi-alternative optimization model. The multi-plan character (taking into account the different ranges) and the multi-criteria of such an experiment require an automatic search for optimizing the scanning at different ranges. The alternative for inclusion in the relevant group, with good result achieved is set in accordance with expert estimations (for the time of the scanning), Boolean variables (alternative for the inclusion of different ranges) or stochastic variables – by nomograms. The six-steps method (for six different ranges) is presented as a sequence of achieved optimums for the different ranges. The optimal set of functions determined by the task solution method is a solution of a stochastic problem including scanning time, pulse duration and ranges.

1 Introduction

The formalized decision-making task assignment (determining the optimal values of R^*, C^* ...) for different ranges is considered as a multi-alternative, multi-criterion optimization model of a complex system [1, 2, 3, 4]. For the individual levels (ranges), whose sequence is determined under T-algorithmically constrained limits, the exact L (scanning) value for "long" ranges depends on the specified L optimum for "small" ranges. The approach is a combination of a priori (expert assessments and different nomograms), axiomatic (person adopting a solution) and adaptive (stochastic) approximation. It allows for alignment within the entire optimization procedure, movement both in algorithmically defined and dependent on one another levels X_i (for different ranges) and optimization of scanning time of the appropriate ranges $L_j = f[T(R^*), T(C^*)]$ at repair,

where: R^* (Setup Resistance) and C^* (Setup Capacity) are to be considered as stochastic quantities.

2 Formulation of the problem

The process of task assignment is done as described below.

Dependence $L = f(T)$ for the different ranges and dependence $T = f(R^*)$, $T = f(C^*)$ given in [5] are experimentally predefined. Here T is pulse duration.

Fig. 1 indicates the respective specified resistant tolerances for the different ranges.

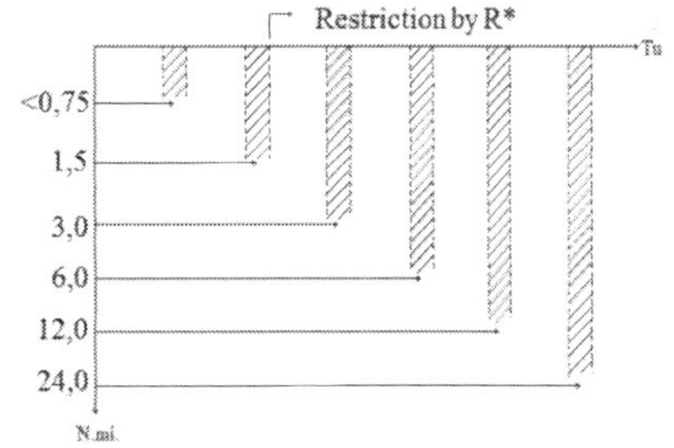

Fig. 1. Resistance tolerance for different ranges.

1. By multi-alternative optimization in Fig. 2 is set the optimization sequence for the different ranges shown in Fig. 1.

2. The accurate limits of the scanning are precised by the person making decision (PMD). This person determines the corresponding actual duration of the scanning required for the procedure in Fig. 3 (structural diagram of multi-criteria optimization).

3. In Fig. 3, using the method described in [5], the L_{opt} and the corresponding T_{opt} are defined by the nomogram $L = f(T)$.

4. In Fig. 3, applying the same method by nomogram $T = f(R^*)$, $T = f(C^*)$ follows determination of the optimal value of R and C.

* Corresponding author: egigugi@abv.bg

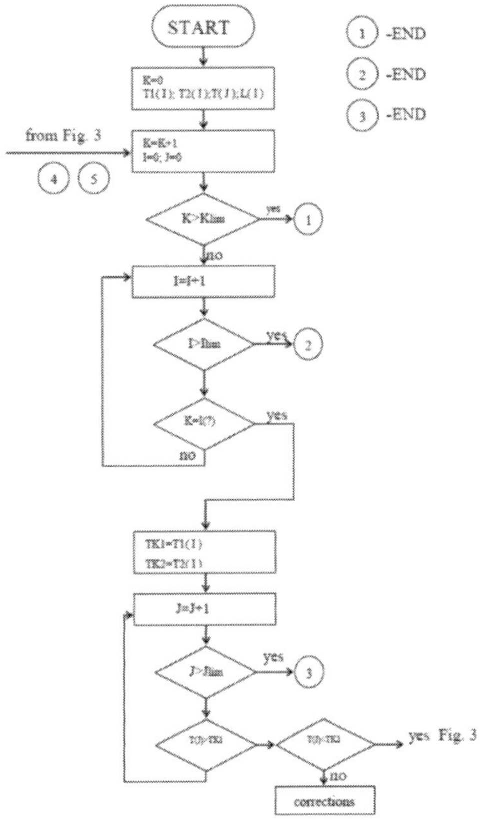

Fig. 2. Multi-alternative optimization.

3 Essentiality of the implemented survey approach

The multi-plan character of such an experiment require takes into account the different ranges and the multi-criteria of such an experiment require an automatic search to optimize the scanning at different ranges providing the extreme value of the target function [6], i.e.:

$$\Psi[X_i, L_j(T_j), T_k(R^*), T_k(C^*)] \rightarrow \min, \quad (1)$$

where X_i is Boolean variable, implementing the appropriate range when the corresponding limitations are met.

1) $$X_i = 1 \quad (2)$$

at:

a) $i \equiv j$ – at a certain range there is a matching of exactly j from a certain combination $L_j(T_j)$, for each radar scale from 1 to 6 (see Fig.1);

b) $$T_1(i) < T_j < T_2(i),$$

where T_j are limitation rows described in [5];

c) Performance of other conditions;

2) $$X_i = 0, \quad (3)$$

in case of non-fulfillment of any of the conditions *a*, *b*, *c*.

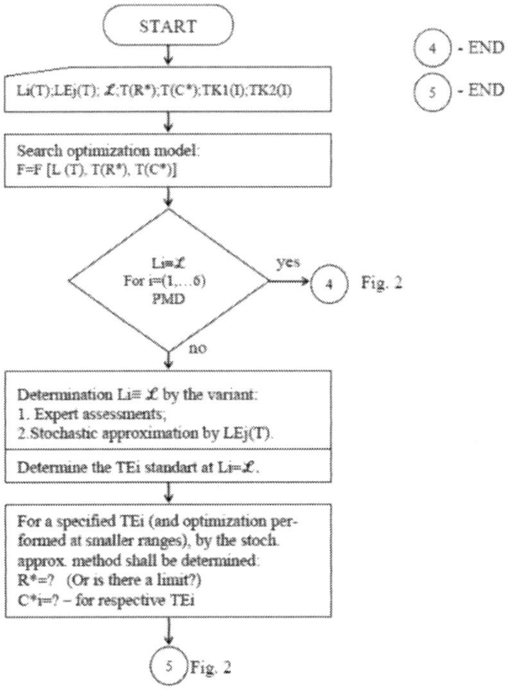

Fig. 3. Structural diagram of multi-criteria optimization.

$L[T(R^*), T(C^*)]$; $L_j(T_j)$ – require multi-criteria optimization;

L_j; T_j; R^*; C^* – Stochastic quantity [5];

L_j; T_j – Current non-optimal values for each separate radar scale (range).

Steps of optimization

Step 1: A priori research

For a certain level, according to the realization of the stochastic quantity L, the $L_i(T)$ is determined either by $T = T(R^*)$ or by $T = T(C^*)$.

R^*, C^* change using the rules: $T = \alpha R^* C^*$ – given in [5], and determining $L_i = L_i(T)$.

Performing multi-alternative optimization (for different levels or corresponding ranges) and multi-criteria optimization (for each level separately) lead to a global optimum [7].

Step 2: It is done by using decision-makers (PMD)

The question set to the PMD is, "Are all the parameters of the L_i scanning satisfactory?"

In case of a negative response, the possibility is analyzed:

1. $$\Psi_i L = f(\Delta L) = f[\Delta L(T)] \rightarrow \min. \quad (4)$$

After reaching an optimum by L, T_{opt} is determined and is accepted that $T_{standatd} = T_{opt}$.

2. $$\Psi_i T = f'(\Delta T) = f'[\Delta T(R^*), \Delta T(C^*)] \rightarrow \min. \quad (5)$$

After reaching an optimum by T, R_{opt}, C_{opt} are determined to achieve optimal scanning.
ΔL – deviation from the standard value of the scanning – $\mathcal{L}$,
ΔT – deviation from T_{opt} defined in [5].

Step 3: Going to an optimum

A setting of the values is done:

1. $$L^{r+1} = L^{r} + \Delta L(T) \quad (6)$$

$L^* = \mathcal{L}$ is determined at

$$\Delta\Psi_{iL} \to \min, \quad (7)$$

where L is scanning,
$\mathcal{L}$ – set reference,
L^* – optimal scanning, which determines the corresponding T_{opt} on the nomogram $L(T)$.
2. An exact (optimal) reference is defined $T_E = T_{opt}$

$$T^{r+1} = T^{r} + \Delta T(R^*) + \Delta T(C^*) \quad (8)$$

$T^* = T_E$ is determined at

$$\Delta\Psi_{iT} \to \min. \quad (9)$$

Next follows the determination of R^*, C^*.

$$R^{r+1} = R^{r} + \Delta R \quad (10)$$

The optimal R^* is defined using the method of [5].

$$C^{r+1} = C^{r} + \Delta C \quad (11)$$

The optimal C^* is defined using the same method.

First procedure – Task 1

It is demonstrated in Fig. 2.

If the pulse duration does not fall within any of the restriction intervals for the different ranges, then a correction is made by the PMD.

Second procedure – Task 2

After a successful test of the first procedure in Fig. 2 a second procedure is performed – the multi-criteria optimization of Fig. 3. This procedure leads to determination of the optimum, which is the above mentioned Step 3.

It is possible to increase the speed of convergence at the expense of the management of the choice by means of objective (by degree of change of the target function) and subjective (according to PMD information) local-prognostic estimations of the direction of movement to extremum.

The described dialogue procedure uses the experience of the expert, his ability to work with abstract and fuzzy concepts in conducting automated analysis with multiple alternative optimization models [8].

4 Experiments

The experiments are organized under the following conditions:
1. The nomograms $L_i(T_i)$ (i = 1÷6 for each radar scale) and $T(R^*)$, $T(C^*)$ are determined in advance.
2. The current measured L_i, T_i (i = 1÷6 for each radar scale) have to be tuned. (see Fig.1).
3. Using the expert assessments or PMD, L_i & $\mathcal{L}$ are compared and the path of decision is determined.
4. Solution of Task 1: "Multi-alternative optimization at algorithmically constrained limitations" – according to Fig. 2.

Fig. 1 gives the respective permissible intervals.

Rows $T_1(i)$, $T_2(i)$ are pre-defined under the Technical Conditions (the corresponding pairs give the margin for the respective levels).
5. Solution of Task 2 ("Multi-criteria optimization" – according to Fig. 3.):

$$\Delta\Psi_{iL} \to \min \text{, at } L_i \equiv \mathcal{L} \text{ we get } T_{ie} = T_{opt}, \quad (12)$$

where T_{ie} is the reference;

$$\Delta\Psi_{iT} \to \min \text{, at } R^* \equiv R_{opt} \quad (13)$$

or $C^* \equiv C_{opt}$ or both at $T_i = T_{ie}$.

6. After each successful test in Fig. 3, the specified and the actual $L_i(T_i)$ are compared by PMD and the decision is taken.
7. Additional optimization (if required):
• The dependence $L = f(T)$ is determined experimentally.
• There is an a priori data on the dependences $T = f(R^*)$, $T = f(C^*)$ given in [5].

5 Conclusions

The algorithms described in this research make it possible to completely exclude the "active" participation of a person in complete setup of the radar repairing process.

The research provides a general solution to the task of obtaining optimal values of the radar parameters.
By using information obtained from the optimization of the radar setup process, real applicability is ensured.

The developed algorithms, i.e. "instrumentation" can be used to repair ship radars.

This study has been carried out under project BG05M2OP001-2.009-0037-C01 "Support for the development of PhD students, post-doctoral students, young scientists and lecturers at the Nikola Vaptsarov Naval Academy", with Nikola Vaptsarov Naval Academy, Varna as beneficiary. The project is funded by the Intelligent Growth Science and Education Operational Program, co-financed by the European Union through the European structural and investment funds.

References

1. A. Donets, Y. Lvovich, V. Florov, *Automatic analysis and optimization of radio equipment constructions and technologies*, in Russian, Moscow (1983)
2. C. Floudas, P. Pardalos, *Encyclopedia of optimization*, Dordlecht (Kluwer Acad. Pub., 2001)
3. Z. Nikolov, M. Tsvetkov, T. Tsanev, Experimental research of a jamming over a direct sequence spread spectrum system, Journal of Marine Technology and Environment, **2**, 83-88 (2009)
4. Y. Sivkov, Building maritime data hub by using Arduino IoT platform, *IAMU AGA 2017*, Varna, **I**, 533-540 (2017)
5. E. Guglev, Y. Dachev, Stochastic approach to optimize the setting of radiolocation stations at repair, Journal of Marine Technology and Environment, Constanta, **2**, 33 - 37 (2017)
6. N. Madjarov, L. Michailova, Adaptive multiple model algorithms for monitoring and control of stochastic systems, Journal of Automatics and Informatics, in Bulgarian, **2** (1998)
7. S. Stoyanov, *Methods and algorithms for optimization*, in Bulgarian, Sofia (Technica, 1990)
8. P. Pardalos, R. Horst, *Introduction to global optimization*, Norwood (Kluwer Acad. Pub., 1995)

Optimal designs of electric vehicles for long-range mobility

Zlatina Dimitrova[1,*]

[1]PSA Groupe, Research and Innovation Department, 78943 Vélizy-Villacoublay, France

Abstract. This paper introduces the major priorities of the automotive industry of reducing energy consumption and emissions of the passenger cars of the future and of delivering an efficient mobility service for customers. The number of electric vehicles and hybrid electric vehicles is increasing in the mobility market. The problems of the range and the energy storage of the vehicle on board are important. This paper studies the energy system of electric vehicles for different ranges and mobility usages. A multi-objective optimization method is applied to estimate the optimal vehicle energy system designs for urban mobility and for long way electric mobility (> 500 km). Optimal designs considering technical, economic and environmental criteria are presented. The relations between the vehicle ranges and the energy densities of high voltage batteries are illustrated. The boundary of the system is extended to the vehicles and the grid system, including the charging infrastructure. The vehicle energy systems and recharging needs are analysed for a range of 500 km on electric drive.

1 Introduction

The electrification of vehicles is driven by the trend for decarbonisation and emission reduction of road transport.

Around 2030, electric vehicles are predicted to increase their market penetration and to bring evolution concerning the main technologies for energy storage and conversion, powertrain components and energy management [1]. The industrialisation of those components on a high scale and volume reduces the cost of the electrification and the democratisation of the electrification on all vehicle segments. Hybrid electric vehicles with different level of hybridisation are proposed for several vehicle segments. They are designed for urban and peri-urban drives, and allow for zero emissions drives from thank-to-wheels perspective for 25 km or 50 km. Hybrid electric vehicles with "zero emission vehicle" modes are supported by incentives for circulation in big cities centres. According to the ERTRAC Report [1] it is possible to distinguish between two types of electric vehicles: user friendly affordable vehicles and high performance electric vehicles. The first category will develop in small and middle segments (A/B/C) and they represent a more optimized range/cost solution enabling a very big market penetration with considerable impact on greenhouse emissions reduction. The ranges of these vehicles are adapted for ranges of 200 km and are suited for urban and sub-urban use. The major challenge is to achieve significant cost reduction of the battery and the propulsion system components. The introduction of these vehicles is related to the support of the governmental policies (transportation energy taxes, city centre driving), the infrastructure deployment and the cost of electricity. High performances electric vehicles are deployed in the C/D/E segments. They will maximize the range in the performances offering at least 600 km on a single charge. The cost of these vehicles is high because of the high battery capacity on board. The cell energy density needs to be improved to allow for the deployment of these vehicles. The significant battery capacity can be rapidly recharged with high power chargers. These cars are designed for long distances, on the highways and their deployment is related to the deployment of the high speed chargers infrastructure all along the highways.

Customers' expectations and considerations for buying electrified vehicles are acceptable prices, range, reliability, durability and re-sale value similar to conventional vehicles. Some advantages may also be achievable through synergies of the electrification with the connectivity and the automation of the vehicles. This will provide benefits for the users through easy access to the charging infrastructure, optimization of the traffic and minimization of energy consumption. Charging is a crucial topic for the success of electrification. Purely electric vehicles are pushed to compete with internal combustion engine vehicles one-on-one, where refilling takes place around once every week. With the right mix of infrastructure, electric vehicles could compete with vehicles powered by internal combustion engines, but with the added dimension of the refilling comfort. Daily e-charging would mean that drivers never have to stop at filling stations, except at rapid charging stations for long way drives. Information and connectivity solutions that inform, in real time, the drivers about the availability and distribution of the charging infrastructure, including the possibility to book and pay for charge points in different places, is an important part of the solution. Combined with fast, i.e. high power charging capacities, pure electric vehicles would be able to serve long range needs.

* Corresponding author: zlatina.dimitrova@mpsa.com

Zenginis and al. propose in [2] the waiting time in the queue of the customers for recharging their batteries. The model considers the stochastic recharging of the battery capacities and the class of the vehicles. Furthermore, a charging strategy is proposed according to which the drivers are motivated to limit their energy demands. Karakitsios and al. introduce in [3] a methodology to analyse the impact of the fast inductive charging technologies to the grid. Levinson and al. [4] explore the potential of public charging infrastructure to spur US battery electric vehicle. They used market analysis and simulations. Zhang and al. develop in [5] a multi-period flow refuelling location problem for electric vehicles (EVs) and charging infrastructure. The optimization model determines the optimal location of chargers for every station. Sun and al. explore in [6] the choice of the users to fast charging of battery electric vehicles as a function of the available charging stations and the distance between them, in Japan. Su and Sioshansi develop in [7] a model to optimize the location of public fast charging stations for electric vehicles (EVs). They use a stochastic flow-capturing location model to solve the uncertainty about where EV charging demands appear. Schroeder and Traber provide in [8] the business case of fast charging technology in a case study for Germany. They estimated that the Return on Investment (ROI) of a public fast charging station constitutes the main contribution. The general EV adoption rate is estimated to be the main risk factor for investment in public charging infrastructure.

The novelty of the present study is the application of a systematic multi-objective optimization methodology for defining the optimal design and operation parameters of the electric vehicle energy system. The optimization considers optimal vehicle energy systems designs, in terms of technical, economic and environmental criteria. This three-objective optimization is also called environomic optimization [9], [10]. The methods and techniques of analysing, improving and optimizing energy systems have to deal not only with energy consumption and economics, but also with environmental impact. The multi-objective environomic optimization includes all these activities.

2 Methodology

The developed methodology combines flowsheeting models of technologies, energy integration techniques, evaluation of the economics and life cycle assessment in a computational platform. Multi- objective optimization techniques are used to explore the superstructure possibilities and to find the optimal design solution. The master optimization is done by using a genetic algorithm and mixed integer nonlinear programming. Such an integrated approach is innovative as compared to the traditional heuristic design engineering method, based on iterations of designs and their cost evaluation. Fig. 1 presents the generic computational framework for environomic design of a vehicle energy system. The superstructure contains a vehicle simulation model, with dynamic layouts. The cost equations are implemented in the model of economics. A Genetic algorithm based on optimizer is used. The set of decision variables includes the types and the size of powertrain components. The master optimization is solved by an evolutionary algorithm with 3 objectives: the minimization of the electric consumption, the minimization of the investment cost and the environmental impact of the technologies (Fig.1). The multi-objective optimization results converge to the Pareto frontier curve. In this study energy integration is not applied.

2.1 Electric vehicle dynamic model

Electric vehicles are characterized by a powertrain that stores and converts electric energy. This energy conversion chain consists of a battery (or another electricity storage system) and the electric motor with its controller. The size of the high voltage battery and its density affects directly the autonomy of electric vehicles. Permanent magnets synchronous machines are mostly used as traction motors. The peak power ranges are from 50 to 80 kW. Lithium-ion battery energy typically ranges from 15 to 25 kWh. The top speed and range of such cars reaches 135 km/h and 160 km autonomy on combined driving cycles. Electric consumption is measured in kWh/100 km or Wh/km. The electric consumption depends on the efficiency of the powertrain components: the electric machine, the battery and the transmission. The electric machine acts in motor and generator mode and

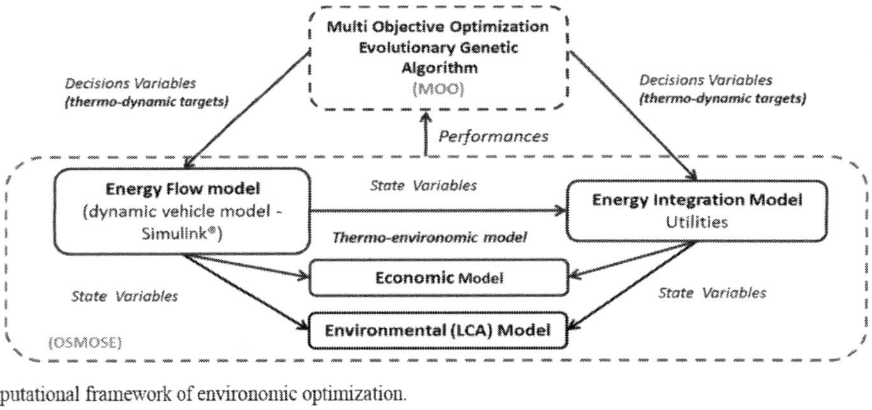

Fig. 1. Computational framework of environomic optimization.

during braking the kinetic energy is recovered and stored in the high voltage battery as electricity. Typical values of the energy consumption of recent electric vehicles are 10-15 kWh/100 km, characterized on combined driving cycles. The quasi-static approach is used for the modelling. The voltage and the current are the constituents ("power factors") of electric power. Since the former can only have positive values, it plays an analogous role to that of speed in mechanical power. The current can switch direction to represent both power flows from (positive current) or to the storage system (negative current). The negative current occurs during the regenerative braking. Thus, current is the electric analogue of torque and force. Figure 2 lists the causality of power factors of an electric vehicle.

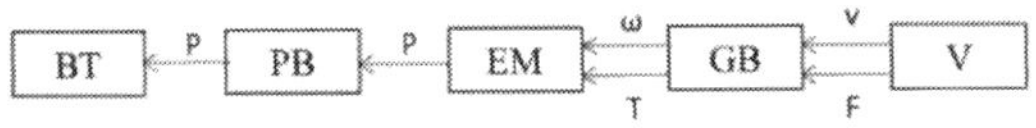

Fig. 2. Electric vehicle quasi-static model: BT – battery, PB – power link, EM – electric machine, GB – gear box, V– vehicle.

The main characteristics of the electric vehicle model are given in Table 1.

Table 1. Electric vehicle main characteristics [11].

Sub-System	Characteristics	Value
Vehicle	Nominal mass, kg	1195
Gear box	IVT efficiency, -	0,9
Energy	Electricity	
Electric motor	P_{EM}, kW	125
Battery	Type	Li-ion
	Capacity, kWh	19

2.2 Vehicle cost model

The size and efficiency of energy converters and energy storage devices impact the economic evaluation of the vehicle. The cost is calculated for every run. The economic equations of the components come from the literature and are related to the size of the components. Table 2 presents the cost equations.

Table 2. Equations for the economic model.

Components	Cost, €
Electric motor [12]	30 €/kW × P_{EM} , kW (1)
Thermal engine [12]	15 €/kW × P_{TE} , kW (2)
Battery [12]	600 €/kWh × 0,2477 × log($battery_{specific\_mass}$ (bat_type) + 0,5126q_{batt}), kWh (3)
Supercapacitor [12]	15 €/kW × P_{SC} , kW (4)
Nominal cost (car shell)	17,3×car_shell_mass kg – 3905,4 € (5)
Electricity household [12]	0,14269 €/kWh
Diesel [12]	1,451 €/L

The cost of the electric machine includes the cost of the power unit. The battery cost is sensitive to the battery type and the energy storage capability of the material. The nominal cost represents the vehicle shell cost, without the powertrain components. The linear correlation (5), (Table 3) takes into account the price of the parts, as well the manufacturing cost of the vehicle shell. The margin of the carmaker is included as well. A simplified vehicle objective economics function is constructed (7). It takes into account the vehicle powertrain cost (production) (6) and vehicle nominal cost (5).

$$Cost_{powertrain} = Cost_{EM} + Cost_{battery} + Cost_{SC}, €, \quad (6)$$

$$Cost_{vehicle} = Cost_{powertrain} + Cost_{car\_shell}\ , €. \quad (7)$$

2.3 Environmental model

In this work the Life Cycle Assessment is applied as an indicator for the vehicle energy system design. In this study one category from the CML short impact is used. - the 100-year Global Warming Potential (GWP). The life cycle of a product, a system or a service has usually three distinct successive phases: the production phase, the use phase and the end-of-life phase. The inventory of the unitary processes and the raw materials for the production of the parts is built from the Eco Invent® database. The vehicle is divided into seven substructures. They allow identifying the powertrain: electric machine, low voltage battery, high voltage battery, power unit, thermal engine, gearbox, vehicle body (car shell). The electric consumption of the vehicle corresponds to the use phase. The inventory for the corresponding electricity production comes from the Eco Invent® database. The end-of-life phase is represented by the average vehicle disposal process. This process also comes from the Eco Invent® database. We consider that the vehicles are operated in France with the French electricity mix. The current normalized driving cycle for characterization of the commercial vehicles is the New European Driving Cycle (NEDC). Table 3 presents the characteristics of the driving cycles used in this study.

Table 3. Drive cycles characteristics.

Cycle	Distance, km	Duration, s	Average speed, km/h
NEDC	11.023	1180	32.26

3 Results – multi-objective optimization

3.1 Problem definition

A multi-objective optimization is defined for electric vehicles. The environmental category is chosen as optimization function, in addition to the cost and the efficiency. Environomic optimization is researched and evaluated for electric vehicles. Based on the Pareto curves obtained with the results from the multi-objective optimization, optimal vehicle configurations of the cars are discussed and decided.

The environmental impact category (GWP), coming from the life cycle assessment model as performance indicator in the superstructure, is used as an objective function. The variables for the decision are defined in Table 4.

Table 4. Decision variables for design – multi-objective optimization problem.

Electric propulsion system components	Range	Unit
Electric machine	15-50	kW
High voltage battery	10-100	kWh
Number of super capacitors in a parallel structure	0-5	-
Max Power of supercapacitors	87,5	kW

3.2 Multi-objective techno-economic optimization

The vehicle range is the main technical performance parameter of the electric vehicle. The range depends on the battery capacity on board, and on the energy density of the battery cells.

The cost of the battery is also related to these two parameters and especially to the energy density (equation 3). The battery specific energy, which depends on the technology (cathode material, conductivity, etc.), takes considerate time and resources to evolve. The battery capacity can be more easily increased for a given battery type.

In this section a conceptual point-of-view design is adopted by varying the battery specific energy during the two-objective techno-economic optimization. The optimization problem is defined in Table 5.

The techno-economic optimization converges on the Pareto curve presented in Figure 3. The results indicate the effect of the battery specific energy on the autonomy and cost, for current and future battery technologies. The cost and the ranges of the electric vehicle are related to the battery capacity and to the battery density.

The autonomy ranges vary between 100 km and 500 km. The vehicle cost increases also with the autonomy and the high specific energy. The minimal cost is obtained for batteries with lowest specific energy of 90 Wh/kg. The maximal autonomy of 500 km is obtained for batteries with 90 kWh of capacity and 400 Wh/kg of specific energy of the battery cells. Figure 4 presents the evolution of the battery cost as a function of the specific energy for a battery with 20 kWh of capacity. The battery cost varies between 8000 € and 42000 €.

Table 5: Multi-objective optimization objectives and decision variables

Techno-economic optimization – Objectives		
Autonomy	km	Max
Cost	€	Min
Electric propulsion system components	**Range**	**Unit**
Electric machine	15-50	kW
High voltage battery capacity	10-100	kWh
High voltage battery specific energy	90-400	Wh/kg
Number of supercapacitors in a parallel structure	0-5	-
Max Power of supercapacitors	87,5	kW

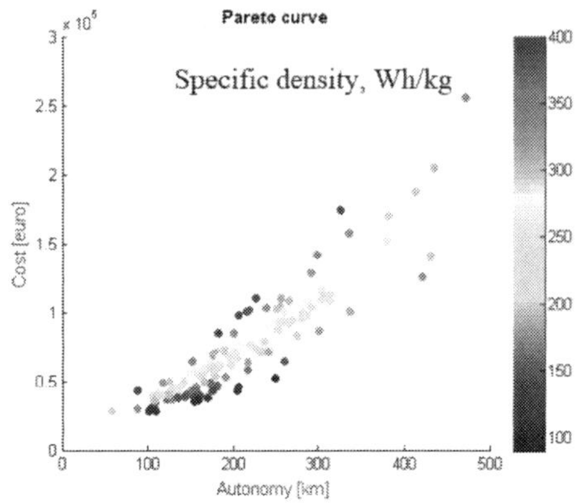

Fig. 3. Pareto for techno-economic optimization, battery specific density in colorbar.

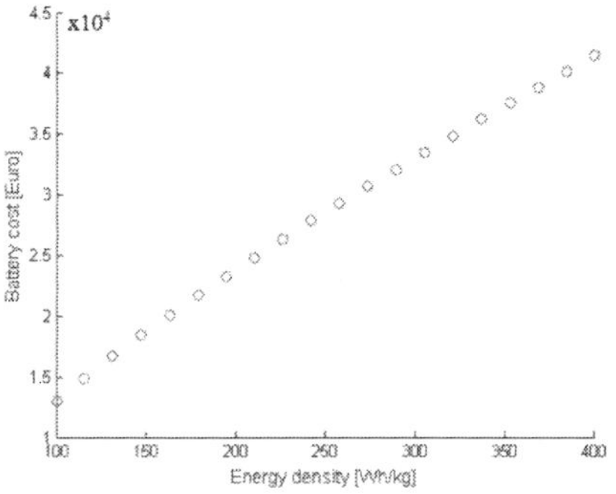

Fig. 4. Evolution of battery cost as a function of battery density, for battery capacity of 20 kWh.

The energy capacity of 20 kWh is a reference for the serial electric vehicles and allows for at least 150 km of autonomy. The autonomy of the electric vehicles needs to be increased and for that purpose the battery capacity on board needs to be increased. Figure 5 presents a simulation of the vehicle range as a function of battery capacity. The simulation is done for the maximal specific energy density of 400 Wh/kg considered in the study. One can see that a range of 550 km could be reached with a 100 kWh of battery capacity of the energy board, which is 5 times more that the state-of-the-art of on-board capacities.

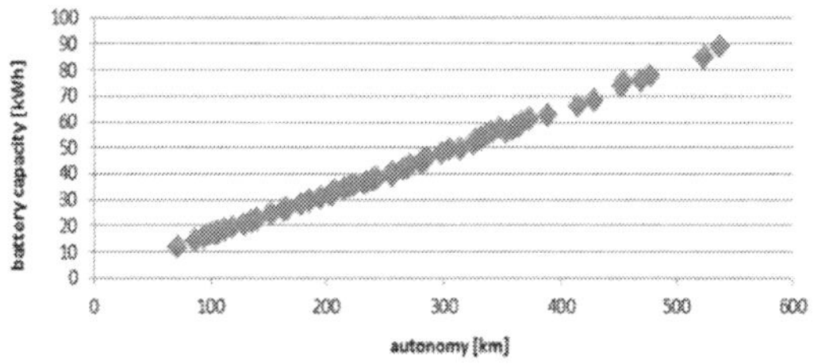

Fig. 5. Evolution of vehicle autonomy as a function of battery capacity, specific energy of 400 Wh/kg.

Ranges higher than 300 km are important for the development of electric vehicles as a solution for long way mobility. Thus, electric vehicles can be used for holiday drives and not just for home to work urban and peri-urban commuting. In addition, the important capacity of 100 kWh sets the question of the vehicle recharging systems and necessitates the development of fast charging systems and their availability on highways.

The time of the recharging of the batteries depends on the battery capacity and also on the power that the recharging infrastructure is capable to deliver. To insure long durability of the battery, the battery optimal use is considered to be for states of the charge between 20% and 80% of the maximal capacity. The computation of the time is considered for these boundaries. Table 6 presents the times in minutes for the charging of the batteries with different powers for the chargers. For example, to charge an electric vehicle with 100 kWh of battery with a charger of 50 kW, will take 96 minutes. If the charging is done on the highway, this time is considered to be too long. The charger of 50 kW can be used in city parking or in shopping centres. An acceptable time for a highway stop is considered to be 15 minutes. To reach this time performance a charger of 350 kW is needed. Thus, the vehicle with 100 kWh of battery capacity can be charged in 14 minutes.

Table 6. Times of charging of the batteries for different powers of the chargers.

Power (kW)	Time for 80% state of charge (min)		
	Battery capacity		
	50 kWh	100 kWh	150 kWh
50	48	96	144
100	24	48	72
150	17	32	51
350	7	14	21

One can conclude that the long way electric mobility with ranges over 500 km, will develop with increasing battery capacities (> 100 kWh) and with the deployment of fast charging infrastructure on highways with chargers capable to deliver at least 350 kW.

3.3 Multi-objective environomic optimization

In the environmental model, one can consider the 100-year GWP as the main impact category and the GWP is considered as environmental objective to be minimized. The multi-objective problem is defined with 3 objective functions: the autonomy to be maximized, the cost and the GWP impact of the electric vehicles to be minimized. The decision variables are the same ones defined in Table 5.

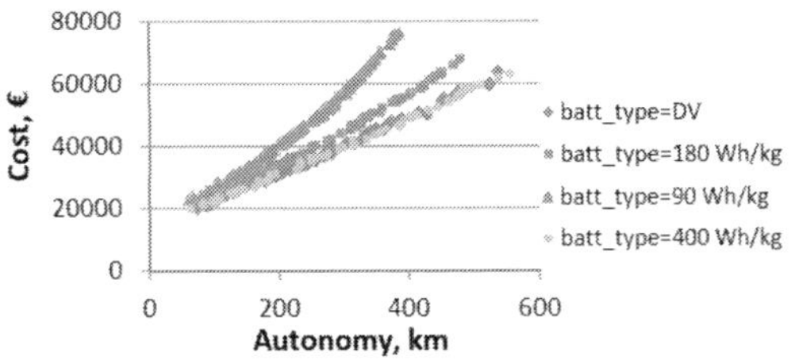

Fig. 6. Pareto of the environomic optimization, 2D plan cost/ autonomy.

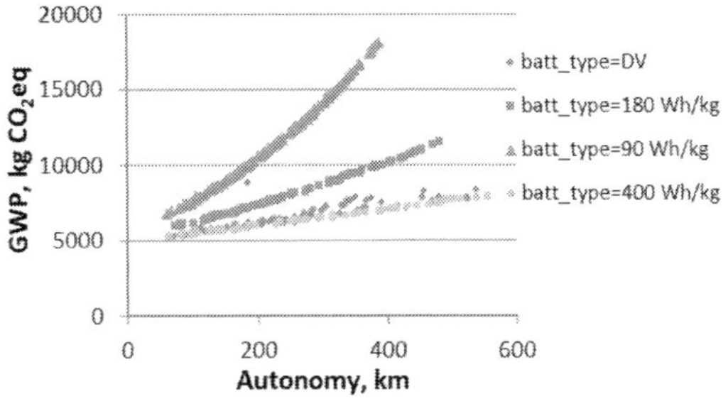

Fig. 7. Pareto of the environomic optimization, 2D plan GWP/ autonomy.

The optimal environomic solutions converge on the Pareto frontiers presented in Figures 6 and 7, as projections in the plans cost/autonomy and GWP/autonomy.

One can see that for battery types with specific efficiency of 90 Wh/kg the maximal range is estimated at 380 km. For this solution the vehicle cost is estimated to be 76000 € and the GWP is estimated to be 18000 kg CO_2 eq. For the battery with specific energy of 180 Wh/kg, the maximal range is 480 km, with a vehicle cost of 68000 € and GWP of 12000 kg CO_2 eq. For the battery with specific energy of 400 Wh/kg, the maximal range is 550 km, with a vehicle cost of 63000 € and GWP of 8000 kg CO_2 eq. The first conclusion is that the increase of the specific energy increases the autonomy, but also decreases the battery capacity that is needed to be stored on board. For the value of autonomy of for example 300 km, the cost of the vehicles with 90 Wh/kg of battery is 58000 €, the GWP is 14400 kg CO_2 eq. For the same range of 300 km the cost of the vehicles with 180 Wh/kg of battery is 44000 €, the GWP is 8600 kg CO_2 eq. For the range of 300 km the cost of the vehicles with 400 Wh/kg of battery is 40000 €, the GWP is 6500 kg CO_2 eq.

The optimizer chooses the highest specific energy (defined by the upper boundary of the variation range). With the cost model, the slope of the Pareto curve decreases when increasing the specific energy, which explains the choice of the optimizer.

The environomic designs are presented in Table 7 for autonomies of 300 and 550 km.

Table 7. Environomic designs for long range electric vehicles, specific energy of 400 Wh/kg.

	Units	ID = 17	ID = 78
Autonomy	km	300	550
Cost	€	40000	63000
GWP impact	kg CO_2 eq.	6500	8000
Power EM	kW	40	47
Battery capacity	kWh	50	90
Specific energy	Wh/kg	400	400
Number of supercapacitors	-	1	2

5 Conclusions

This paper presents an energy system study of electric vehicles. The study considers different vehicle usages through adapted ranges and referred on the NEDC normalized cycle. The optimal environomic configurations are researched by means of multi-objective optimization. The optimization methodology is based on a genetic algorithm and is applied for defining the optimal set of decision variables for energy system design. The relations between the battery capacity, the battery energy density and the range of the electric vehicles are examined. The designs for different mobility services (urban and long ways distance electric mobility) of the electric vehicles are defined and the time interaction with the charging infrastructure is analysed. Information and connectivity solutions that inform in real time the drivers about the availability and the location of the charging infrastructure, including the possibility to book and pay for charge points in different places, is an important part of the solution. Combined with fast, i.e. high power charging capacities, pure electric vehicles would be able to serve long range needs. The analysis of the environomic Pareto curves on NEDC also illustrates the relation between the economic and the environmental performances of the solutions. Electric vehicles can be used for long way mobility (>400 km), if they have a consequent capacity of battery on board (between 100 and 150 kWh). The density of the battery is also important for the autonomy of the electric vehicles. The importance of battery capacity creates the need for high power chargers, which will allow for fast charging of the batteries. The pick power demands induce to investigate adequate links for such electric vehicles with the electricity grid. Multi-objective optimization can be used in an enlarged energy system, including electric vehicles and the electric grid.

Abbreviations

IVT	Electric continuous variable transmission
EV	Electric vehicle
GWP	Global warming potential
TE	Internal combustion engine
F	Force, N
P	Power, kW
T	Torque, Nm
q_{batt}	battery capacity, kWh
w	wheels
Subscripts and superscripts	
TE	thermal engine
SC	supercapacitors

References

1. ERTRAC Report, European road map, electrification of road transport, www.ertrac.org/uploads/documentsearch/id31/electrification_roadmap_june2012_62.pdf, accessed on 23.10.2017
2. I. Zenginis, J. S. Vardakas, N. Zorba, Ch. V. Verikoukis, Analysis and quality of service evaluation of a fast charging station for electric vehicles, In Energy, **112**, 669-678, (2016)
3. I. Karakitsios, E. Karfopoulos, N. Hatziargyriou, Impact of dynamic and static fast inductive charging of electric vehicles on the distribution network, In Electric Power Systems Research, **140**, 107-115, (2016)
4. R. S. Levinson, T. H. West, Impact of public electric vehicle charging infrastructure, Transportation Research Part D: Transport and Environment, Available online 16 October 2017, ISSN 1361-9209, https://doi.org/10.1016/j.trd.2017.10.006.
5. A. Zhang, J. Eun Kang, C. Kwon, Incorporating demand dynamics in multi-period capacitated fast-charging location planning for electric vehicles, In Transportation Research Part B: Methodological, **103**, 5-29, (2017)
6. X. Sun, T. Yamamoto, T. Morikawa, Fast-charging station choice behavior among battery electric vehicle users, In Transportation Research Part D: Transport and Environment, **46**, 26-39, (2016)
7. F. Wu, R. Sioshansi, A stochastic flow-capturing model to optimize the location of fast-charging stations with uncertain electric vehicle flows, Transportation Research Part D: Transport and Environment, **53**, 354-376, (2017)
8. A. Schroeder, T. Traber, The economics of fast charging infrastructure for electric vehicles, In Energy Policy, **43**, 136-144, (2012)
9. Z. Dimitrova, F. Maréchal, Environomic design for electric vehicles with an integrated solid oxide fuel cell (SOFC) unit as a range extender, Renewable energy, **112**, 124-142, (2017)
10. Z. Dimitrova, Vehicle propulsion systems design methods, MATEC Web of conferences, **133**, 02001, (2017)
11. Z. Dimitrova, F. Maréchal, Environomic design of vehicle energy systems for optimal mobility service, Energy Journal, **76**, 1019-1028, (2014)
12. Z. Dimitrova, F. Maréchal, Techno-economic design of hybrid electric vehicles, Energy Journal, **91**, 630-644, (2015)

Dynamics simulation of mechanical gearbox vibration

Jan Furch[1,*], and *Cao Vu* Tran[1]

[1]University of Defence, Faculty of Military Technology, Kounicova 65, 662 10 Brno, Czech Republic

Abstract. This paper focuses on creating a virtual model of mechanical gearbox used in medium-sized terrain vehicle using MSC.Adams software. This software is regarded as the most common and effective tool to simulate the gearbox as multibody system and to record and analyse the vibration signal from the gearbox. The paper makes an overview of modelling and simulation and performs an analysis with frequency spectrum. The paper demonstrates that it is possible to simulate vibration signals through the model of the gearbox created in 3D CAD software and then analyse in multi-body dynamics software MSC.Adams. Successful application of the virtual model not only help us decrease the cost of design work, but also help us identify the patterns of the vibration signal and the relations between the signal and the technical condition of the gearbox. The goal is to create a virtual model of a mechanical gearbox. In MSC.Adams, the vibration values of the rotating components can be detected in different gears. These values are then analysed and evaluated. The result is a simulation of fault states and identification of vibration frequencies for practical applications.

1 Introduction

The mechanical gearbox is the basic part of all mechanical machines. In a vehicle it is one of the most important elements which enables torque to be transferred from the combustion engine to the final gears of single driving axles.

The vibrations of the gear box are transferred by the joints of single shafts. Vibrations occur because of the rotation of an input shaft followed by tooth wheel meshing, and also because of the single bearings placement. That is the reason why strokes are made depending on revolutions and load which stimulate natural vibration signals in a gear box. The operation of the gear box, which is time-dependent, results in the wear of machine parts and failure [1]. Our aim is to design a gear box by modelling so that no critical vibration spectrum would occur and the presumed service life would be preserved. Another important aspect we are going to deal with is the observation of a vibration signal during the gear box service life. In practice it is essential to determine a critical vibration signal which would help to prevent failure occurrence. For this purpose we can use a virtual model which enables us to examine a vibration signal at low cost [2]. The aim is to decide how detailed the model should be to correctly describe the vibrational and acoustic performance of the powertrain [3].

The development of computer technology and software in the area of modelling is quite advanced nowadays, therefore we used modern software MSC.Adams for modelling the mechanical gearbox vibrations as it seems to be the best [4].

2 Theory of modelling of mechanical gearbox vibrations

In practice the aim of modelling is to understand the observed phenomena, imitate the behaviour of the examined system, simulate it on the model and then impact its behaviour in a required way, e.g. by its observed inner states. The model is only an approximation of reality since the real system can be very complex and the model might not fit the system completely. Therefore, we perform verifying experiments which enable us to complete and improve the model parameters or the model itself [5].

Modelling and simulation are the activities connected with making models of real world objects and experimenting with these models.

MSC.Adams (Automatic Dynamic Analysis of Mechanical System) is computing software used for modelling, analysing and optimizing mechanical systems called MBS (Multi-Body System). It supports static, kinematic, and dynamic analyses of the model. In other words, this software is based on a physical model describing the model specifically by relevant ideal objects (mass, rotational inertia, length, kinds of joints, etc.) [6].

MSC. Adams consists of a number of different modules, e.g. Adams/View, Adams/Car, Adams/Chassis, Adams/Driveline, etc. For modelling gear box vibrations and analysing the results we have used a universal module MSC.Adams/View [6].

The mechanical gearbox model has numerous parts. A basic part is the gear box where an input, output and counter shaft are placed. These shafts are supplemented by tooth wheels and bearing structures. In order to

* Corresponding author: jan.furch@unob.cz

describe the gearing box, the input, output and counter shaft more accurately we have chosen program Inventor 2014 (software 3D-CAD), which can be seen in Fig. 1. Then it is exported to MSC.Adams. For the tooth wheels and the bearing structures we have used module MSC. Adams/View.

Fig. 1. Gearing box, the input, output and tooth wheel in program Inventor 2014.

The basic steps in modelling and simulating in the MSC.Adams/View module are as follows:

- create models and bodies – geometrical shapes (length, width, depth, dimension, distance among shafts, etc.),
- put physical characteristics – rotational inertia, mass, dynamic friction coefficient, etc.,
- create structures – kinematics definition (translation, rotation, calm, etc.),
- run a simulation – make motion forces, observe model behaviour,
- record and analyse results in MSC.Adams/ Postprocessing [4], [7].

2.1 Description of the creation of a body model

The mechanical gearbox of vehicle consists of three shafts and five speed gears with the following gear ratios for single speed gears – 1st speed gear – 4,124; 2nd speed gear – 2,641; 3rd speed gear – 1,58; 4th speed gear – 1,00; reverse movement – 5,224).

The shafts and the gear box were created in the Inventor environment in the 3D-CAD module to have accurate dimensions, construction, shapes, and rotational inertias. After that the values were transported to the MSC. Adams environment of the module View where the rest of the parameters were modelled according to the following parameters, see Tables 1 and 2.

The bearing between a propelling shaft and a driven shaft was modelled according to the following information. In the bearing there are 14 rollers and the diameter of the rollers is 5,51 mm. The rotational inertia of the tooth wheels was calculated in the Inventor environment as [8]:

$$I_{xx} = \frac{1}{2}mr^2; \tag{1}$$

$$I_{yy} = \frac{1}{12}m(3r^2 + L^2), \tag{2}$$

where: m – mass of the tooth wheel, r – radius of the tooth wheel, L – width of the tooth wheel, xx – rotating axis, yy, zz – perpendicular axes with rotating axis (Fig. 2).

The values of the selected variables are given in Table 4. The adjustment of the remaining two variables such as the contact exponent and the damping distance was left unchanged.

Table 1. Basic geometrical data of tooth wheels of the mechanical gearbox of a vehicle.

Table Name of the part	Module (mm)	Number of teeth	Width of teeth (mm)	Type of teeth
Drive gear of an axle drive reduction	3	15	20	rectilinear
Driven gear of an axle drive reduction	3	32	20	rectilinear
Drive gear third speed	3	27	20	spiral
Driven gear third speed	3	20	20	spiral
Drive gear second speed	3	21	21	spiral
Driven gear second speed	3	26	21	spiral
Drive gear first speed	3,5	15	20	rectilinear
Driven gear first speed	3,5	29	20	rectilinear
Driven gear of a reverse movement	3,5	19	13	rectilinear
Drive gear of a reverse movement	3,5	15	13	rectilinear

Table 2. Name of bearing.

Table Name of the Bearings	Numbers
Input shaft bearings	SKF 6208
Countershaft bearings	SKF 6305 + SKF 6306
Output shaft bearings	SKF BS2-2207-2CS

Table 3. Calculated rotational inertias in the Inventor environment.

Name of the part	Mass (kg)	I_{xx} (kg.mm^{-2})	I_{yy} (kg.mm^{-2})	I_{zz} (kg.mm^{-2})
Driven gear of an axle drive reduction	1,181	1849,805	1173,785	1173,785
Drive gear third speed	1,134	1203,801	722,55	722,55
Driven gear third speed	0,696	459,411	306,872	306,872
Drive gear second speed	0,502	398,682	216,189	216,189
Driven gear second speed	0,826	981,071	533,479	533,479
Driven gear first speed	1,15	1993,909	1077,648	1077,648
Reverse movement wheels	0,684	466,87	337,549	337,549

Table 4. Variables necessary for defining the contact of the gearing [7].

Variables	Values
Static friction coefficient (μ_s)	0,1
Static transonic speed (v_s)	1 (mm/s)
Dynamic friction coefficient (μ_d)	0,08
Dynamic transonic speed (v_d)	10 (mm/s)

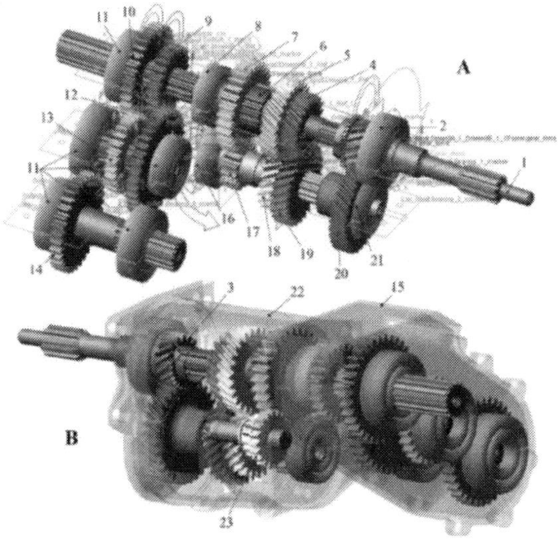

Fig. 2. Virtual model of the mechanical gear box of a vehicle [7, 9]. A – Model with all constraints; B– Model with all cases.

1 – Input shaft with constant-mesh drive gear;
2 – Single row deep groove ball bearing SKF 6208;
3 – Roller bearing SKF NU 2204 between the input shaft 1 and the output shaft 6;
4 – 3rd speed driven gear;
5 – 2nd speed driven gear;
6 – Output shaft of gearbox;
7 – 1st speed driven gear;
8 – Double row angular contact ball bearing SKF 3207BN;
9 – Rear axle and underdrive engaging gear;
10 – Rear axle drive shaft with gear;
11 – Single row deep groove ball bearing SKF 6307;
12 – Layshaft of transfer box with idle gear;
13 – Front axle drive gear;
14 – Front axle drive shaft with gear;
15 – Transfer gearbox case;
16 – Single row deep groove ball bearing SKF 6306;
17 – Layshaft (or countershaft) with 1st speed drive gear;
18 – 2nd speed drive gear;
19 – 3rd speed driver gear;
20 – Constant-mesh driven gear;
21 – Single row deep groove ball bearing SKF 6305;
22 – Gearbox case;
23 – Reverse gears.

2.2 Description of the creation of the joints

The operation of transmission gearings and synchronizer clutches of a real gear box are simulated in the virtual model by joints. The first assumption is that the tooth wheels are fastened on the countershaft with the shaft designated as a "fixed joint". The second assumption is that the tooth wheels are not loaded on the driven shift in the idle position and are fastened only if the speed is engaged. Three shafts in relation to each other are placed by rotation joints designated as "Revolute joint". Single shafts are places in ball-bearings which are connected with the box of the transmission. It enables the vibration signal to be transmitted through these bearings to the gear box (Fig. 2).

2.3 Description of simulation

It is a special process performed in MSC.Adams/View [6] module during which the mechanical gear box of the vehicle is replaced by its model. The aim of this process is to obtain the information about the examined dynamic system and find whether the model is designed properly and whether its behaviour corresponds to reality. To put it simply, simulation is a process during which a virtual gear box model is created. It resembles a real gear box with the properties observed during the operation. In our case this is the simulation of vibrations.

Generally, vibrations consist of vibration signals in three axes – x, y and z. The basic parameters which are calculated for vibrations are trajectory, velocity and angular velocity, acceleration and angular acceleration [10]. All these aspects are simulated and then analysed in the MSC.Adams/Postprocessor module.

The gearbox is used to transfer the engine power and the moment to the axles of the vehicle. In the process of operation it could suffer from heavy stimulation of the engine, the axles, the clutch and other resources such as oil, backlash, and clearance. The stimulation makes the changes of the dynamic load very complex which is the main reason for vibrations, impact, noise and other signs. The selected operating conditions are based on the working characteristics of the gearbox in the field and they are among the huge number of possible choices which can be applied to the virtual gearbox:

a) Rotation speed n = 1500 rpm is applied on the input shaft (Fig. 2, position 1);
b) A position of 3rd speed gear (i = 1,58) is chosen in the gearbox and two-wheel-drive mode is chosen in the transfer box;
c) The load torque can be calculated from the maximum torque of the engine used on the vehicle (166,7 Nm) and the respective ratio (i = 1,58), then the load torques of 263,386 Nm is applied on the output shaft (Fig. 2, position 6);
d) Other parameters are applied as follows:

• simulation time (end time) 1 s,
• simulation step (step size) 0,0001 s.

3 Analysis of the simulation results

3.1 Algorithms for the dynamic simulation

MSC.Adams offers four solvers (the Gstiff, Wstiff, Dstiff and Constant-BDF) to solve the Differential-Algebra Equation (DAE) for the multi-body dynamic simulation. All of them use multi-step, variable order algorithms and apply one of these three integration formats including the Index3 (I3), Stabilized Index 1 (SI1) and Stabilized Index 2 (SI2) [4].

The Gstiff and the Wstiff use a variable step and fixed coefficients. The former helps us to calculate faster and with higher accuracy, but when computing velocity, they can make an error which might excite discontinuities in acceleration. Because of this, the error must be controlled by limiting the maximum step during the simulation. The latter is more useful and stable because it could be modified according to variable steps without any accuracy loss, but it requires more calculated time than those by the Gstiff. The Dstiff algorithm is similar to the Wstiff, but it allows us to choose only the integration format Index3. Unlike them, the Constant-BDF algorithm uses fixed steps, so it is very useful when SI2 format is selected with short step. Although it does not calculate as fast as the Gstiff and Wstiff, it also reaches high accuracy and it is not as sensitive to the discontinuity of the acceleration and force as the Gstiff [4].

Integration formats differ a lot, for instance the SI3 monitors only the error of the displacement and other state variables of the differential equations, but not the velocities and constrained reaction forces. Therefore, its accuracy when calculating velocities, acceleration and constrained reaction forces is not as high as that of the others. The SI1 can monitor all state variables such as displacement, velocity and Lagrange multiplier by introducing the velocity constrained equations instead of acceleration constrained equations. Therefore, it calculates quite accurately but it is very sensitive to the models with friction and contact problems. Unlike the SI1, the SI2 is able to control the errors of the Lagrange multiplier and velocity by considering the velocity constrained equations, so more accurate result could be obtained for the velocity and acceleration computation.

Based on the above information about the solvers and integration formats, the Wstiff solver with SI2 integration is adopted for the dynamic simulation of the gearbox [4].

3.2 Algorithms analysing the contact force

Because the function of contact force is periodic, a Fourier transform (FFT) will be used to decompose it into a sum of simple harmonic functions, namely sines and cosines. Theoretically, the FFT can be defined as follows [11]:

$$x(t) = a_0 + \sum_{n=1}^{\infty} \left(a_n \cos\frac{2n\pi t}{T} + b_n \sin\frac{2n\pi t}{T}\right), \quad (3)$$

where $x(t)$ is the function of contact force with period T, a_n and b_n are constants called the coefficients of the transform and given by the Euler formulas [11]:

$$a_0 = \frac{1}{T}\int_{-T/2}^{T/2} x(t)dt; \quad (4)$$

$$a_n = \frac{2}{T}\int_{-T/2}^{T/2} x(t)\cos\frac{2\pi nt}{T}dt, n = 1, 2, \dots ; \quad (5)$$

$$b_n = \frac{2}{T}\int_{-T/2}^{T/2} x(t)\sin\frac{2\pi nt}{T}dt, n = 1, 2, \dots . \quad (6)$$

However, in practice, the function of contact force is a set of data with discrete and finite values $x_n (n = 1, 2, \dots)$. To perform the analysis using these finite values of discrete data, the discrete Fourier transform (DFT) should be applied [11]:

$$X_n = \frac{1}{N}\sum_{k=0}^{N-1} x_k e^{-j2\pi nk/N} \; nm = 0, 2, 2, \dots N-1, \quad (7)$$

where N is the number of x_n in a constant interval Δt and X_n is called DFT of the discrete values $x_0, x_1 \dots x_{N-1}$. Equation (8) will transfer correspondingly finite values on the time axis to the discrete spectra on the frequency axis. DFT can also use real numbers instead of complex ones [11]:

$$\begin{cases} A_n = \frac{1}{N}\sum_{k=0}^{N-1} x_k \cos\frac{2\pi nk}{N} \\ B_n = \frac{1}{N}\sum_{k=0}^{N-1} x_k \sin\frac{2\pi nk}{N} \end{cases} \; n = 0, 1, 2, \dots N-1, \quad (8)$$

where $X_n = A_n + jB_n$.

We apply a window function in the work in order to provide the discrete values that appear to be continuous and periodic. Discontinuities are "filled in" by forcing the function of contact force to be equal to zero at the beginning and the end of the calculated period.

There are many available windowing functions such as Rectangular (it is equivalent to saying that no window was used), Gaussian, Hamming, Blackman-Harris and Hanning.

$$w[n] = 0{,}723 - 0{,}498\cos\frac{2\pi n}{N} + 0{,}0792\cos\frac{4\pi n}{N} \quad (9)$$

After many trial calculations, Blackman-Harris windowing function was chosen to be used in this work.

4 Results and discussion

4.1 Gear mesh frequencies (GMF)

The gear mesh frequency also called "tooth mesh frequency" is the rate at which gear teeth mesh together in a gearbox. It is equal to the number of teeth on the gear times the rotation speed of the gear [8]

$$f_m = fZ, \quad (10)$$

where f_m is the gear mesh frequency (Hz), f is rotational frequency of the gear (Hz) and Z is the number of teeth.

The number of teeth on the drive gear multiplied by the speed of the drive gear must equal the number of teeth on the driven gear times the speed of the driven gear. As the pinion rotates against the driven gear, the individual cycles of the frequency generated are a profile of the individual teeth meshing. Gear mesh frequencies of gears in the gearbox were calculated as follows (see Table 5):

Table 5. The GMF in 3rd speed gear, two-wheel-drive mode and 1500 rpm on the input shaft [7].

Description	Gear	Number of teeth	Values (Hz)
The constant-mesh gears	Drive gear	$Z_{0R} = 15$	$f_{m0} = 375$
	Driven gear	$Z_{0N} = 32$	
1st speed gear	Drive gear	$Z_{1R} = 15$	$f_{m1} = 175{,}78$
	Driven gear	$Z_{1N} = 29$	
2nd speed gear	Drive gear	$Z_{2R} = 21$	$f_{m2} = 246{,}094$
	Driven gear	$Z_{2N} = 26$	
3rd speed gear	Drive gear	$Z_{3R} = 27$	$f_{m3} = 316{,}406$
	Driven gear	$Z_{3N} = 20$	

4.2 Impact of input revolution on the vibrational signal

In order to analyse the dependence of a vibration signal on the number of revolutions we selected only three input shaft revolutions (1000, 2000 and 3000 rpm^{-1}) shifting into second gear. This number of results is adequate for expressing the dependence. In order to express the simulation results we chose the acceleration values in vertical axis y which is perpendicular to the axis of mechanical gear box rotation (Fig. 3).

On the basis of the above results (Fig. 3) we can hypothesize that the higher the input shaft revolutions, the higher the amplitude in the frequency spectrum. This specific trend is probably the result of the assumption that the higher the number of input shaft revolutions, the higher the gear box oscillation. This analysis proves that higher revolutions of parts result in higher operation wear and later in failure occurrence.

4.3 Impact of shifting speed gear on vibrational signal

In order to verify the impact of shifting single speed gears on the vibration signal we selected 2000 rpm^{-1} on the input shaft (Fig. 4). The graphs showing single speed gears designated as CW1 for first gear up to CW4 for fourth gear present the difference in a vibration signal. Since the tooth wheels have different number of cogs, when connected together on the input shaft, countershaft and output shaft, they cause the difference in the vibration signal. The biggest difference in the vibration signal is at fourth speed gear because of the direct connection of the input and the output shaft. The frequency spectrum at first, second and third speed gear is similar since the countershaft and its tooth wheels are in contact.

4.4 Impact of gear box construction on vibrational signal

The vibration signal is influenced by the number of teeth, shafts and their positioning. In this chapter we selected two possible types of bearings and the direction of the gearing pair of first gear. In another step we simulated the change of the ball-bearing SKF 6305/6306 (6 balls) on the countershaft into the ball-bearing SKF 6305/6306 ETN9 (5 balls) (Fig. 5). The figure shows that the smaller the number of bearing balls, the higher the oscillation amplitude of a gear box.

4.5 Example of an experimental mechanical vibration measurement of the mechanical gearbox

In order to verify the impact of the damaged outer ring on the vibration signal of the tested roller bearing, the outer ring was artificially damaged. We cut a groove parallel to the axis of rotation and perpendicular to the track of rolling elements, see Fig. 6. For the experimental analysis, we selected the measurement at 1200 RPM of the shaft [12].

Table 6 shows the calculation results of the frequencies of typical bearing failures for the shaft revolving at 1200 RPM. The value of *BPFO* corresponds to the result of the frequency analysis, which is given in the upper right hand corner in Fig. 7. Small errors may be caused by the inaccurate measurement of bearing shaft revolutions. Fig. 7 was created after using the envelope detection which, on the basis of a band-elimination filter, enables us to filter the lower and upper undesirable vibrations which may have a negative impact on the envelope detection analysis [12].

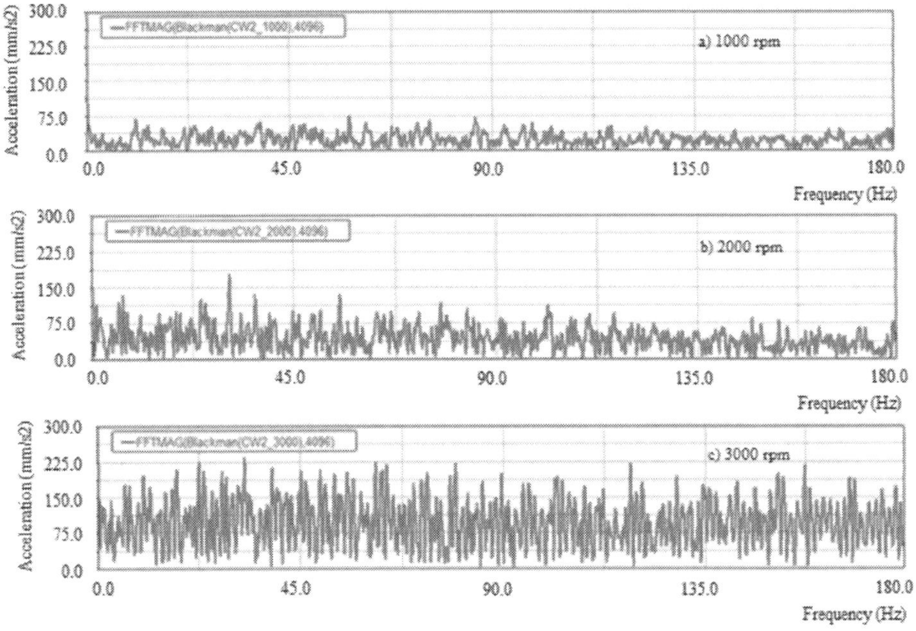

Fig. 3. Record of a frequency spectrum by the number of revolutions, Gearbox UAZ469.

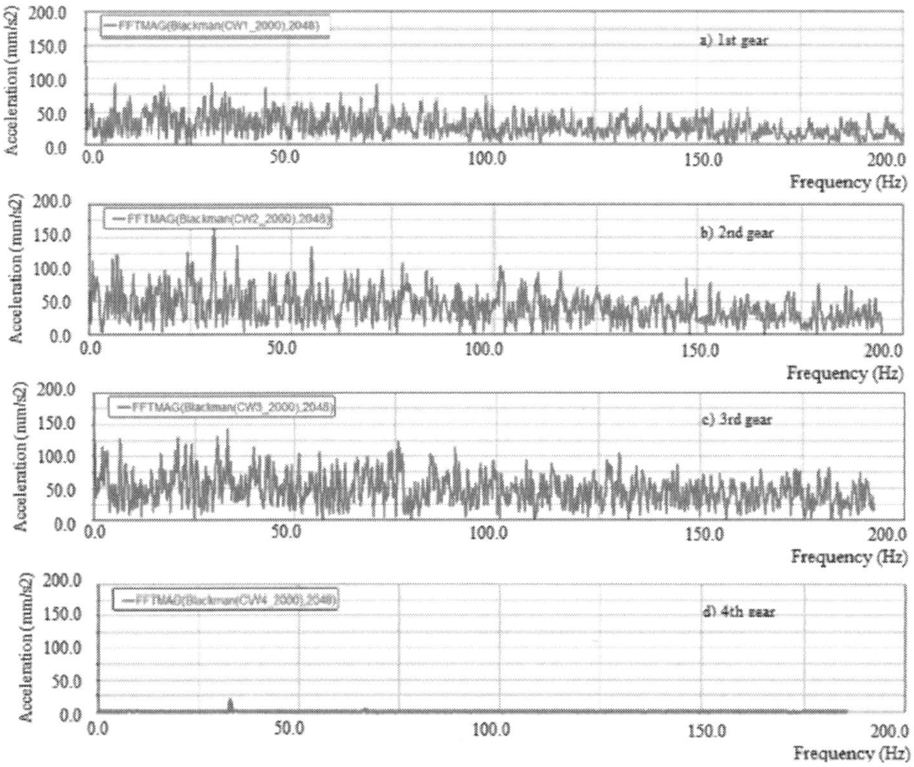

Fig. 4. Frequency spectrum record based on shifting a speed gear, Gearbox UAZ469.

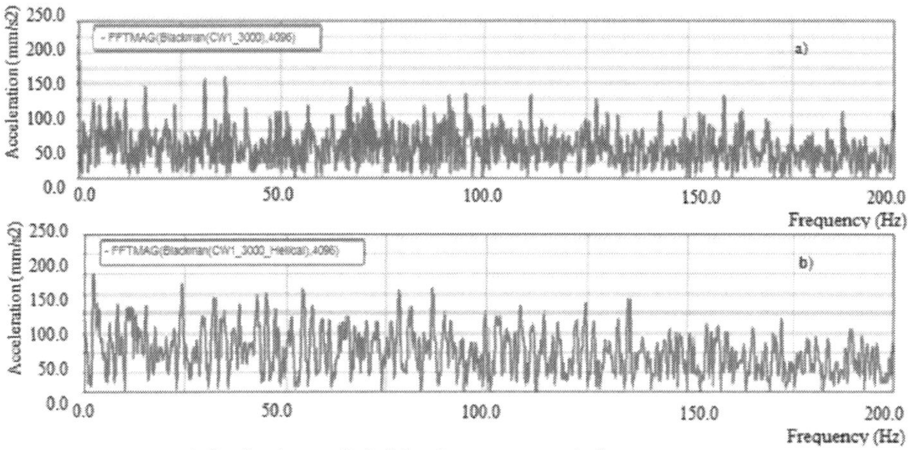

Fig. 5. Frequency spectrum record after the change of a ball-bearing on a countershaft:
a) ball-bearing SKF 6305/6306 (6 balls), *b*) ball-bearing SKF 6305/6306 ETN9 (5 balls).

Fig. 6. A groove created on the outer ring [12].

Table 6. The values calculated for bearing fault frequencies labelled FAG 31308A [12].

Designations of fault frequency	f_r frequency 20 Hz
Snap ring defect – *FTF*	8,42
Bearing defect – *BPF*	107,22
Outer race – *BPFO*	136,64
Inner race – *BPFI*	185,36

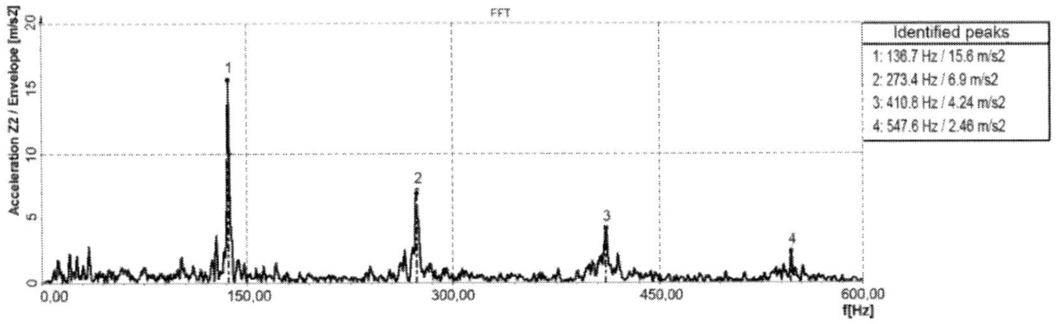

Fig. 7. The frequency spectrum of the damaged outer ring obtained by means of envelope detection [12].

5 Conclusion

The aim of the paper is to establish whether it is possible to simulate vibrodiagnostic signal to monitor the technical condition of rotating components in a mechanical gearbox, which is one of the most important elements in a combat vehicle. On the basis of the process we can state that the vibration signal is a very effective tool when tracking the technical condition of the mechanical gearbox. In the process of modelling, it is important to solve some problems such as creating accurate bodies, determining physical values (mass, rotational inertia, etc.), creating proper motions, setting software for simulation, presenting the simulation results. For modelling and simulation we have used the software of ADAMS/View of the MSC company, which seems to be the most suitable. Together with the MSC Adams/View module, we have used the Inventor software of the Autodesk company for building bodies and determining their physical values. This software enables the bodies to be created more accurately. Next, the virtual bodies were transformed into the MSC Adams/View software where they were simulated and the frequency spectra were recorded. This software was used to make some changes in the mechanical gearbox structure influencing the vibration. Simulations are an effective tool for making different designs which are difficult to make in practice. We established the impact of the different design modifications on the vibration signal which can help us to monitor the technical condition and predict failure in the future.

The present work has been prepared with the support of the Ministry of Defence of the Czech Republic, Partial Project for Institutional Development and Specific Research, Department of Combat and Special Vehicles, University of Defence, Brno.

References

1. D. Wu, J. Meagher, C. Xu, Nonlinear contact analysis of gear teeth for malfunction diagnostics, *IMAC XXVI Conference and Exposition on Structural Dynamics* (2008)
2. Q. Wilson, I. Fatimah, G. Farid, Assessment of gear damage monitoring techniques using vibration measurements, Mechanical Systems and Signal Processing, **15**(5), 905-922 (2001)
3. P. Novotný, A. Prokop, M. Zubík, K. Řehák, Investigating the impact of computational model complexity on noise and vibration modeling of powertrain, Journal of Vibroengineering, **22**(4), 277-392 (2016)
4. Mechanical dynamics. *Building Models in ADAMS/View* (Michigan, Ann Arbor, 2000)
5. X. Wu, A. Sommer, J. Meagher, Spectrum Diagnostics of a Damaged Differential Planetary Gear during Various Operating Conditions, Physical Science International Journal, **9**(3), 1-13 (2016)
6. Adams team at MSC Software. *Adams tutorial kit for mechanical engineering courses*, (Newport Beach, CA, 2003)
7. J. Furch, T. T. Nguyen, Simulation of Failure in Gearbox Using MSC.ADAMS, Acta Universitatis Agriculturae et Silviculturae Mendelianae Brunensis, **65**(2), 419-428, (2017)
8. M. Norton, D. Karczub, *Fundamentals of Noise and Vibration Analysis for Engineers*, 2nd ed. (Cambridge, Cambridge University Press, 2003)
9. J. Furch, J. Glos, T. T. Nguyen, Modelling and simulation of mechanical gearbox vibrations, International scientific conference Transport means, 133-139 (2016)
10. D. Hroncová, M. Binda, P. Šarga, F. Kičák, Kinematical Analysis of Crank Slider Mechanism Using MSC Adams/View. Procedia Engineering, **48**, 213-222 (2012)
11. P. Bakir, *Vibration based structural health monitoring*, (Berlin, Technische Universität Berlin, 2008)
12. J. Furch, J. Glos, T. T. Nguyen, Identification of the Technical Condition of Roller Bearings by Means of Vibrodiagnostics and Tribodiagnostics. Transactions of FAMENA, **41**(3), 99-113, (2017)

Methodology of testing and optimal setting of an asynchronous traction drive for electric vehicles

Vasil Dimitrov[1,*]

[1]Todor Kableshkov University of Transport – Sofia, Faculty of Telecommunications and Electrical Equipment in Transport, Department of Power Engineering and Electrical Equipment in Transport, 1574 Sofia, 158 Geo Milev St., Bulgaria

Abstract. This paper presents a methodology of tests of traction asynchronous drives applied in electric vehicles. It is similar to Urban Driving Cycle ECE-15 (UDC). The objective of the research is to verify the developed methodology through tests performed on a laboratory simulator, which consists of energy-saving asynchronous motor controlled by a frequency inverter (converter). Vector Control on the drive is used for smooth speed regulation over a wide range at high values of efficiency and power factor. Contemporary measuring and control devices and special software products are used for capturing the dynamic characteristics when starting, stopping and changing the speed set point of the drive. Some of tests are presented and an experimental verification of the developed methodology is made. An optimal setting of the drive parameters can be made to minimize the electric energy consumption.

1 Introduction

Contemporary electric vehicles are designed on the basis of power drives controlled by highly efficient devices and microprocessor control and safety systems [1, 2]. Asynchronous motors with improved efficiency and performance are introduced into many modern trams, trolleybuses, underground trains and locomotives, as well as in hybrid and fully electric cars. These motors possess extremely long life and high reliability, offer considerable energy savings and protect the environment, their weight-optimized design has a positive effect on the stability of the equipment unit. Motors with aluminium housing have particular low weight and they are preferred for applications in vehicles. There are also motors with increased output and compact construction which could be used to advantage in confined spaces. They are also optimized for efficiency and designed for inverter-fed operation with a pulse width modulation (PWM). They feature an insulation system consisting of high-quality enamel wires and insulating sheeting in conjunction with solvent-free resin impregnation. A smooth speed control from 0 to several times over the rated one is limited only by the limits for the roller bearings, critical rotor speed and rigidity of the rotating parts.

In this paper, a methodology of testing traction asynchronous drives applied in electric vehicles is developed. It is similar to Urban Driving Cycle ECE-15 (UDC) used to assess the emission levels and fuel economy consumption of car engines, as well as for the measurement of electric energy consumption and electric range in hybrid and fully electric vehicles [3-6]. A laboratory system has been designed and built and contemporary measuring and devices are used for monitoring and control. The objective of the research is to verify the developed methodology through tests performed on the simulator using special software products to capture the dynamic characteristics when starting, stopping and changing the speed set point of the drive.

2 Possibilities of speed and torque control on the traction drive

In electric vehicles, load torque varies significantly and speed should be adjusted smoothly from zero to the maximum value. In this case, the primary law of frequency controlled asynchronous drives (established by M. P. Kostenko) must be used [1, 2]:

$$\frac{U_1}{U_{1N}} = \frac{f_1}{f_{1N}} \sqrt{\frac{M}{M_N}}\,, \qquad (1)$$

where U_1 and U_{1N} are the adjustable and nominal voltage; f_1 and f_{1N} – adjustable and nominal frequency; M and M_N – load torque and rated motor torque.

This method is known as "Vector Control" (Field-orientated closed-loop control). It ensures reliable and flexible motor operation and provides a stable speed irrespective of any disturbances (like the supply pressure fluctuations, climatic conditions, load mass, road profile, track condition etc.). The Vector Control (VC) has also many other advantages: stable parameters when changing loads and setting point values, the motor and braking torque are controlled independently of the speed, accelerating and braking are possible with a maximum adjustable torque, high values of the power factor and the efficiency can be achieved. Therefore, the VC must be used in all applications where the load varies widely, as in

* Corresponding author: vdimitroff@abv.bg

electric vehicles. The use of maximum adhesion, if necessary (e.g. starting or climbing a slope), and the realization of skid resistance is also possible.

3 Methodology of testing

3.1 Driving cycles

A driving cycle is a series of data points representing the speed of a vehicle versus time. Different driving cycles are developed to assess the performance of vehicles in various ways, as for example fuel consumption, polluting emissions, performance of internal combustion engines, transmissions, electric drive systems, batteries, fuel cell systems, and similar components [3-6]. The Urban Driving Cycle ECE-15 (or just UDC) has been designed to represent typical driving conditions in busy European cities (defined by ECE R83, R84 and R101). It is characterized by low engine load and a maximum speed of 50 km/h (Fig. 1).

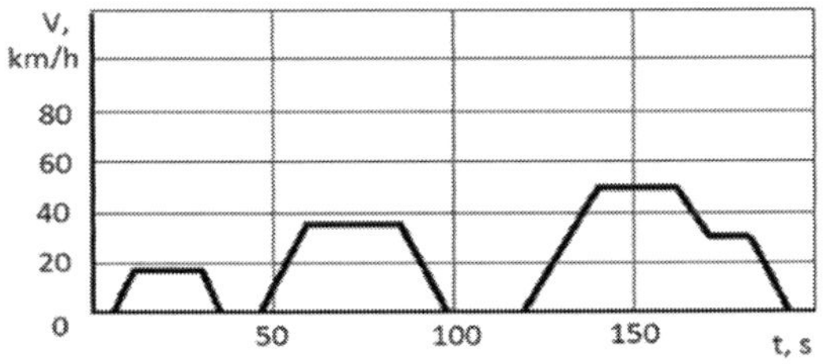

Fig. 1. Scheme of the Urban Driving Cycle (UDC).

3.2 Laboratory simulator

The laboratory simulator is built on contemporary devices, based on the newest technologies (Fig. 2) [7]. The main elements involved in the configuration of the simulator are developed by Siemens: three phase energy-saving asynchronous motor (AM – 2,2 kW) with lightweight construction; Sinamics G120 frequency converter (includes three phase IGBT inverter and Microprocessor Control Unit – CU); Simatic S7-200 Programmable Logic Controller (PLC); TP-177micro Touch Panel. An encoder is mounted on the motor shaft and is used for closed loop motion control on the drive. A synchronous generator (SG – 30V, 60A) with a bridge rectifier and resistors simulates the load of the motor. It can be changed smoothly using the Flux current Regulator or resistors. A WINDOWS-based computer system is used for configuring and data storage. The necessary commissioning software STARTER is installed on this computer and is used for setting-up the Sinamics G120. A subprogram TRACE is developed for real-time tracing of the inverter parameters. It can trace up to 30 parameters simultaneously, and then export the obtained data into an Excel sheet and create graphs. The intelligent energy meter EMPS is connected to the drive and allows for monitoring the operation and analyzing the current consumption and power flows.

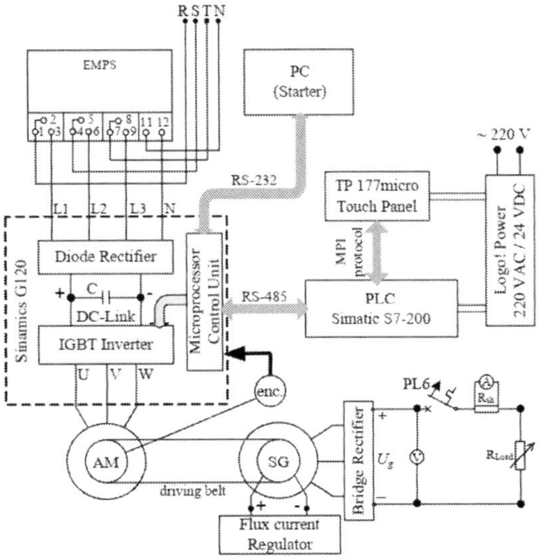

Fig. 2. Scheme of the laboratory simulator.

3.3 Experimental research

The software STARTER and the subprogram TRACE have to be run. VC is used and the load is constant during the tests (equal to 75% of the rated motor torque). The developed methodology consists of the following tests:

- The motor slowly accelerates to 450 rpm (equivalent to 30% of the nominal speed – 15 km/h), runs at constant speed, and then slowly stops.
- After a short break the motor slowly accelerates to 960 rpm (equivalent to 64% of the nominal speed – 32 km/h), runs at constant speed, and then slowly brakes to a full stop.
- After a new short break the motor slowly accelerates to 1500 rpm (equivalent to the nominal speed of 50 km/h), runs at constant speed, and then slowly stops (according to SAE J227a driving schedule the last element of the UDC is not necessary to be included when electric vehicles are tested - it could be replaced by a 'coast' phase, during which the tractive effort is set to zero but the speed is not specified, and the energy consumption is negligible [3]).

The obtained graphs are shown in Fig. 3.

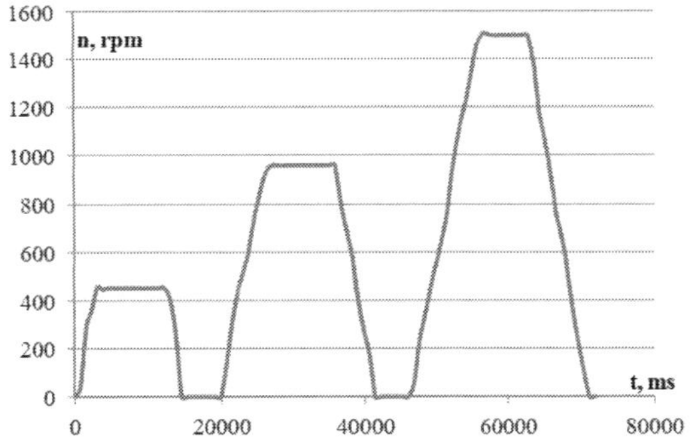

Fig. 3. Graph for the developed methodology.

3.3.1 Test results

The speed set points, as well as the “Start” and “Stop” commands, are put by Touch panel and PLC [7]. Except for the actual rotor speed (shown in Fig. 3), the frequency, torque, power, voltage, power factor and current (as well as its components) have been traced by TRACE subprogram and STARTER commissioning software. Some of test results are shown in Figures 4, 5, and 6. The energy consumption has been measured by EMPS.

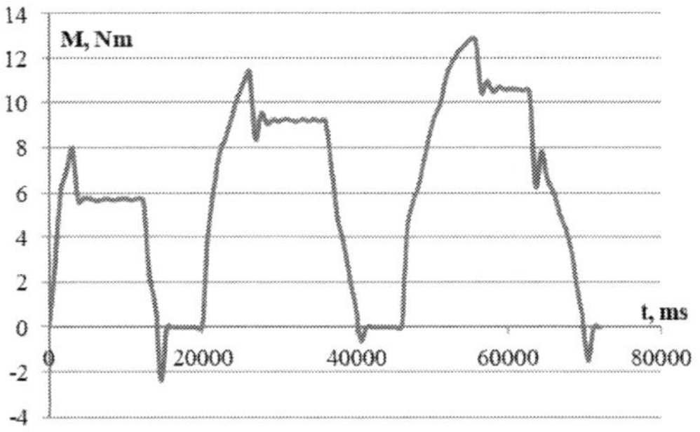

Fig. 4. The motor torque during the tests.

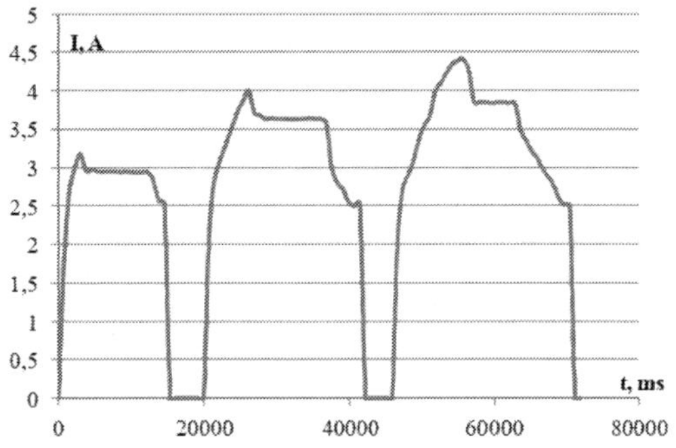

Fig. 5. The motor current during the tests.

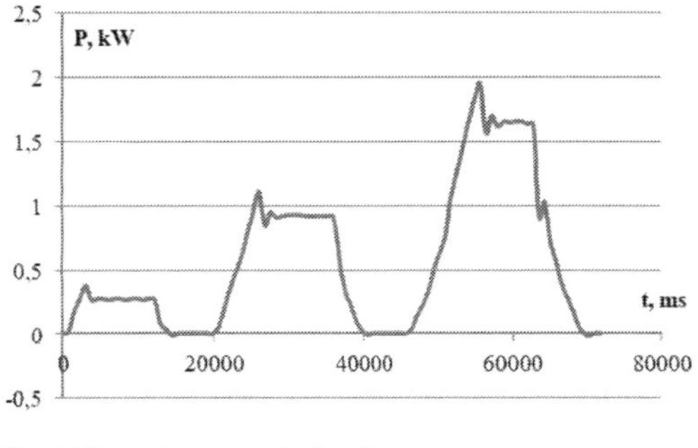

Fig. 6. The active power during the tests.

3.3.2 Study of the impact of the acceleration time on the drive performance

As opposed to the UDC, the acceleration and deceleration time can be changed and their impact on the drive performance and energy consumption can be examined. The ramp-function is used to limit the acceleration when the speed set point n_{sp} (proportional to f_{sp}) changes according to a step function. This helps to reduce the stress on the mechanical system of the drive. An acceleration ramp can be set using parameter $T_{ramp\text{-}up}$ of Sinamics G120. This is the time it takes the motor to accelerate from standstill up to maximum motor frequency f_{max} when no rounding is used. This function allows also for a controlled transition when the set point is changed. The acceleration time T_{acc} from a standstill up to a frequency set point f_{sp} can be calculated by the following equation:

$$T_{acc} = \frac{f_{sp}}{f_{\max}} T_{ramp-up} . \tag{2}$$

The synchronous speed is 1500 rpm at the nominal frequency 50 Hz. The maximum frequency is set to 100 Hz. Some examinations were performed at different acceleration time and the same load. Subprogram TRACE and VC on the drive are used again and the obtained dynamic characteristics are drawn in Fig. 7.

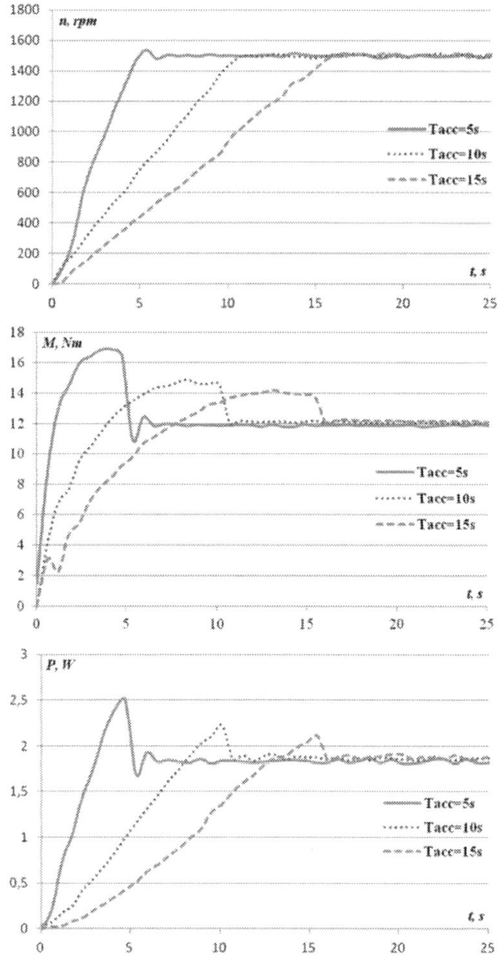

Fig. 7. Dynamic characteristics when starting at different acceleration time.

The main dynamic indicator is the overshoot σ – this is the maximum excess of any parameter y over the set-point value, expressed in percentage:

$$\sigma = \frac{y_{M1} - y_{sp}}{y_{sp}}.100,\% \quad (3)$$

where: y_{M1} is the maximum value (exceedance) of the parameter y reached during the transient mode; y_{sp} is the set-point value of this parameter defined by the operator or calculated by the microprocessor CU.

Using (3) the speed, torque, current and power overshoots have been calculated and their dependence on the T_{acc} is shown in Fig. 8. Analyzing Fig. 7 and Fig. 8 it can be seen that over-speed is negligible (under than 3%) but could be totally reduced at longer T_{acc}. Shorter acceleration time requires bigger current and torque surges and leads to increased jerk impact on the mechanical part and the power supply (accumulator batteries).

The power consumption also depends on the acceleration time (Fig. 9). Longer T_{acc} results in lower consumption.

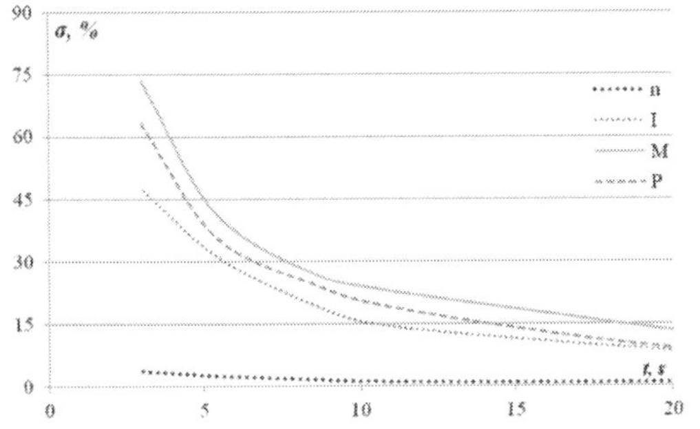

Fig. 8. Graphs representing overshoots dependence on acceleration time.

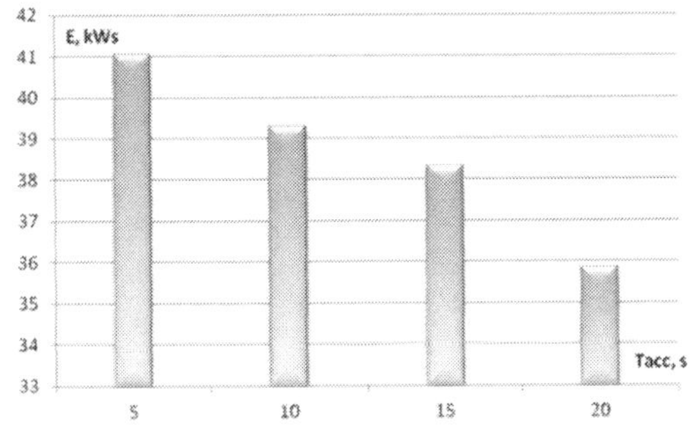

Fig. 9. Graphs representing energy consumption dependence on acceleration time.

In order to avoid torque surges at the transitions from constant velocity phase to accelerating phase, additional rounding-off times could be programmed (parameters "Ramp-up initial rounding time" and "Ramp-up final rounding time"). This is important for electric vehicles, which especially require "soft" (jerk-free) acceleration. In this case, the speed and torque are changed more slightly and smoothly. This leads to increased comfort of the passengers during vehicle acceleration.

Similar tests can be performed to determine the parameters dependence on the deceleration time T_{dcc}. A part of the energy can be recuperated (see Fig. 6) and used to recharge the batteries. Shorter T_{dcc} requires bigger over-voltage and breaking torque surges and leads to increased jerk impact and big recuperate current that could damage the batteries or shorten the service life.

The impact of many parameters (PWM frequency, current, torque and speed limitations and many others) on the drive performance could be defined using the developed methodology and the optimal setting of the control system can generally be achieved.

4 Conclusion

The implementation of asynchronous motors as a traction machine in electric vehicles provides good performance. Using VC, an infinitely variable speed and torque in a wide range can be realized. The torque and speed fluctuations could be reduced. Very good energetic performance and ride comfort can be achieved.

The developed methodology provides the opportunity of an optimal setting of the drive and combines the advantages of the UDC and SAE J227a driving schedules. The coasting phase is ignored - during it the tractive effort is set to zero but recuperation is not possible , the speed and the energy consumption are not specified, that leads to somewhat of a nuisance to model. The methodology is based on the possibility of examining the influence of the inverter parameters on the energetic drive performance. The optimal ratio of the dynamic features of the car to the energy consumption can be accomplished. The verification has been done using a laboratory simulator – the impact of the acceleration time on the drive performance and energy consumption has been defined. This methodology and the obtained results can be used in the design of drives for electric vehicles and their optimal setting – a longest run and maximum passengers comfort can be achieved.

References

1. L. Balgaranov, *Electric transport*, Sofia, (2009)
2. U. Bakshi, M. Bakshi, *Electrical Drives and Control*, (Technical Publications Pune, India, 2009)
3. J. Larminie, J. Lowry, *Electric Vehicle Technology Explained*, (John Wiley&Sons Ltd., England 2003)
4. P. Mock, Discrepancies between type-approval and "real-world" fuel-consumption and CO_2 values, *ICCT Working paper 2012–02* (2012)
5. E/ECE/TRANS/505/Rev.2/Add.100/Rev.3 (12.04.2013)
6. Vehicle Regulations, UNECE Transport Division, http://www.unece.org/trans/main/welcwp29.html
7. V. Dimitrov, E. Dimitrova, Using PLC for control on asynchronous drives – laboratory simulator, *ICEST 2016*, Ohrid, Macedonia, 471-474 (2016)

Experimental study of electric vehicle gearbox efficiency

Josef Břoušek[1,*], and *Tomáš* Zvolský[1]

[1]Technical University of Liberec, Department of Vehicles and Engines, 461 17 Liberec, Czech Republic

Abstract. This paper presents an experimental analysis of the overall efficiency of the gearbox for an electric vehicle with a Dual Motor Drive System (DMDS) propulsion system. The efficiency was measured in the powertrain testbed where the gearbox was placed between two dynamometers. The main aim of the experiment was to get relevant values of an overall efficiency as a function of input speed and torque. All these measured values of the gearbox efficiency are included in graphs and efficiency maps. Tests were run with two different amounts of oil charge. The experiment confirmed the expected high values of the overall mechanical gearbox efficiency. The results can be used further for advanced simulations and other electric vehicle experiments.

1 Introduction

Nowadays, researchers and automobile producers come together in search of a way to increase the efficiency of vehicles by perfecting or changing the technology and concepts incorporating the idea of modern transportation [1]. The whole world has been focused in recent years on vehicles with alternative vehicle propulsion such as electric (EV) and hybrid (HEV) vehicles because of emission restrictions and fuel consumption of vehicles related to global warming and limited energy resources. Research and development are focused on new drive concepts and systems that allow for greater availability of known technologies in terms of affordability and reliability of these drives [2]. The main topics in the field of EV are drive range extension per one charge, performance increase and drive efficiency [3].

The overall efficiency of the EV propulsion system is shown in Fig. 1 and described in formula (1). The overall efficiency η_{EV} includes battery efficiency η_1, power converter and electric motor efficiency η_2 and gearbox efficiency η_3 [4]. P_1 is input power, P_2 is output power. P_{LOSS1} are losses in battery, P_{LOSS2} are losses in power converter and electric motor, P_{LOSS3} are losses in gearbox. If we want to focus on improving the overall efficiency we need to deal with all of these individual components. This article focuses on the last mentioned component – the gearbox efficiency.

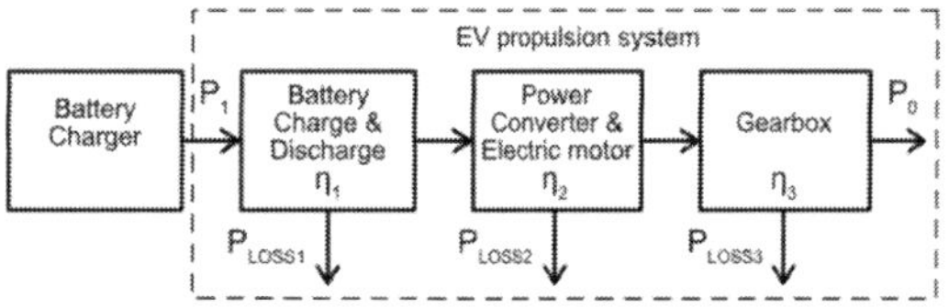

Fig. 1. The individual components of the efficiency of the EV propulsion system [4].

$$\eta_{EV} = \frac{P_0}{P_1} = \frac{P_0}{P_0 + \sum_{i=1}^{3} P_{LOSSi}} = \eta_1\eta_2\eta_3 \quad (1)$$

Overall gearbox efficiency is determined by individual components loss or equivalently the power losses of the individual components of the powertrain: gearing losses, bearing losses, sealing losses, synchronizing losses, clutch losses, torque converter losses, auxiliary units [5]. Gear power losses are strongly related to lubrication. Load dependent losses are caused by the frictional effects in the lubricant film and load independent losses are from squeezing, churning and windage effects [6]. These hydraulic losses are well analysed in [6].

In scientific publications certain values of gearbox efficiency are often assumed. These values are often based on calculations and simulations and then they are used for other calculations. For example, in book [4] on page 12 are listed values 95-98 %. The authors of paper [7] work with values 86-94 % during the load values 5 Nm. Tehrani in his thesis [8] works with gearbox efficiency values above 98% in some areas of speed and load.

At the Technical University of Liberec current development is focused on a Dual Motor Drive System (DMDS) electric propulsion system shown in Fig. 2. DMDS consists of two electric motors and one single-speed gearbox with a mechanical differential. Current research shows that this system can provide high drive efficiency under all loads, without limiting the driving dynamics [9]. All components of the propulsion system we test in our lightweight battery electric vehicle for educational purposes, which has a completely open frame from aluminium profiles. This solution gives an excellent access to all parts of the vehicle and allows easy modification [10].

* Corresponding author: josef.brousek@tul.cz

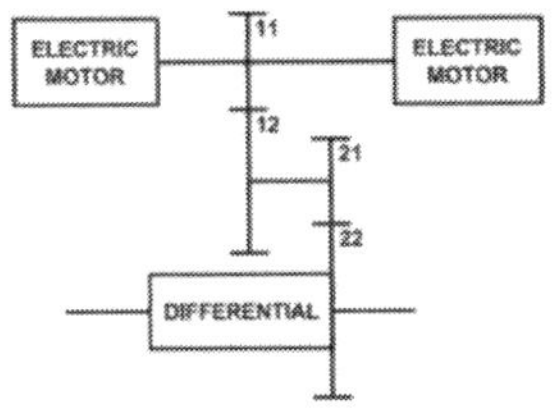

Fig. 2. Dual Motor Drive System components arrangement.

For our next steps in the research, we need to consider the exact values of each component of the overall efficiency. The main aim of our experiment was to measure the efficiency of an EV gearbox in our powertrain testbed.

2 Methods and Materials

2.1 Prototype DMDS gearbox

The subject of measurement was our prototype DMDS gearbox (shown in Fig. 3) which we designed and made for the purposes of our DMDS research. This prototype gearbox allows for the connection of any two electric motors with a standardized flange NEMA C-face with the limitations of overall maximum input torque at 200 Nm and maximum input speed of 7000 rpm. The overall ratio is 5,857.

Fig. 3. Prototype DMDS gearbox used for the experiment.

Each of the motors is placed on one side of the gearbox and the torque of both motors is brought to the first shaft of the gearbox. First gear set and the final drive with differential are all used from the third gear of Skoda MQ200 02T gearbox. All bearings are also used from this gearbox. Internal components are placed in our designed gearbox housing. The material of the housing is aluminium alloy EN AW 7021 [AlZn5,5Mg1,5].

The main aim of the experiment was to get relevant values of the overall efficiency as a function of input speed and torque for our gearbox. We plan to continue to use these values for subsequent simulations and other experiments.

2.2 Powertrain testbed

We decided to use a special testing mode of our powertrain testbed (shown in Fig. 4) for this gearbox efficiency experiment. This testbed is mainly designed for complete powertrain testing. It means the assembly of the engine-clutch-gearbox-drive shafts.

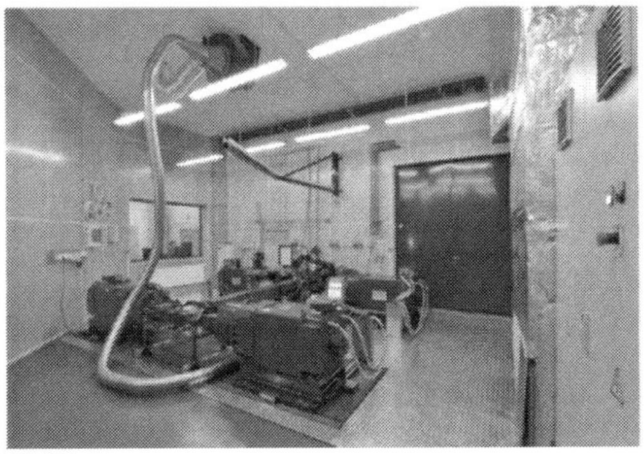

Fig. 4. Powertrain testbed room.

The testbed consists of four asynchronous dynamometers. One pair of dynamometers is composed of Siemens 136 ADG 288WP (each one 120 kW / 2600 min^{-1} / 440 Nm; max. speed 3000 min^{-1}). The second pair is composed of Siemens 111 ADG 286WP (each one 80 kW / 2600 min^{-1} / 294 Nm; max. speed 3000 min^{-1}). This solution provides testing of 2wheel drive and also 4wheel drive and hybrid powertrain where one axis can be power driven by the combustion engine and the second one by the electric motor shown in Fig. 5. Manual shifting is provided by a shifting robot, which can control the shift lever of manual gearboxes. The testing room is further equipped with the advanced control system, air condition, and extraction of exhaust gases. As fuels, it is possible to use petrol, diesel, LPG, CNG, and hydrogen. We use this testbed mainly for gearbox testing, efficiency measuring and for long-term tests of complete powertrains. During these tests, we normally use combustion engines as propulsion of the testing assembly and the braking torque is generated by a different number of dynamometers as required.

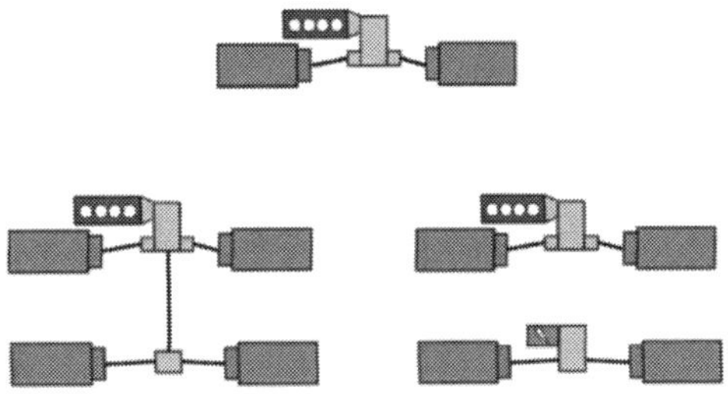

Fig. 5. Powertrain testbed layout possibilities.

2.3 Experimental setup

The special mode mentioned in the previous subsection makes it possible to use one dynamometer as a motor and the second one on the other side as a dynamometer as shown in Fig. 6. Each dynamometer is equipped with very precise measurement of speed and torque. Specifically, the torque is measured by torque flange HBP T40B with accuracy class 0,05. The torque flange is combined with the integrated speed measuring system.

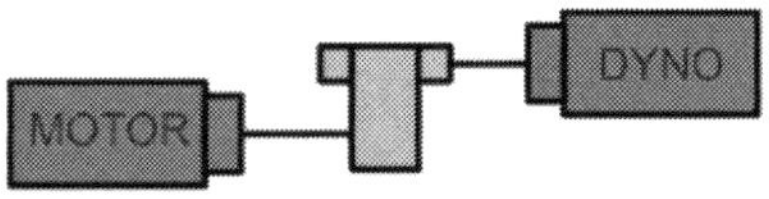

Fig. 6. Powertrain testbed layout for the gearbox efficiency measurement.

The gearbox was placed between two dynamometers on the experimental frame from aluminium profiles at the powertrain testbed shown in Fig. 7. Mounting points were retained as they are used in the vehicle. The gearbox input shaft was connected to the dynamometer which was used in a motor mode. The output shaft was connected to the opposite dynamometer. To evaluate gearbox efficiency we measured and logged values of input speed and torque and the same variables for the output. During testing, we controlled input speed and output torque by the testbed control system.

Fig. 7. Experimental setup of the gearbox efficiency measurement.

For the purpose of this experiment, we had to lock the differential, because we could use only one output shaft of the gearbox. Locking was made by using a special element which was put instead of the differential gear (shown in Fig. 8).

Fig. 8. A special element used to lock the differential

The oil temperature was measured by thermocouple OMEGA Engineering TJC200-CAIN-M025U-150 placed at the inner bottom of the gearbox housing.

2.4 Experiment details

As we mentioned in the previous subsection we controlled the values of input speed and output torque. The measurements were performed at the points with the required values shown in Table 1. It was measured for each speed at all loads.

Table 1. Required values for the experiment.

Controlled		Calculated
Input speed	Output torque	Input torque
rpm	Nm	Nm
500	117,1	20,0
1000	234,3	40,0
1500	351,4	60,0
2000	468,6	80,0
2500	500	85,4
3000	-	-

Note: We couldn't measure in the full range of gearbox input torque (max. 200 Nm) because we used only one input point of the gearbox which is designed for max. input torque 100 Nm. The control system of the powertrain testbed allowed us to use max. output torque 500 Nm. That is the reason why the last required value of output torque is 500 Nm and not 585,7 Nm, which would correspond to the max. usable input torque 100 Nm for only one input point.

The amount of oil charge was determined on the basis of a 3D model during the development process of the experimental gearbox. We decided to run two complete efficiency measurements with two different amounts of oil charge to determine the effect of the oil charge on the efficiency. For the first experiment, we filled the gearbox with 500 g of oil. With this amount of oil charge the oil level reached the bottom part of the final drive gear. For the second experiment, we used 820 g of oil. In this case the oil level reached the bottom of the gear on the input shaft. The type of used gear oil was Mogul Syntrans 75W-90 with the specification API GL-4+ for both measurements. This type of oil is normally used in conventional Skoda MQ200 vehicle gearbox. We used this type of oil mainly because almost all internal components come from this gearbox as we described in subsection 2.1.

Each measured point was measured for 30 s with the recording frequency of 100 Hz. During the experiment, we logged the following values of input speed ω_{in} and torque Min, output speed ω_{out} and torque M_{out} and oil temperature. From the logged values were calculated the mean values for each measured point by Matlab. Then we calculated the input power P_{in} and output power P_{out} on the basis of the obtained mean values and calculated the overall gearbox efficiency $\eta_{gearbox}$ according to the equation (2).

$$\eta_{gearbox} = \frac{P_{out}}{P_{in}} = \frac{\omega_{out}\, M_{out}}{\omega_{in}\, M_{in}} \quad (2)$$

3 Results

The calculated overall gearbox efficiency from the measurement with 500 g oil charge is shown in Fig. 9 and with 820 g oil charge in Fig. 10. Each symbol corresponds to a different input speed of the gearbox. For each input speed can be found a point close to the input torque values 20 Nm, 40 Nm, 60 Nm, 80 Nm and 85,4 Nm.

The resulting values of the overall gearbox efficiency for 820 g oil charge in Fig. 10 are lower at every point than the other results for 500 g oil charge.

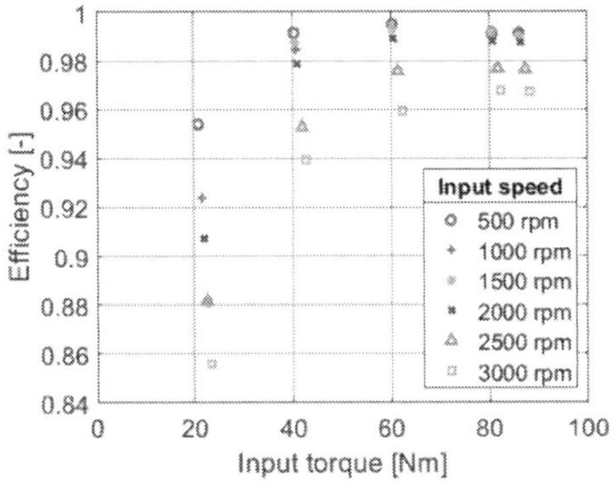

Fig. 9. Results of the overall gearbox efficiency for 500 g oil charge.

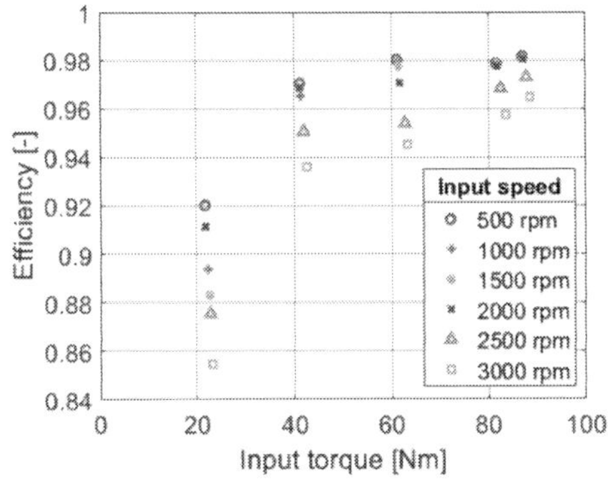

Fig. 10. Results of the overall gearbox efficiency for 820 g oil charge.

We used the obtained resulting values from both previous graphs to build efficiency maps in Matlab shown in Fig. 11 and Fig. 12. Mesh grids for both maps were created by the linearization with 30 evenly spaced points between the minimal values and maximal values of the input speeds and input torques.

The oil temperature values measured at the inner bottom of the gearbox housing were during the measurement with 820 g oil charge in the range 40,3-51,2 °C with average value of 44,5 °C and with 500 g oil charge in the range 38,1-51,2 °C with average value of 44,0 °C.

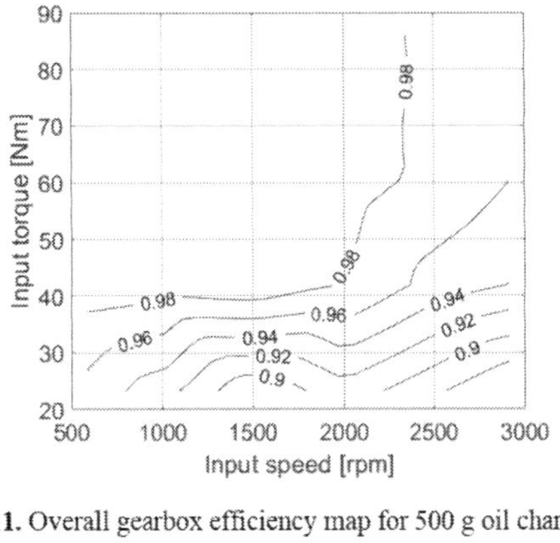

Fig. 11. Overall gearbox efficiency map for 500 g oil charge.

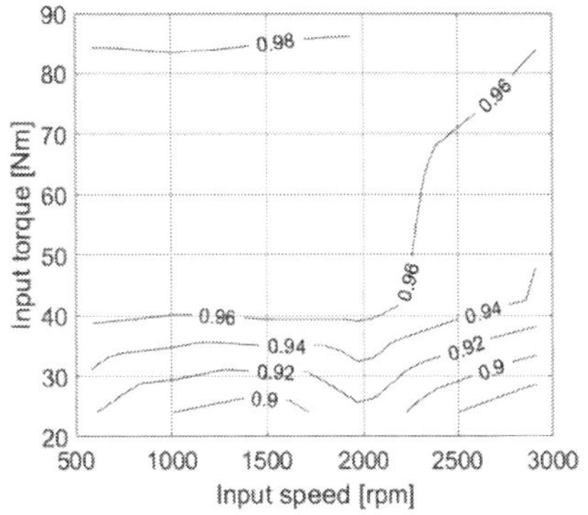

Fig. 12. Overall gearbox efficiency map for 820 g oil charge.

4 Discussion

The results of the experiment provided detailed data on the efficiency of our prototype DMDS gearbox depending on the input speed and the input torque in the range of input speed of 500-3000 rpm and in the range of input torque of 20-85,4 Nm.

In the introduction section of the paper, we mentioned some efficiency values stated in books and other scientific articles. There we can find values of gearbox efficiency around 86% to more than 98%. A lot of them are based on calculations and simulations. The resulting values of our experiment with our prototype DMDS gearbox are almost in the same range. Values for 500 g oil charge at input speeds of 500 rpm, 1000 rpm and 1500 rpm in some areas of input torque of 40 Nm, 60 Nm, 80 Nm, 86 Nm are even higher than 99 %. On the other hand, in the areas of the whole measured input speeds of 500-3000 rpm and at a low load of around 20 Nm, the values are significantly lower – in the range of 85,6-95,4 %.

A lot of the inside components of the gearbox come from Skoda MQ200 02T gearbox commonly used in today's vehicles. The results of our experiment may also indicate efficiency values in some conventional vehicle transmissions.

The experiment was performed with two different amounts of oil charge. The results show that the amount of charge in the measured gearbox has a relatively significant effect on its overall efficiency. The resulting values are on average by 1,2% lower with the oil filling 820 g than with 500 g. The maximum difference is 3,4% at the point 500 rpm of input speed and 21 Nm of input torque. We plan to investigate more precisely the choice of oil type and its right amount in the next steps to improve the efficiency of our prototype DMDS gearbox.

In our opinion, the results from our experimental analysis of the overall efficiency of our prototype DMDS gearbox in the powertrain testbed confirmed the quality design and the technological process of its production. The resulting overall gearbox efficiency maps can be used for detailed advanced simulations of new EV propulsion systems because they are built from efficiency values measured in specific values of input speed and input torque.

For further work on analyzing the efficiency of this gearbox it would be appropriate to analyze the efficiency in the rest of input torque range 100 Nm to 200 Nm and in the rest of input speed range 3000 rpm to 7000 rpm. For such measurement it would be necessary to use another special testbed with required parameters. Using the resulting efficiency values it certainly would be interesting to measure the efficiency in the smaller intervals of input speed and input torque in the areas of lower input torque than 40 Nm. This would allow us to build more accurate efficiency maps in these areas where there are significant changes in efficiency values. Also it would be appropriate to select such controlled output torque values so that the input torque values would be closer to decimal values. Further work could also be done on efficiency analysis at higher oil temperature values.

This publication was written at the Technical University of Liberec as part of project 21127 with the support of the Specific University Research Grant, as provided by the Ministry of Education, Youth and Sports of the Czech Republic in the year 2018.

References

1. T. Gechev, M. Mruzek, D. Barta, Comparison of real driving cycles and consumed braking power in suburban Slovakian driving, *MATEC Web of Conferences – BulTrans 2017*, **133** (2017)
2. C. Chan, The state of the art of electric, hybrid, and fuel cell vehicles, *Proceedings of the IEEE*, **95**, 704-718 (2007)
3. K. Rajashekara, Present status and future trends in electric vehicle propulsion technologies, *IEEE Journal of Emerging and Selected Topics in Power Electronics*, **1**, 3-10 (2013)
4. I. Husain, *Electric and hybrid vehicles design fundamentals* (CRC Press, Florida, 2003)
5. H. Naunheimer, B. Bertsche, J. Ryborz, W. Novak, *Automotive transmissions*, (Springer-Verlag Berlin Heidelberg, Berlin, 2011)
6. C. Gorla, F. Concli, K. Stahl, B. Höhn, K. Michaelis, H. Schultheiß, J. Stemplinger, Hydraulic losses of a gearbox: CFD analysis and experiments, *Tribology International*, **66**, 337-344 (2013)
7. J. Li, C. Li, J. Huang, Transmission efficiency analysis and experiment research of gear box, Advances in Engineering Research (AER), **105**, 298-303 (2017)
8. M. Tehrani, Energy efficiency consideration in electric vehicle transmissions. Lappeenranta University of Technology, (2013) https://www.doria.fi/bitstream/handle/10024/90781/Final%20thesis%20last%20revision%20END.pdf?sequence=2 (April 23, 2018)
9. P. Jandura, J. Břoušek, M. Bukvic, The concept of a highly efficient powertrain for an electric vehicle with respect to vehicle driving dynamics. *International Conference on Electrical Drives and Power Electronics*, Tatranska Lomnica, Slovakia, 422-429 (2015)
10. P. Jandura, M. Bukvic, Lightweight battery electric vehicle for educational purposes, Applied Mechanics and Materials, **390**, 281-285 (2013)

Study of the vibrational behaviour of the components of a car suspension

Zlatin Georgiev[1,*], and *Lilo* Kunchev[1]

[1]Technical University – Sofia, Faculty of Transport, 8 Kliment Ohridski Blvd., 1000 Sofia, Bulgaria

Abstract. The paper presents a study of the vibrational behaviour of a car suspension in the frequency range from 50 to 200 Hz. The vehicle has a torsion spring suspension. The aim of the study is to provide a numerical approach, which evaluates the impact of the tyre vibration characteristics and the elastic properties of suspension elements on the vibrational behaviour of the "tyre-suspension-body" system. The object of the study are the elements of the suspension – stub-axle, upper and lower suspension arms, strut rod, torsion bar and others. The suspension is equipped with pneumatic tyres with different vibration characteristics. Laboratory experiments are done. The tyre contact patch is disturbed by harmonic force. The accelerations of the wheel axle and specific points of the suspension are obtained. A numerical model of the system is created by using the Finite Element Method available in SolidWorks Simulation software. The model allows to determinate the accelerations at different points of the suspension. The results of the experimental and theoretical studies are compared. The values obtained in the two studies are similar.

1 Introduction

The pneumatic tyre perceives the road unevenness and roughness and transmits them to the suspension elements during the movement of a vehicle. The generated vibrations enter into the passenger compartment through connections between the suspension and the vehicle body. Correctly selected elastic and damping characteristics of the suspension elements (shock absorber, spring, rubber mounts), and modal parameters of the metal elements (rods, shells, plates) can improve car comfort in terms of noise and vibrations. This reduces driver fatigue. Depending on the vehicle class, the vibration and noise levels in its cabin are different.

The driver feels these vibrations in the seat, the floor panels and the steering wheel. It is known from the physiological tests made on a person on a vibrating platform that he perceives vibrations in a different way. The vibrations up to 5 Hz have effect on the entire human body. Those of up to 10 Hz affect the limbs and the torso, up to 20 Hz - the lower torso and limbs, up to 60 Hz - the limbs. A person perceives vibrations of over 60 Hz only with the receptors of the skin.

2 Overview of the problem and setting the aim

The interaction between the tyre and the road surface is a source of vibrations. These vibrations affect the ride and the vibro-acoustic comfort as well as the road holding [1, 2]. They depend on the tyre characteristics, which are similar to those of air-springs [3] for frequencies below 30 Hz. The natural frequencies of the pneumatic tyre depend on its construction. In [4, 5], the authors conduct theoretical and experimental studies, showing that the sidewall stiffness has the greatest impact on the tyre natural frequencies. In [6, 7], the authors study the effect of circular velocity on their natural frequencies. They determine that the first natural frequency is the same for both unloaded, non-rotating tyre, and for loaded, rotating tyre. There is a splitting of modes associated with higher frequencies [8] and in case of a loaded rotating tyre, the natural frequency decreases as the circular velocity increases. Kindt et al. [9] conduct a theoretical and experimental study showing that the transmission ratio between the road excitations and the wheel axis as well as the tyre dynamic transfer stiffness depend on its natural frequencies.

Despite the small amplitude of high frequency vibrations, the vibrations generated by the interaction between the tyre and the road are not absorbed by the elastic elements and the shock absorbers. Through the contact points of the suspension, they are transmitted to the vehicle body without noticeably lowering their levels. Klerk and Ossipov [10] conduct experiments with six different types of radial tyres (from slicks to off-road profile, as well as various seasonal tyres). The analysis shows that, up to 650 Hz, structure-borne noise is critical for vibro-acoustic comfort. They point out about the given suspension (McPherson type), that the transmission ratios between the wheel axle acceleration and structure-borne noise in the passenger compartment are similar for different tyre types and for different road surfaces.

In work [11], the authors examine the dampening effect of the tyre, the elastic element and the shock

* Corresponding author: zlaty.georgiev@gmail.com

absorber in vertical direction. At frequencies above 25 Hz, the vibrations are absorbed only by the pneumatic tyre and the upper insulator mounted between the McPherson strut and the car body (the coil spring and the shock absorber can be considered as rigid bodies with a specific mass). In semi-active suspensions, it is noted that even with the change of the damping coefficient, for frequencies over 50 Hz, it does not affect the vibrations of the sprung mass [12].

The vibration behaviour is specific for each type of suspension. Many authors focus on describing different methods of study of structural vibrations. One of them is the method of transmission ratios, in which a specific connection (ball joint, bushing, etc.) as well as a whole system, which contains several such connections, are analyzed [13]. The transmission ratio shows the quantitative relation between the output and the input vibrations of the connection (system). It should be noted that the source transfer formulations are expressed in the form of spectra and frequency response functions. Another way to determine the parameters of the high frequency vibrations in the connection points between the suspension elements and the vehicle body is by using an equivalent model and applying the dynamic stiffness method [14]. The frequencies and the amplitudes of the vibrations are used to assess the impact of the suspension elements on the frame at the different connection points. On the other hand, changing the stiffness of the rubber insulators at the connection points affects the natural frequencies of metal elements of the suspension [15].

The acoustic noise created by the interaction between the tyre and the road is a subject of new and more restrictive regulations of the European Union [16]. Reduction of the acoustic noise can be achieved by developing new tyre design strategies [17, 18] or by installing reflective noise barriers [19].

The aim of the study is to provide a numerical approach, which evaluates the impact of the tyre vibration characteristics and the modal parameters of suspension elements on the vibrational behaviour of the "tyre-suspension-body" system in the frequency range from 50 to 200 Hz. This frequency range contains the first five natural frequencies of the tyre, so it gives good representation of tyre vibrational behaviour [20].

3 Method of the experimental investigation

The study uses the method of transmission ratios for investigation and evaluation of high frequency vibrations in the vehicle suspension elements. For the measurements, the ratio of the output acceleration to the input acceleration is chosen. The accelerations are measured in a vertical direction. The measurement points are shown in Fig. 1.

The study is focused on the following steps:

- transmission of vibrations to the wheel axle when the tyre is being disturbed;
- transmission of vibrations through the insulators, joints, dampers, etc.;
- determining the natural frequencies of the suspension elements – from the measured frequency response can be determined the maximum amplitudes, corresponding to their own natural frequencies. These amplitudes change for the different points and by tracking their change, the mode shapes of the suspension elements (torsion bar, strut rod) can be obtained. The measured frequency responses also show the influence of the tyre natural frequencies on the suspension elements.

The study is carried out on a stand, constructed from a front independent suspension, complete with a pneumatic tyre (Fig. 1). The suspension elements are mounted on a foundation with the help of suitable stands, which make it possible to change the stiffness at the mounting points of the supports. The tyre is disturbed in the frequency range 50-200 Hz, by an electrodynamic shaker. The shaker exerts harmonic force in front of the contact patch of the tyre.

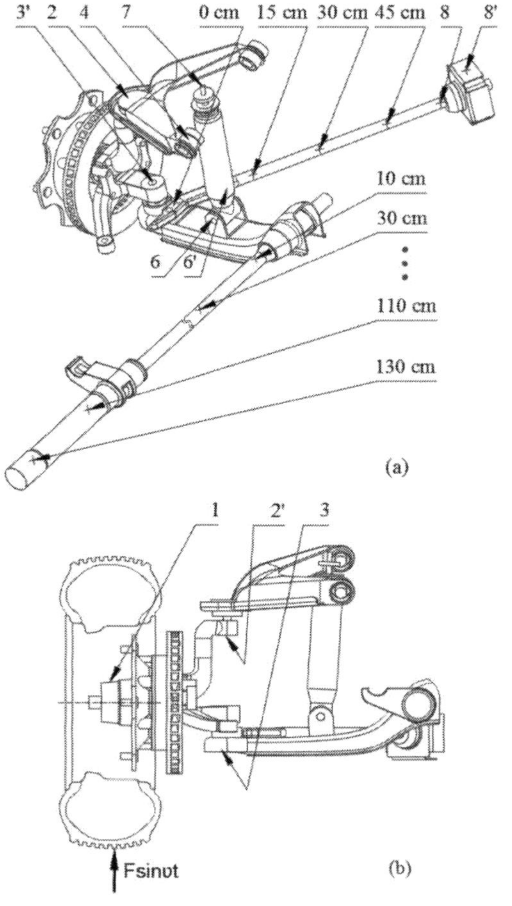

Fig. 1. Points of study for the suspension elements: (a) – isometric view, (b) – back view.

The layout of the testing system is shown in Fig. 2. It is an electrodynamic excitation system with feedback control built on standard Bruel&Keer equipment. The shaker head 6 has the potential to maintain steady acceleration. In this case it is 10 *g*. Accelerometers 2 and 9 are piezoelectric.

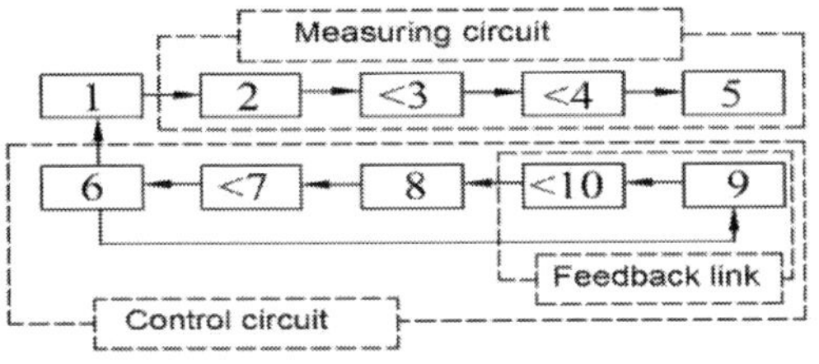

Fig. 2. Layout of the testing system (1 – test object, 2 and 9 – accelerometer, 3 and 10 – preamplifier, 4 – amplifier, 5 – read out, 6 – shaker, 7 – amplifier of the disturbing signal, 8 – generator).

The experiment is carried out in the following manner. First, a pre-check of the pneumatic tyre pressure is performed. After that, the force of the shaker head on the pneumatic tyre is checked (≈16 kg). Preliminary tests are made to determine the scale of measurements. Three consecutive measurements are performed in which the accelerometer 2 (Fig. 2) is placed on the wheel axle (point 1 in Fig. 1) in a position to measure the accelerations in the vertical direction. Successive tests are made to obtain the transmission ratios of the ball joints (positions 2-2', 3-3' in Fig. 1), of the mounting points of the shock absorber (positions 6, 6', 7 in Fig. 1) and of the strut rod insulator (positions 8, 8' in Fig. 1). Measurements at different points, marked along the torsion bar and the strut rod (Fig. 1), are performed. The corresponding mode shapes are determined.

4 Method of the numerical investigation

For the purpose of the study, it has been used Frequency response analysis called Harmonic study in the SolidWorks Simulation software. A Harmonic analysis assumes that the load is a function of the frequency rather than being directly dependent on time as is the case with Time response analysis [21]:

$$[M]\ddot{\mathbf{d}} + [C]\dot{\mathbf{d}} + [K]\mathbf{d} = \mathbf{F}[A\sin(vt) + B\cos(vt)], \quad (1)$$

where $[M]$ – mass matrix; $[C]$ – damping matrix; $[K]$ – stiffness matrix; F – vector of nodal loads (this vector is a function of frequency); d – unknown vector of nodal displacements; v – excitation frequency; t – time.

A model based on Finite Element Method (FEM) is built in SolidWorks Simulation software. The construction of the suspension is shown in Fig. 1. The metal elements are made of Plain Carbon Steel. The Elastic modulus for this steel is $E = 2{,}1\times10^{11}$ N/m^2, the Poisson's ratio is $\mu = 0{,}28$ and the mass density is $\rho = 7800$ kg/m^3. The rubber bushings and the connections between the suspension elements and the stand supports (vehicle body) are set by means of elastic elements without their own mass and damping – Spring Connection in SolidWorks Simulation software. The corresponding values of the spring connections' stiffness are presented in Table 1. The friction in the spline connection of the torsion bar is represented by radially distributed springs (Spring Connection).

The bearings between the hub and the wheel axle are set using the Bearing Connection in the SolidWorks Simulation software. The ball joints between the wheel axle (wheel spindle) and the upper arm (2'-2 in Fig. 1), the wheel axle and the lower arm (3'-3 in Fig. 1) are represented by Pin Connection (link without any play). For frequencies over 50 Hz, the shock absorber is modelled as a rigid body (hollow cylinder with a piston) with mass equal to that of the actual shock absorber.

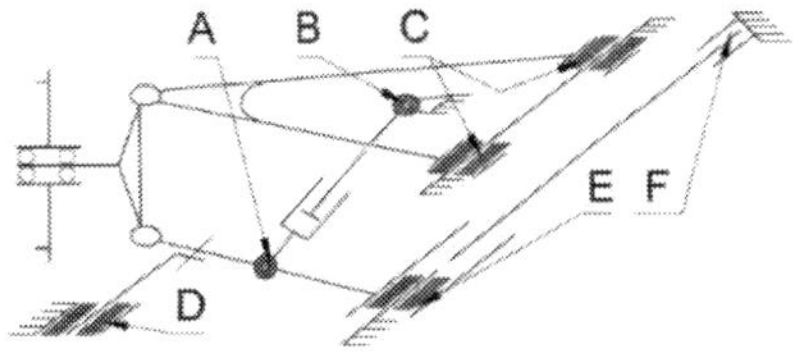

Fig. 3. Positions of the rubber bushings and connections in the vehicle suspension represented as springs in the FEM model.

Table 1. Stiffness values for rubber bushings and connections in the vehicle suspension.

Position in Fig. 3	Stiffness, N/m	Description
A	220000	Rubber bushing between the lower arm and the shock absorber
B	220000	Rubber bushing between the shock absorber and stand supports
C	5100000	Rubber bushing between the upper arm and the stand supports
D	800000	Rubber bushing between the strut rod and the stand supports
E	5100000	Rubber bushing between the lower arm and the stand supports
F	50000	Friction in the spline connection of the torsion bar

The mesh is constructed by linear tetrahedral solid elements with a total size of 10 mm and tolerance of 0,5 mm.

Input parameters are different vertical accelerations for the wheel hub, simulating the excitation from the road surface. For this purpose, it has been used the experimentally obtained transmission ratio of accelerations between the response in the wheel hub and the excitation from the electrodynamic shaker. Besides a standard radial tyre, a tyre with different vibrational behaviour is also tested. In this way, the influence of the tyre natural frequencies on the suspension elements can be evaluated. The wheel hub is subjected to harmonic motion (Selected Base Excitation in SolidWorks Simulation) with amplitude and frequency from the measured transmission ratio (Fig. 4).

Fig. 4 shows that the first two natural frequencies of the standard radial tyre are at 96 and 115 Hz, and for the modified tyre – at 150 and 175 Hz.

A modal analysis of the individual suspension elements is performed. The natural frequencies and mode shapes of vibration of the suspension elements, falling within the frequency range of the study (50-200 Hz), are determined. From the performed modal analysis, it is established that the natural frequencies of the torsion bar and the strut rod fall within this frequency range (Table 2).

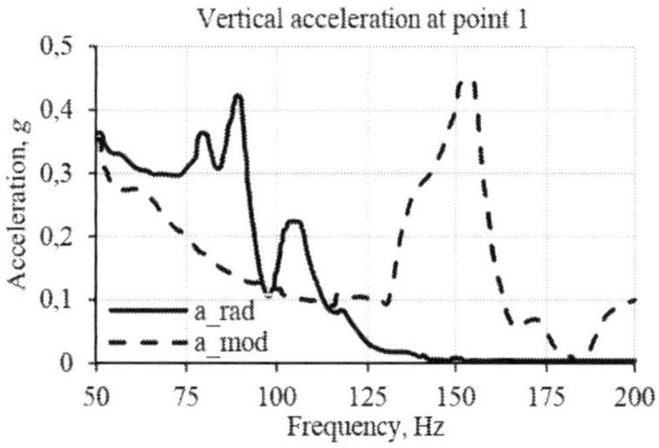

Fig.4. Experimental transmission ratio of accelerations between the response in the wheel hub and the excitation from the electrodynamic shaker (a_rad – acceleration for a standard radial tyre, a_mod - acceleration for a modified tyre).

Table 2. Natural frequencies of the suspension elements within the frequency range 50-200 Hz.

Mode	Natural frequencies ω, Hz			
	torsion bar		strut rod	
	experimental value	calculated value	experimental value	calculated value
I	55-58	56	90-95	91
II	145-155	158	125-130	121

For the numerical study the suspension is divided into individual subsystems. The separation is made at points of connection, where we assume that the transmission ratio between the subsystems is one (rigid connections). In this case, this is the bolt connection between the lower arm and the strut rod as well as the spline connection between the lower arm and the torsion bar. For the following subsystems, the input parameter is the frequency response of vertical acceleration at the point of connection. It is determined from the previous subsystem.

It is necessary to perform a numerical solution with smaller element size of the mesh to eliminate the impact of so-called artificial stiffness (stiffness which depends on the element size of the mesh) on the results. The transmission ratios (experimental and numerical) allow us to determine the contribution of the tyre natural frequencies on the vibrations of each suspension component. Also critical points with maximum amplitude of vibration can be located.

The FEM model makes it possible to determine the numerical values of the vibrations (displacement, velocity or acceleration) at the selected specific points (Fig. 1). In this case, the frequency responses of acceleration are best suited for the study, the values of which are compared to those of the experiment.

5 Results

In the following figures, the experimental and numerical results are denoted by (exp) or (num) respectively. The results for the standard radial tyre are denoted by (r) and for the modified tyre by (m).

Figures 5 and 6 illustrate the vertical accelerations before and after the ball joint connection between the wheel axle and the upper arm (position 2' and 2 in Fig. 1). From these curves it can be determined that the transmission ratio for the ball joint connection is approximately one.

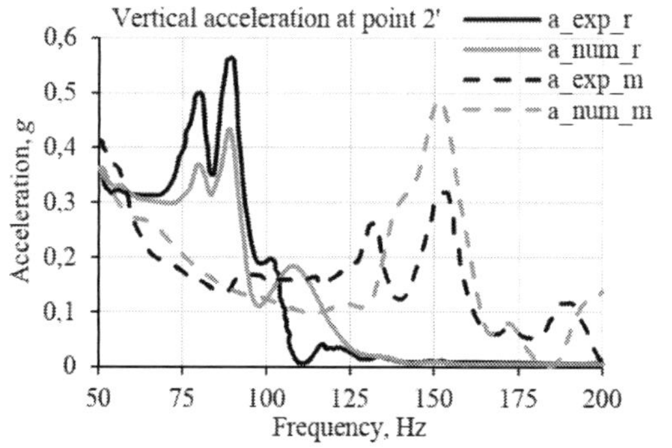

Fig. 5. Frequency response of acceleration at point 2' (a_exp_r – experimentally measured acceleration for a standard radial tyre, a_num_r – numerically calculated acceleration for a standard radial tyre, a_exp_m – experimentally measured acceleration for a modified tyre, a_num_m – numerically calculated acceleration for a modified tyre).

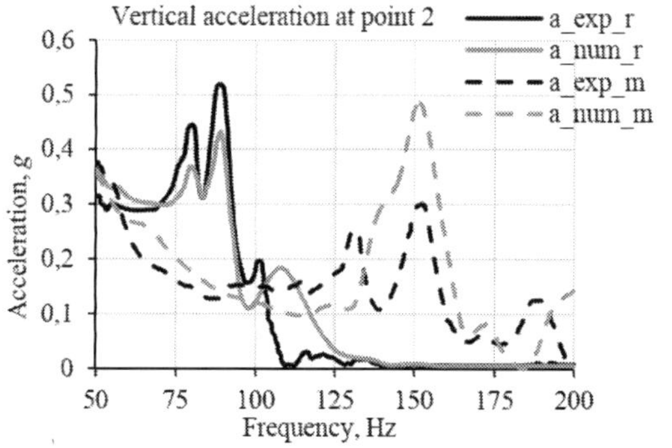

Fig. 6. Frequency response of acceleration at point 2.

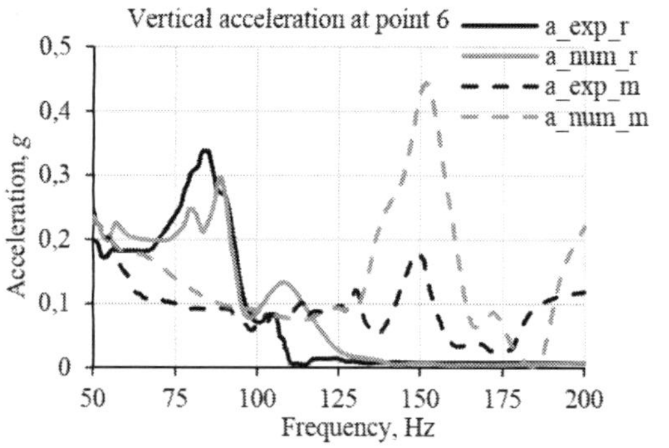

Fig. 7. Frequency response of acceleration at point 6.

Fig. 8 shows the vertical accelerations at the upper end of the shock absorber piston rod. The calculated values have a peak at 55 Hz due to the simplification of the shock absorber. This value is the natural frequency of a rigid body (with mass equal to that of the shock absorber) attached to springs with stiffness equal to that of the rubber elements (discrete system).

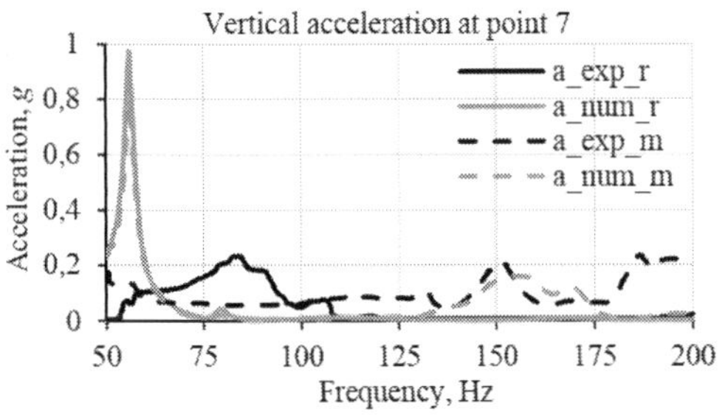

Fig. 8. Frequency response of acceleration at point 7.

Figures 9 and 10 show the vertical accelerations of the 10th and 130th cm of the torsion bar. The peak at 100 Hz is maintained for the standard radial tyre as well as for the modified tyre. This indicates that the torsion bar vibration is not affected by the tyre natural frequencies. Its natural frequencies depend on the boundary conditions (the friction in the spline connection).

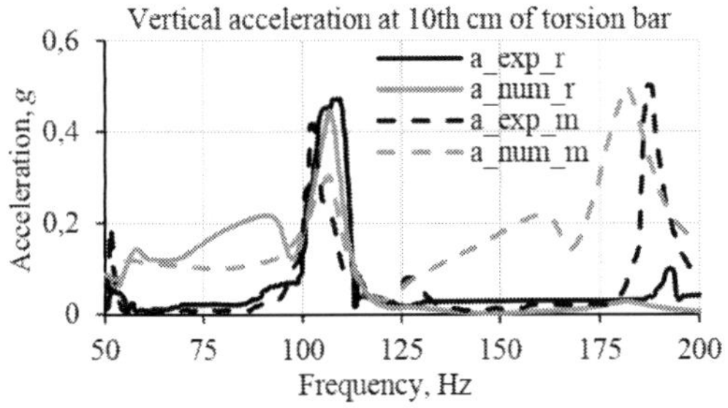

Fig. 9. Frequency response of acceleration at the 10th cm of the torsion bar.

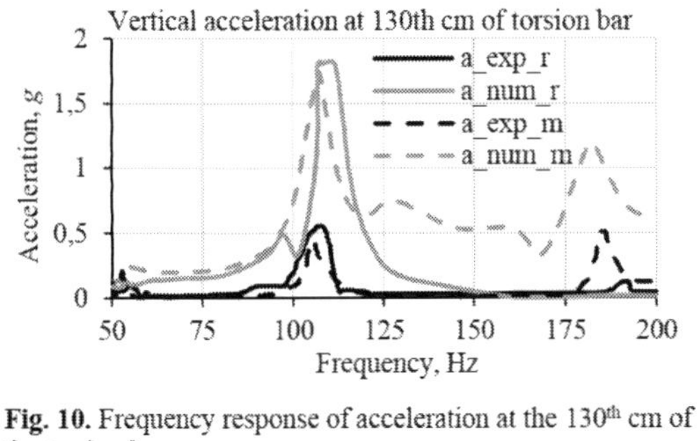

Fig. 10. Frequency response of acceleration at the 130th cm of the torsion bar.

Figures 11 and 12 show the vertical accelerations at the beginning and the end of the strut rod. Amplitude peaks are observed at frequencies, corresponding to the different natural frequencies of the two pneumatic tyres. It can be concluded, that the natural frequency of the standard radial tyre at 115 Hz coincides with the natural frequency of the strut rod and a resonance occurs.

This resonance amplifies the vibrations that pass through the strut rod to the vehicle body. One way to overcome the problem is to shift the strut rod natural frequencies by varying the stiffness of the insulator at its end. The impact of the insulator stiffness on the strut rod natural frequencies and the acceleration values is shown in Fig. 13.

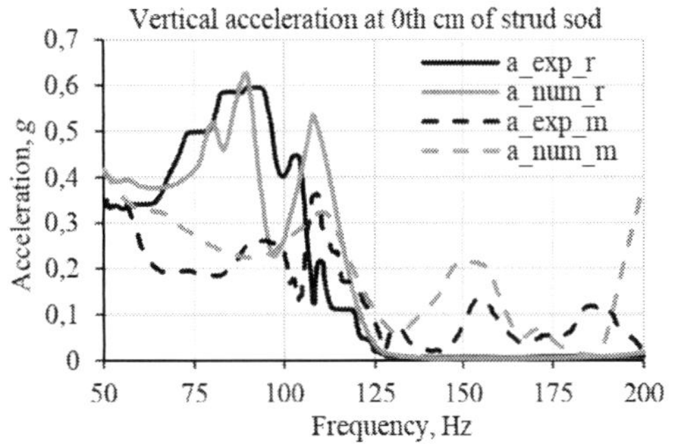

Fig. 11. Frequency response of acceleration at the 0th cm of the strut rod.

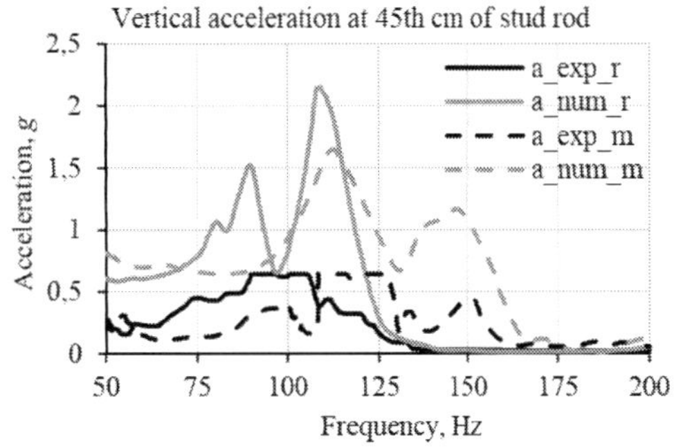

Fig. 12. Frequency response of acceleration at the 45th cm of the strut rod

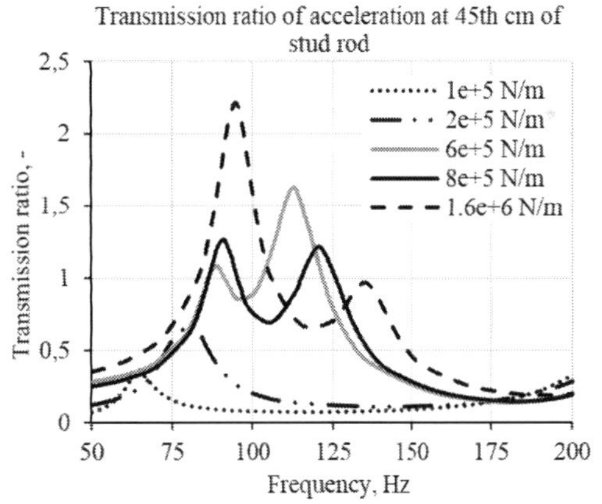

Fig. 13. Transmission ratio of resultant acceleration at the 45th cm of the strut rod for different stiffness of the insulator.

6 Conclusion

The work presents a numerical model of a vehicle suspension built using SolidWorks Simulation software and the Finite Element Method. It describes the vibrational behaviour of the suspension elements in the 50-200 Hz frequency range. It is assumed that the rubber insulators can be represented as springs without their own mass and damping. Also, it is assumed that the frequency response of acceleration, at a connection point between two subsystems, can be used as input excitation for the sequent subsystem. The work presents methods for conducting the natural and the numerical experiments. The methods in the study are used to determine the influence of the vibrational behaviour of a pneumatic tyre on the suspension elements. The model is confirmed to be fairly accurate, showing a good correlation between the measured and calculated values despite the assumptions made. The comparability of the experimental and numerical results by frequency is by up to 9% and the difference in amplitudes is by up to 70%. The model can be used to predict the effects of structural changes such as changing the input forces, different transmission ratios of connections or other characteristics of the suspension elements. In this way, it is possible to evaluate the effects of structural changes on the facelift of a car model or to set the correct objectives for the next model.

This work is part of a research project № 182ПД0009-04 in support of PhD students funded by the Research Centre at the Technical University – Sofia.

References

1. N. Pavlov, Influence of the shock absorber temperature on the vehicle ride comfort and road holding, *BulTrans-2017*, MATEC Web of Conferences, **133** (2017)
2. J. Genov, S. Tashkov, Multi-objective synthesis of vehicle semi-active suspension control represented in a vertical plane: Part 1 Model and algorithm for multi-objective synthesis, *BulTrans-2015 Proceedings*, 77-83 (2015)
3. N. Pavlov, E. Sokolov, V. Dimitrova, D. Cekov, A method and test equipment for obtaining characteristics of air springs (in Bulgarian), 54-th Annual Scientific Conference of Angel Kanchev University of Ruse and Union of Scientists – Ruse, *Proceedings Transport and Machine Science*, **54**, 54-58 (2015)
4. N. Kang, Prediction of Tire Natural Frequency with Consideration of the Enveloping Property, International Journal of Automotive Technology, **10**, 65–71 (2009)
5. P. Petkov, Analytical determine the parameters is characterized the vibration activity of pneumatic tire (in Bulgarian), Mechanics Transport Communications Academic journal, **14**, 45-51 (2016)
6. I. Kozhevnikov, Vibrations of a rolling tyre, Journal of Sound and Vibration, **331**, 1669-1685 (2012)
7. C. Diaz, P. Kindt, J. Middelberg, S. Vercammen, C. Thiry, R. Close, J. Leyssens, Dynamic behaviour of a rolling tyre: Experimental and numerical analyses, Journal of Sound and Vibration, **364**, 147-164 (2016)
8. T. Vu, D. Duhamel, Z. Abbadi, H. Yin, A. Gaudin, A nonlinear circular ring model with rotating effects for tire vibrations, Journal of Sound and Vibration, **388**, 245-271 (2017)
9. P. Kindt, F. De Coninck, P. Sas, W. Desmet, Measurement of the tire dynamic transfer stiffness at operational excitation levels, *Proceedings of ISMA 2010 /International Conference on Noise and Vibration Engineering/*, 3991-4002 (2010)
10. D. de Klerk, A. Ossipov, Operational transfer path analysis: Theory, guidelines & tire noise application, *Proceedings of ISMA 2010 /International Conference on Noise and Vibration Engineering/*, 3961-3976 (2010)
11. A. Chiesa, L. Oberto, L. Tamburini, Transmission of tyre vibrations, Automobile Engineer, **54**, 520-530 (1964)
12. N. Pavlov, E. Sokolov, Vibration characteristics of quarter car semi-active suspension model - Numerical simulations and indoor testing, Trans Motauto World, **1**, 11-16 (2017)
13. P. van der Linden, K. Wyckaert, Module target setting and substructuring for vehicle NVH development, *Proceedings of ISMA 2002 /International Conference on Noise and Vibration Engineering/*, **5**, 1921-1938 (2002)
14. P. Petkov, Studying the vibro-insulators parameters impact of automobile suspension on high vibrations transmissions (in Bulgarian), Mechanics Transport Communications Academic journal, **13**, 23-29 (2015)
15. I. Kralov, P. Sinapov, K. Nedelchev, I Ignatov, Influence of the stiffness of the support on the cantilever beam on its natural frequencies (in Bulgarian), *BulTrans-2013 Proceedings*, **5**(1), 206-211 (2013)
16. European Parliament and Council, Regulation (EU) No 540/2014 on the Sound Level of Motor Vehicles and of Replacement Silencing Systems and Amending Directive 2007/46/EC and Repealing Directive 70/157/EEC, Official Journal of the European Union, **L 158**, 131-195 (2014)
17. A. Ossipov, W. Gnoerich, Tire/Road Noise Reduction, *Proceedings of ISMA 2004 /International Conference on Noise and Vibration Engineering/*, 3483-3486 (2004)
18. Société de Technologie Michelin, *The tyre – Mechanical and acoustic comfort*, (2002)
19. I. Kralov, New solution for transport and industrial noise protection through reflective noise barriers, *BulTrans-2017*, MATEC Web of Conferences, **133** (2017)
20. P. Kindt, F. de Coninck, P. Sas, W. Desmet, Analysis of tire/road noise caused by road impact excitations, SAE Technical Paper 2007-01-2248 (2007)
21. P. Kurowski, *Vibration Analysis with SolidWorks Simulation 2016* (2016)

Cascade control algorithm of test bench for studying load-sensing electrohydraulic steering systems

Alexander Mitov[1,*], *Jordan* Kralev[2], and *Ilcho* Angelov[1]

[1]Technical University – Sofia, Department of Hydroaerodynamics & Hydraulic machines, Bulgaria
[2]Technical University – Sofia, Department of Systems & Control, Bulgaria

Abstract. The paper presents the design and synthesis of a three-loop cascade control system with a digital PI controller intended for installation in a test bench for load-sensing electrohydraulic steering units. The structure and layout of both the hydraulic system and the control system are shown. The cascade control system consists of three controllers for measurable variables: proportional spool valve position, steering velocity and steering cylinder position. The developed control algorithm is embedded into a 32-bit microcontroller for mobile applications. Experimental studies have been carried out to present the performance of the entire electrohydraulic system. The developed control algorithm will serve as a basis for further study of the system with advanced control techniques implementable in low speed mobile machines.

1 Introduction

The need for mobile machines with automated control of the direction of movement is a determining factor for the development of built-in electrohydraulic steering systems. A basic component in these systems is the electrohydraulic steering unit (EHSU). Modern EHSUs enable the reconciliation of two modes of steering depending on the control action: mechanical - through the steering wheel and digital – by electronic joystick or GPS signal. In this way, besides meeting the requirements of the safety standards, new advantages are gained with regard to precise remote control and provision of variable steering ratio between the steering wheel and the steered wheels [1].

In addition, the implementation of modern load-sensing (LS) steering systems on the executive servo cylinder significantly influences the energy balance of the entire hydraulic drive system, providing the ability to adjust the input hydraulic power by two parameters – flow rate and pressure. This adjustment capability combined with the relatively high efficiency of individual displacement volume machines results in a high structural efficiency of the entire hydraulic system [2].

It should be noted that in this type of steering systems there is no mechanical connection between the steering wheel and the steering axle of the machine, such as the conventional steering system of cars [3]. The transmission of energy is carried out by means of a hydraulic working fluid. This makes them applicable to mobile machines whose speed does not exceed 60 km/h [4].

On the other hand, in order to meet the requirements for the compactness of the steering devices, related mainly to the limitations of the construction of the mobile machinery, the conversion of the electrical signal into a hydraulic flow is achieved through a digital electro-hydraulic control module (EHCM). The continuous development of this type of modules is an example for one of the applications of the so called digital fluid power. An example of this is the well-known type of Danfoss – PVE [1]. It has an electrohydraulic system consisting of four micro valves, connected in parallel, which serve for pilot hydraulic control of a proportional spool valve (with position feedback) integrated in the steering unit. The proportional hydraulic control valve determines the direction of movement of the servo cylinder which steers the wheels. The parallel connected valves in the module are switched by solenoids, working in pairs according to pulse width modulation (PWM) and the electric control signal is generated by a microcontroller (Fig.1). The embedded control system has to ensure the quality of the entire electrohydraulic system [5].

The main objective of the paper is to present a developed three-loop cascade system for automatic control of a load-sensing electro-hydraulic steering system that is implementable in a low speed mobile machinery. The hydraulic steering system is evaluated on a developed test bench of the EHSU. The control algorithm uses a classic PI law that is embedded in a 32-bit microcontroller for mobile applications. Experimentally obtained dynamic characteristics in the form of transients processes during a reference tracking confirm the performance of the entire electrohydraulic system.

2 Laboratory test bench layout

A hydraulic diagram of a rig for testing the hydraulic system for the load-sensing steering system, shown in Fig. 1, has been developed.

* Corresponding author: a_mitov@tu-sofia.bg

 https://doi.org/10.1051/matecconf/201823402006

2.1 Hydraulic system

A hydraulic drive system is composed of: a tank (1), a variable displacement axial-piston pump type A10VSO (3), displacement volume 18 cm^3 with pressure controller type DR (4), an electrical motor (2), a main pressure-relief valve (6), a pressure filter (9), a gear type flow rate meter (10), a hydrostatic steering device (11), a hydraulic control block (12) and EHCM (13) type PVE. The plant under control is a double-acting symmetrically servo cylinder (22).

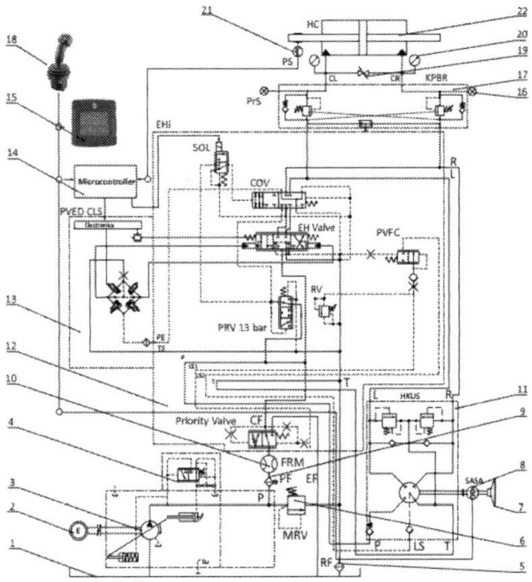

Fig. 1. Internal structure of the target EHSU.

Fig. 2 shows the hydraulic diagram of the pressure controller of the axial-piston pump [6].

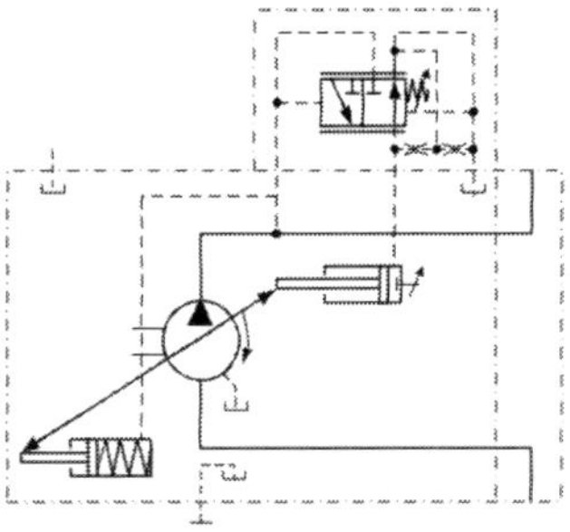

Fig. 2. Hydraulic diagram of a pressure controller – type "DR".

The inclusion of a variable displacement axial-piston pump in the open-loop hydraulic systems makes it possible to adjust the flow rate delivery to the steering system by changing the pump displacement volume from a maximum to a minimum level. The pump is equipped with a DR type pressure control which allows for the reduction of the pump displacement volume, and hence of the supplied flow when the pressure increases (up to the maximum limitation) in the steering system due to the increase in the external force acting on the piston of the servo cylinder.

The developed schematic solution of the test bench complies with the requirement for operation in two modes: mechanical by steering wheel (7) and digital by electronic joystick (18).

The laboratory bench meets also the testing requirements for electrohydraulic steering devices under varying loads. The pressure load is realized by a designed hydraulic block with adjustable over-center valves (17) which are connected to the two chambers of the hydraulic cylinder (22).

2.2 Control system

The purpose of the system is to control the stroke of the piston of the servo cylinder (22) by receiving a signal from an electronic joystick (18). The embedded in microcontroller (14) cascade controller calculates a control voltage for an electrohydraulic control module (EHUM, 13) that switches the hydraulically proportional EH valve. The proportional valve determines the direction of the movement of the piston of the servo cylinder (22), supplying working fluid to one of the two chambers (CL and CR). EHCM (13) is constructed of small two-way two-position valves connected in parallel switch by PWM signal.

The developed bench provides the ability to record the experimental dynamic processes, such as starting and stopping the cylinder piston, and monitoring a certain reference (driving law) formed by turning the steering wheel, electronic joystick or pre-programmed into the microcontroller. For this purpose, the rig is equipped with a potentiometric measuring transducer (21) sensing the steering cylinder position.

In addition, pressure sensors (16) are included for measuring the dynamic pressure variation in the two chambers of the cylinder, and in order to allow for operator observation of the values, they are duplicated with high-accuracy pressure gauges (20).

The inclusion of a gear flowmeter (10) allows for a dynamic record of the flow rate supplied from the pump to the hydraulic control block (12).

In the implementations of modern steering systems with architecture according to ISO13849-1, it is necessary to incorporate a sensor (8) for recording the angular position of the steering wheel, respectively [7].

Fig. 3 shows an electrohydraulic steering system model for the purpose of control.

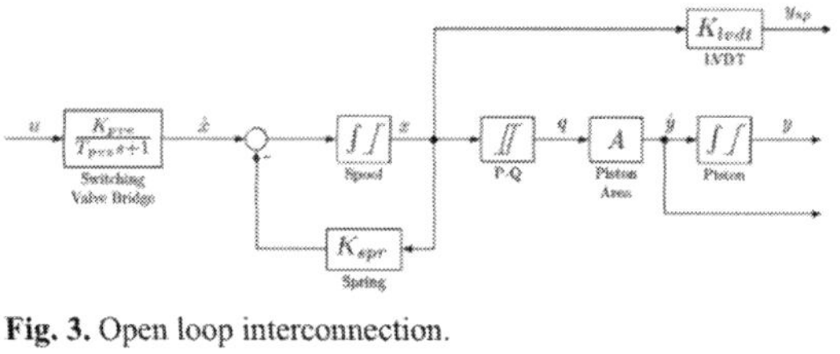

Fig. 3. Open loop interconnection.

The system is designed with dominating linear dynamics, however, some nonlinearities arise – due to naturally nonlinear pressure-flow characteristics and artificially introduced hysteresis for safety reasons. The control signal (manipulated variable) u is a generated voltage from the microcontroller [8]. It is converted to a hydraulic signal with the help of a switching valve module (PVE) which can be modelled as a first order dynamical system with gain K_{pve} and time constant T_{pve}. Switching valves generate a flow, which proportionally moves a directional spool with a velocity $\dot{x}$ such that

$$\ddot{x}(t) = -\frac{1}{T_{pve}}\left(-\dot{x}(t) + K_{pve}u(t)\right). \quad (1)$$

The motion of the directional spool is constrained within the range of ±3.5 mm. The spool is returned to a neutral position with the help of equivalent to a spring-like action K_{spr} accounting for actual mechanical spring and hydraulic reaction force arising from jet flow through the spool valve. The position of the directional spool is sensed by a LVDT (linear variable differential transducer) sensor with a gain K_{lvdt} giving a voltage signal y_{sp} to the microcontroller within a range ±2500 mV.

Every left or right shift of the spool causes a flow q to the chambers of the steering cylinder, which proportionally to its piston area A translated it in a corresponding direction. Usually, there is a certain dead-band in flow response due to the presence of a positive covering in the directional spool valve to protect the device from unintended leaks which can lead to steering drifts. The velocity of the steering cylinder A is integrated to give the piston position measured by a resistive sensor. So the plant has one input port - u and three output ports - $\vec{y} = \left(y_{sp}, y, \dot{y}\right)^T$. Since the transfer function from u to y_{sp} is

$$y_{sp} = \frac{K_{pve}}{(s+K_{spr})(T_{pve}s+1)}u \quad (2)$$

the dynamics of the direction spool will be dependent on elastic components in the systems accounted in K_{spr} and pulse-width modulation period related to T_{pve}. The system is constructed so that $T_{pve}K_{spr} > 10$ and the dominating pole will be $-K_{spr}$ determining the transient response. The subordinated pole will be $-1/T_{pve}$, which will cause some high frequency oscillations around the current spool position due to the switching nature of the PVE. With increasing mechanical spring stiffness and increasing the pressure in the system parameter K_{spr} will rise, which will make the response faster. Hence, smaller amount of control power will be required to compensate the spool dynamics. Also when decreasing PWM period the time constant T_{pve} will decrease the amplitude of the steady state oscillations but will increase their frequency. Maximal frequency of the spool oscillations is limited by the mechanical parameters of the switching valves, mainly the mass and friction. Lighter valves will allow faster switching transient and increased frequency.

3 Design of cascade control algorithm

Fig. 4 shows the designed cascade control system structure.

It is easy to note the dependability of the plant outputs since a change in the directional spool position will cause a change in steering cylinder velocity which in turn will cause a change in steering cylinder position. Hence, the cascade control technique to achieve steering position tracking is a natural choice. Three regulators are designed: K_{sp} (for spool position), K_v (for steering velocity) and K_y (for steering cylinder position). The output of an external regulator is seen as a reference signal for the next regulator.

3.1 Spool position regulator

The requirements for the spool position controller K_{sp} are for fast rising time and short transients in y_{sp}. The steady state error and tracking error are of secondary importance for this control loop. It also has a linearizing effect on the plant, compensating to some extend the dead-band. The controller is implemented as a pure *PI* with dominating *P* component [5] ($K_{p,sp} > K_{i,sp}$)

$$K_{sp} = K_{p,sp} + K_{i,sp}\frac{1}{s} = K_{p,sp}\frac{s+K_{i,sp}/K_{p,sp}}{s}, \quad (3)$$

where $K_{p,sp}$ is the proportional gain and $K_{i,sp}$ is the integral gain. As a transfer function this regulator has one zero at $-\frac{K_{i,sp}}{K_{p,sp}}$ which can be used to compensate for the dominating pole of the plant at $-K_{spr}$. So first the proportional gain is increased until the control signal reaches its constraint. Next the integral gain is tuned to compensate the dominating pole and minimize the overshot. The integral gain should not be chosen too high due to possible wind up effects. In the time domain the regulator works at

$$u(kT_S) = K_{p,sp}e_{sp}(kT_S) + K_{i,sp}x_{i,sp}(kT_S), \quad (4)$$

where $e_{sp}(kT_S) = r_{sp}(kT_S) - y_{sp}(kT_S)$ is the spool position error calculated every sample k with sample period $T_S = 0.1s$. The state of the integrator $x_{i,sp}(kT_S)$ is approximated by the *backward Euler difference*

$$x_{i,sp}(kT_S) = x_{i,sp}\left((k-1)T_S\right) + T_S e_{sp}(kT_S). \quad (5)$$

To limit the wind up effect in the integrator its output is limited to a certain percentage of the control range such as $\left|K_{i,sp}x_{i,sp}(kT_S)\right| < \gamma U_{max}$, where $U_{max} = 5000mV$ is the maximal amplitude of the control signal u.

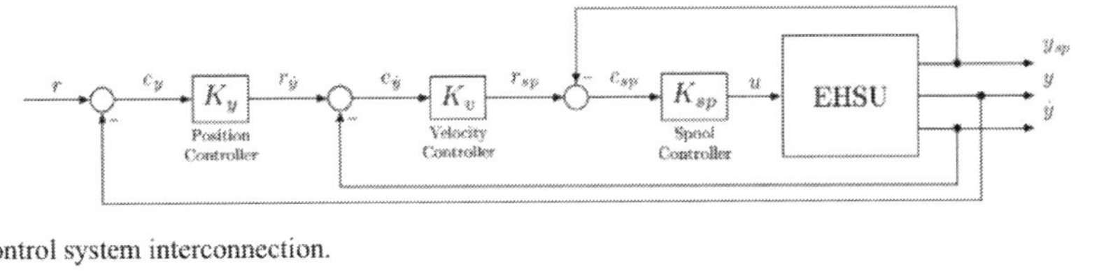

Fig. 4. Cascade control system interconnection.

3.2 Steering velocity regulator

The second control loop takes the translational velocity $\dot{y}$ of the steering cylinder as an input and gives the reference for the spool regulator as its control action r_{sp}. The velocity $\dot{y}$ is estimated by calculating the first difference of the measured position signal y by the resistive sensor

$$\dot{y}_{est}(kT_S) = \frac{y(kT_S)-y((k-1)T_S)}{T_S} \quad (6)$$

Regulator K_v has to react fast to changes in its reference signal $r_{\dot{y}}$ in order to start the steering process as soon as the operator demands it. Such a regulator introduces a derivative like action in the closed loop steering control however minimizing the amplification of the noise in the control signal. Regulator K_v is implemented as a high gain P regulator with dead band compensation

$$r_{sp}(kT_S) = K_{p,v}e_{\dot{y}}(kT_S) + \eta\left(e_{\dot{y}}(kT_S)\right), \quad (7)$$

where $e_{\dot{y}}(kT_S) = r_{\dot{y}}(kT_S) - \dot{y}(kT_S)$ is the loop error signal and $\eta(e)$ is a compensation term calculated on the basis of dead band range $[-\eta_2, \eta_1]$ as

$$\eta(e) = \begin{cases} \eta_1, e > 0 \\ 0, e = 0 \\ -\eta_2, e < 0 \end{cases}. \quad (8)$$

3.3 Steering cylinder position

The outer regulator K_y is designed to achieve the required trajectory tracking performance in the presence of variable load disturbances. Since accuracy is the main requirement this regulator is implemented as a *PI* regulator with increased integral action while keeping the stability margin

$$r_{\dot{y}}(kT_S) = K_{p,y}e_y(kT_S) + K_{i,y}x_{i,y}(kT_S), \quad (9)$$

where $e_y(kT_S) = r(kT_S) - y(kT_S)$ is the error between the desired piston position and the one measured by the resistive sensor. The integrator term $x_{i,y}$ is implemented as a *backward Euler difference* with included backtracking anti-wind up term $w(kT_S)$

$$x_{y,sp}(kT_S) = x_{i,sp}((k-1)T_S) + T_S e_{sp}(kT_S) + T_S K_t w(kT_S). \quad (10)$$

The amount of anti-wind up action is tuned with the parameter K_t. If it is higher the overshot will be attenuated but if it is too high the integrator will not work effectively. The $w(kT_S)$ term is calculated as

$$w(kT_S) = u_{sat}(kT_S) - u_{unsat}(kT_S), \quad (11)$$

where u_{sat} is the saturated control action and u_{unsat} is the unsaturated control action.

4 Experimental results

Reference tracking experiments with typical for steering applications trajectory are used to confirm the prescribed closed-loop performance. They are presented in Figures 5 to 7 when the reference signal is generated with a joystick and in Figures 8 to 10 when the reference signal is pre-programmed in the microcontroller.

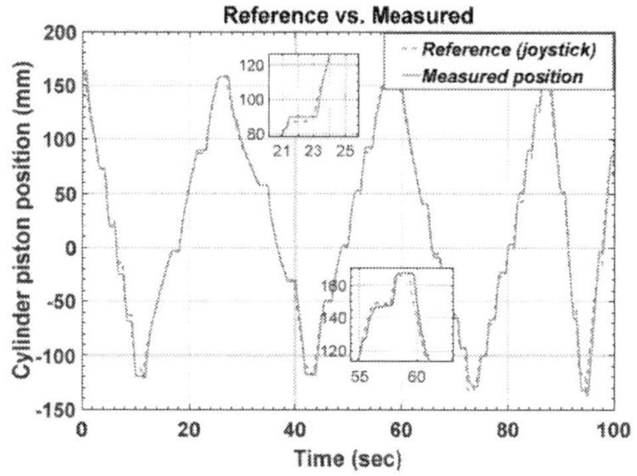

Fig.5. Reference tracking with joystick generated trajectory.

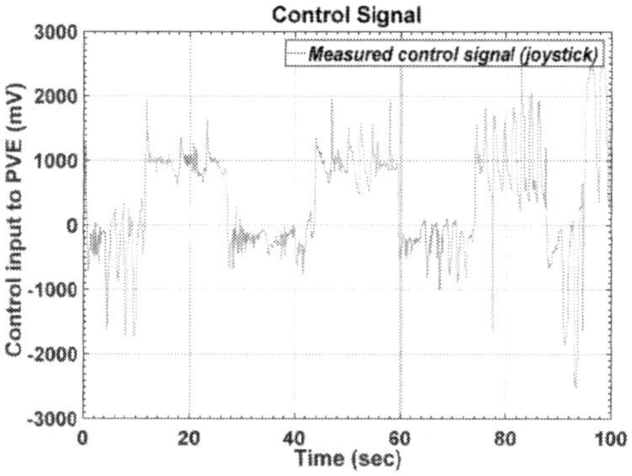

Fig. 6. Control signal to the PVE with joystick generated trajectory.

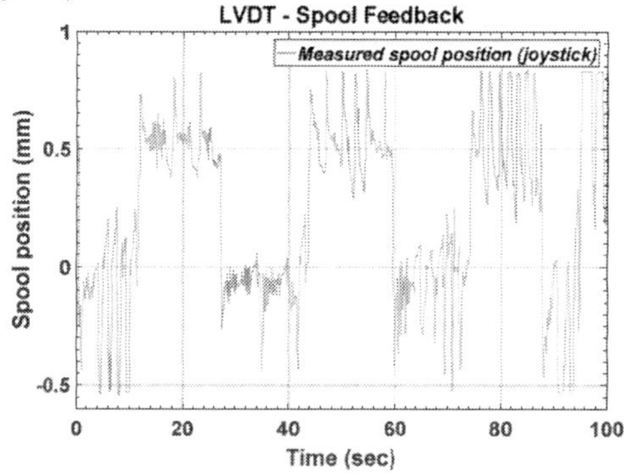

Fig.7. Spool position controller by the PVE with joystick reference trajectory.

The cascade control algorithm is programmed in the microcontroller MC012 with the help of Danfoss IDE – PLUS 1 GUIDE [1]. It has standard capabilities similar to programming environments used in PLC (programmable logic controller) of other vendors. The resultant program is in the form of graphically connected blocks which comes from dedicated library and custom blocks. Custom blocks are described in structured text (STL) language.

4.1 Joystick generated reference signal

The test-bench is equipped with single axis JS6000 Danfoss joystick. It can be programmed through CAN to filter the operator commands and outputs two PWM mutually exclusive sequences each corresponding to the position set point in each direction. When the sequence for the one direction is present the sequence for the other direction is not present. In this way the operator can send the commands to the microcontroller safely. The duty of the PWM signal is proportional to the desired position. When the stick of the joystick is moved by the operator the corresponding output signal begins to increase proportionally to the angle of the stick. So actually the operator commanded the desired velocity. As a result the generated reference signal can be described as

$$r(t) = \begin{cases} r_1(t), d_{joystick} > 0 \\ r_2(t), d_{joystick} < 0' \end{cases} \quad (12)$$

where the $d_{joystick}$ is the displacement of the joystick introduced by the operator and r_1, r_2 are corresponding PWM sequences read by the microcontroller

$$r_k(t) = \begin{cases} 0, t \; mod \; T_{j,pwm} < \gamma T_{j,pwm} \\ 1, t \; mod \; T_{j,pwm} > \gamma T_{j,pwm}' \end{cases} \quad (13)$$

and the PWM duty γ is calculated by integrating the displacement of the joystick as

$$\begin{cases} \dot{\gamma}(t) = a\left|d_{joystick}(t)\right|, \gamma < 1 \\ \dot{\gamma}(t) = 0, \gamma = 1 \end{cases} \quad (14)$$

Fig. 5 shows the tracking performance when the joystick signal is used. The generated trajectory by the operator is a piecewise linear with quasi-periodic form spanning the whole stroke range of the servo cylinder. As can be observed from the figure the tracking error (a difference between desired trajectory and measured position) is very small (around 1 to 3 mm). The largest value of the tracking error is during the transients when the operator desires to suddenly start or stop the steering. However, such value of the tracking error is acceptable for the steering applications of heavy mobile machinery.

Corresponding control action send from the microcontroller to the switching valve module (PVE) is shown in Fig. 6. As can be seen its amplitude stays in the prescribed limits of ± 5000 mV. This is an indication that the system can tolerate additional load disturbances or higher rates of reference change. However, the inherent dead-band nonlinearity present in the plant model limits the achievable tracking error by causing oscillatory behavior in steady state if the control action is further increased. Another thing evident from the control signal is the low levels of noises introduced by the position sensor.

Also the spool position used in the most internal control loop is presented in Fig. 7. It is measured in millimeter units. The maximal shift of the spool is around 3mm in each direction. As can be seen, the EHSU system is highly sensitive to this signal with small displacement of the spool leading to immediate increase in the steering cylinder velocity.

Another interesting thing is that the spool position is slightly biased from the neutral position despite of the evident symmetry in the reference trajectory. This can be explained by some mechanical misfit in the unit which can be further studied so that it is reduced. However, the closed loop system functions normally even when such misfit is present.

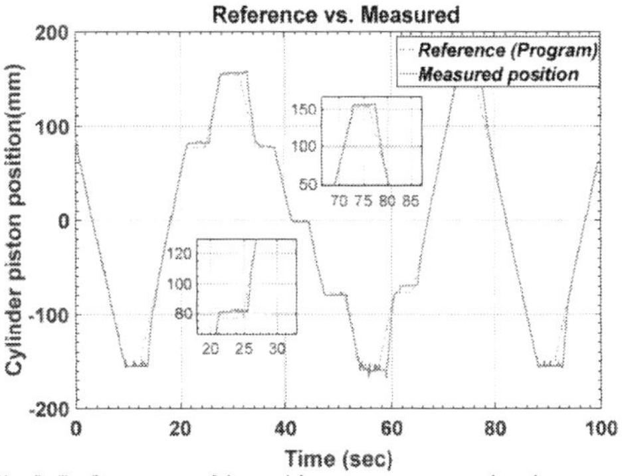

Fig.8. Reference tracking with pre-programmed trajectory.

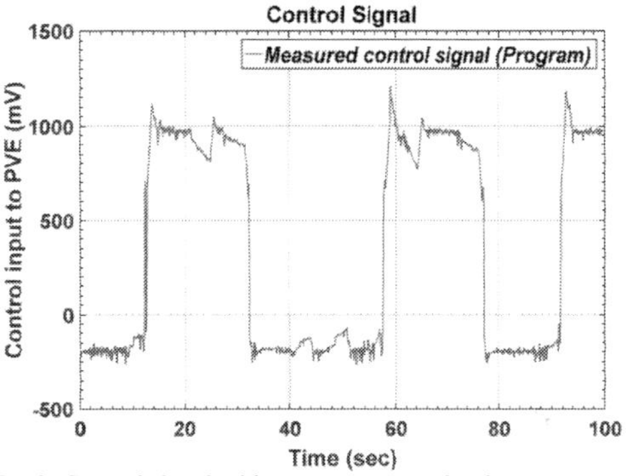

Fig.9. Control signal with pre-programmed trajectory.

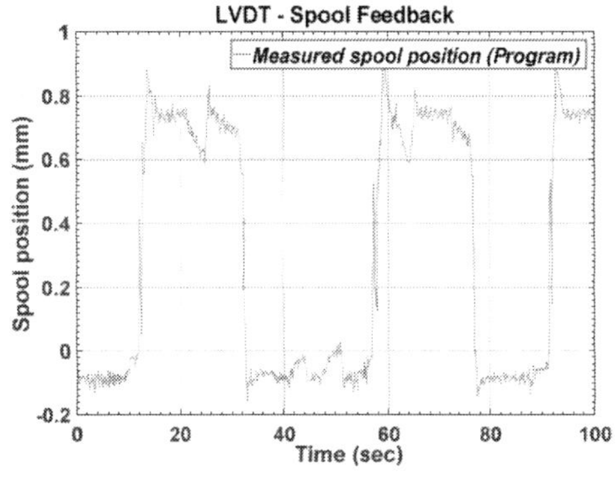

Fig. 10. Spool position controller by the PVE with pre-programmed trajectory.

4.2 Pre-programmed reference signal

In the second experiment the reference trajectory (Fig. 8) is pre-programmed in the microcontroller as a sequence of set point values evenly spaced in time. These values are generated with MATLAB® and recorded in

CSV file which can be loaded in the microcontroller memory as one dimensional array [9]. As can be seen from Fig.8, the results are similar to the case when a joystick generates the reference signal.

The corresponding control action sent from the microcontroller to the switching valve module (PVE) [10] is shown on Fig. 9. The spool position used in the most internal control loop is presented in Fig. 10.

5 Conclusion

The experimental results confirm closed-loop performance which is tested with two different reference trajectories typical for steering applications – joystick generated and pre-programmed. In both cases the tracking error is within the range of 1 to 3 mm, the control signal stays within the prescribed limits of ± 5000 mV and the spool position is within the limits of ± 3 mm. The closed loop system has uniform performance for the whole stroke range of the cylinder ± 150 mm. The system was tested with various loading pressures within the range between 2.5 to 7.5 MPa.

The developed and implemented laboratory bench for studying the load-sensing power steering system enables functional testing of modern electro-hydraulic steering devices designed for a wide range of mobile machinery.

The equipment of a variable displacement axial-piston pump with a pressure controller, as well as the incorporation of modern technical measuring instruments, enable the real-time record of a number of mechanical and hydraulic parameters. This in turn contributes to the study of the energy efficiency of the system with the inclusion of various control devices.

The present research was supported by funding contract № 181PR0020-02 of the Technical University of Sofia, Scientific and Research Sector (SRS) in the field of "Prospective Leaders". Also the authors are thankful to the conference organizing committee for their invitation.

References

1. Danfoss Inc., Steering, EHi steering valve, Technical Information, BC00000379en-US0202 (2018)
2. R. Isermann, *Fahrdynamik-Regelung. Modellbildung, Fahrerassistenzsysteme, Mechatronik*, ISBN 103-8348-0109-7, (Friedr. Vieweg & Sohn Verlag, Germany, 2006).
3. H. Murrenhoff, L. Eckstein, *Fluidtechnik für mobile Anwendungen*, ISBN 978-3-8440-0515-8, (Shaker-Verlag, 2011)
4. K. Åström, T. Hagglund, *PID Controllers: Theory, Design and Tunning*, Second Edition, ISBN 1-55617-516-7, United States of America (1995)
5. Rexroth Bosch Group, Axial Piston Variable Pump, A10VSO, serie 32, RE-A 92714 (2011)
6. J. Weber, Th. Neubert, E. Lautner, *Arbeitsblätter zur Vorlesung: Steurungen, Softwareentwicklung und Sicherheit in mobile Anwendungen,* Institut für Fluidtechnik (IFD), TU-Dresden (2016)
7. J. Weber, *Arbeitsblätter zur Vorlesung: Mobilhydraulik*, Institut für Fluidtechnik (IFD), TU-Dresden (2017)
8. A. Mitov, J. Kralev, Il. Angelov, Ts. Slavov, Identification and synthesis of linear-quadratic regulator for digital control of electrohydraulic steering system, *11th International Fluid Power Conference*, **1**, 120-129, ISBN 978-3-95886-215-9, Germany (2018)
9. D. Gu, P. Petkov, M. Konstantinov, *Robust Control Design with MATLAB®*. Springer-Verlag, London, second edition, ISBN 978-1-4471-4681-0 (2013)
10. A. Mitov, I. Angelov, J. Kralev, Embedded electro-hydraulic controller with digital valve actuation for steering of heavy duty machines, *7th Mediterranean Conference on Embedded Computing (MECO)*, **1**, ISBN 978-1-5386-5684-6, Montenegro (2018)

Life cycle assessment of electric and conventional cars energy consumption and CO_2 emissions

Ivan Evtimov[1], *Rosen* Ivanov[1,*], *Georgi* Kadikyanov[1], and *Gergana* Staneva[1]

[1]University of Ruse, Department of Engines and Vehicles, Bulgaria

Abstract. This paper presents an analysis concerning the effectiveness of electric traction in comparison with conventional cars. The Life Cycle Assessment method is used. It estimates the energy spent for the extraction of the raw materials/sources, manufacturing and transportation of the components and the vehicle, motion, maintenance and repair during exploitation period and the recycling process. The impact of the production technology of the electric energy, needed for charging the battery, is taken into account. The energy consumption and CO_2 emissions for the life cycle of electric and conventional cars are presented on graphs. Examples for Bulgaria and EU countries are given. The exploitation conditions in which the electric car is more effective regarding CO_2 equivalent emissions are shown. The main influence on the effectiveness of electric cars has the structure of the energy mix of the country where the electric car is produced and is used in exploitation.

1 Introduction

In view of decreasing the impact of vehicles on global warming in recent years, more and more electric cars replace the conventional ones. Following this trend, a lot of companies direct their efforts at producing cars with electric propulsion using Li-ion battery. An electric car at appropriate running conditions can be more effective than the conventional one in terms of environment safety. The general structure of an electric car is presented on Fig. 1. This kind of vehicles use the electric energy accumulated in the traction battery, which supplies the electric motor. Different kinds of batteries (Li-ion, $LiFePO_4$ etc.) and electric motors (PMDC, BLDC, AC etc.) exist.

There are studies [1-4] of the effectiveness of electric car vs. conventional ones in terms of emissions of greenhouse gases adjusted to carbon dioxide equivalent (CO_2). Usually, this type of studies are done using the Life Cycle Assessment (LCA) method [5] and the comparison is made for the energy consumption and/or CO_2 emissions. LCA is used to assess the environmental impact during all stages of vehicle life including extraction of raw materials and energy source, materials processing, vehicle manufacture, distribution (transport), use (motion) including maintenance and repair, and finally recycling or disposal [5]. The interest of researchers [1-4, 6-9], and our interest is focused mainly on the results for energy consumption and CO_2 emissions, obtained through LCA.

In [1] different models of cars are studied. The main conclusion is that all the BEVs researched have lower CO_2 emissions than ICE vehicles when the electricity comes from the European mix. The well-to-wheel CO_2 emissions are reduced by approximately 50% as compared to a similar internal combustion engine vehicle. It is not clear what is the energy spent for battery production.

Study [2] calculates the energy inputs and CO_2 equivalent emissions of a conventional gasoline vehicle, a hybrid vehicle, and a battery electric vehicle. The aim is to determine the lifecycle environmental costs of each vehicle type in conditions of California. The main purpose of the study is to examine the environmental impact of each vehicle type, taking into account lifecycle energy usage and both CO_2 equivalents and air pollution emitted. The models are developed and the impact of a variety of factors, including carbon intensity of gasoline and electricity, varied electricity mixes, battery lifetime, and fuel economy is studied. The cost effectiveness for each vehicle type was also calculated.

Study [3] models the relative impact of new BEVs and ICEVs in the US for the year 2015, and it projects the economic and environmental impact of BEVs and ICEVs over the entire assumed twenty-year lifetime of a US passenger vehicle.

A lot of sources [6, 7, 9-16] contain particular data about the elements and processes included in LCA of cars, but some of them are fragmentary and contradictory, which does not permit appropriate use for comparative analysis.

The results from the above mentioned studies show that the average electric energy mix of the respective country has the main impact on CO_2 emissions. A full comparative analysis, based on the LCA method and concerning used energy and generated CO_2 emissions of electric and conventional cars in Bulgaria does not exist. The present paper is concerned with that problem. LCA of electric and conventional cars, based on data about the

* Corresponding author: rossen@uni-ruse.bg

electric energy mix in some specific EU countries (like Norway, Poland and Bulgaria) and in EU-28, is made.

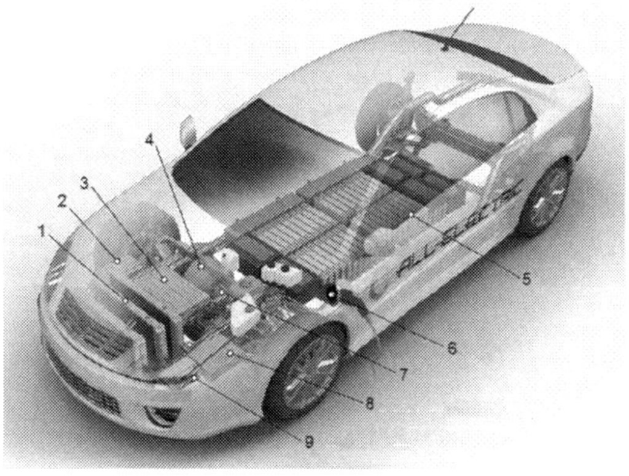

Fig. 1. Structure of an electric car:
1 – cooling system; 2 – DC/DC convertor;
3 – power electronics; 4 – electric motor; 5 – traction battery;
6 – charging contact; 7 – transmission; 8 – charging device;
9 – operational battery.

2 Comparative Life Cycle assessment for the conventional and battery electric cars

The generated emissions in CO_2 equivalent for production of 1 kWh electric energy depend on the electric energy mix for the respective country. In Europe, the larger part of electricity is produced by thermal power stations using coal (TPS), nuclear power stations (NPS) and stations using renewable energy sources (RES). In some countries like Holland, the main part of the electricity is produced using natural gas (NG). In Table 1 the electric energy mix for European counties (EU-28) is presented [10, 11, 17]. In the last row, similar information about Norway is given [10, 11]. In some of the countries, the total percentage is not a full 100% because of using small local electric generators, which is not significant for the statistics.

During the production process, power stations' direct and indirect emissions are generated. The volume of that emission depends on the life cycle of the power station. For example, production of electricity from NPS and from RES has no direct emissions and this is the reason to use electricity produced in such stations for charging electric cars.

The summarized information for the countries of EU-28 concerning the emissions of CO_2 generated for the production of 1 *kWh* electric energy is given in the Table 2 [8, 11]. The whole life cycle of the used primary energy source is taken into account. In different countries, even with the same type of primary energy source, the volume of emissions may not be equal. Many factors impact these emissions (like needed energy for production and transport of the fuel, using the innovative technologies in production process etc.), but they will not be analyzed in this paper.

For a correct LCE of the electric and conventional cars all spent energy, including energy for production the primary energy source, for car and battery production, for using the car and finally for utilizing the old components, have to be considered.

Table 1. Electric energy production (mix) of the EU-28 countries and Norway [10,11,17].

Country	Share of total production, %				
	Nuclear energy	Thermal power station			Renewable energy
		Solid fuels	Natural gas	Crude oil	
Austria	0,0	0,0	8,7	7,3	78,0
Belgium	65,0	0,0	0,0	0,0	28,5
Bulgaria	33,2	48,7	0,7	0,2	17,0
Croatia	0,0	0,0	33,5	15,6	50,7
Cyprus	0,0	0,0	0,0	0,0	97,4
Czech Republic	24,2	58,6	0,7	0,7	14,9
Denmark	0,0	0,0	26,4	48,7	22,5
Estonia	0,0	75,6	0,0	0,0	23,2
Finland	34,2	4,8	0,0	0,4	59,3
France	82,5	0,0	0,0	0,8	15,7
Germany	19,8	35,9	5,3	3,0	32,5
Greece	0,0	67,0	0,1	0,7	31,2
Hungary	36,7	13,6	12,2	7,6	29,0
Ireland	0,0	39,8	5,6	0,0	51,3
Italy	0,0	0,1	15,3	16,1	65,2
Latvia	0,0	0,0	0,0	0,0	99,6
Lithuania	0,0	1,3	0,0	4,8	92,5
Luxembourg	0,0	0,0	0,0	0,0	76,9
Malta	0,0	0,0	0,0	0,0	100,0
Netherlands	2,2	0,0	82,0	4,3	10,1
Poland	0,0	79,6	5,5	1,4	12,8
Portugal	0,0	0,0	0,0	0,0	97,7
Romania	11,3	17,7	33,0	15,6	22,3
Slovakia	62,6	7,8	1,2	0,2	25,2
Slovenia	43,0	25,4	0,1	0,0	30,2
Spain	44,2	3,7	0,2	0,7	50,5
Sweden	43,2	0,3	0,0	0,0	54,6
United Kingdom	15,3	4,3	30,1	39,3	10,0
EU-28	**28,9**	**18,9**	**14,0**	**9,8**	**26,7**
Norway	0,0	1,4	0,0	0,0	98,6

In this study, the following values of the quantities and assumptions are used:

- equal mass of the battery electric (BEV) and conventional gasoline (CV) cars;
- energy consumption of the BEV – 0,210 kWh/km;
- fuel consumption of the CV – 7,6 l/100 km of gasoline;
- capacity of the battery of BEV – 40 kWh;
- equal range of the life cycle of the two types of vehicles – 290 000 km;
- energy for car production – 11 900 kWh [2];
- efficiency of NPS – 29,5% [12];
- efficiency of TPS using coal – 26% [13];
- efficiency of TPS using natural gas – 40% [14];
- efficiency of water power station (WPS) – 60% [15];
- efficiency of wind power station (WiPS) – 40%;
- average efficiency of power stations using renewable energy sources – 50%;
- losses for transport and distribution of the electricity – 5% [1].

Table 2. Emissions of CO_2 in the production of electricity for EU-28 Member States [8] and Norway [11], g/kWh.

Country	Gross electricity production (combustion only)	Gross electricity production (with upstream)	Net electricity production (with upstream)	Electricity consumed at HV (with upstream)	Electricity consumed at LV (with upstream)
Austria	133	151	156	322	334
Belgium	188	224	233	261	267
Bulgaria	507	532	585	618	669
Croatia	231	273	282	487	524
Cyprus	646	737	773	787	810
Czech Republic	518	545	587	657	685
Denmark	316	368	386	364	377
Estonia	1020	1022	1152	878	944
Finland	171	200	209	207	211
France	66	88	92	100	105
Germany	485	534	567	599	615
Greece	655	695	755	732	767
Hungary	310	340	368	383	407
Ireland	459	533	555	588	617
Italy	358	427	444	413	431
Latvia	134	173	185	1110	1168
Lithuania	204	246	262	370	390
Luxembourg	236	288	283	508	513
Malta	731	831	868	954	1032
Netherlands	479	559	582	555	569
Poland	770	847	929	937	980
Portugal	295	346	355	372	400
Romania	356	379	413	449	492
Slovakia	173	199	211	412	420
Slovenia	315	329	351	309	321
Spain	248	295	305	321	341
Sweden	16	24	25	45	47
United Kingdom	469	555	584	593	623
EU-28	**340**	**387**	**407**	**428**	**447**
Norway	–	–	–	–	17

The fuel consumption of the CV is determined based on the specific energy and efficiency of its internal combustion engine, and with the assumption to have the same volume of energy as that one used for motion of the BEV [2]. In the values of efficiency of different power stations, the losses for extraction of the primary energy sources (coal, natural gas etc.) are taken into account.

The efficiency for fuel production is accepted to be 79,6% [6, 7, 16].

The performance of the BEV basically depends on the type of traction battery. The production technology of Lithium-ion battery for electric cars is not so cheap in comparison with traditional lead-acid battery.

The energy spent for the production, transport, recycling etc. of the most popular types of battery was calculated on the basis of data from [9] and is shown in Fig. 2. Our study about battery recycling confirms the popular opinion that this process is not economically effective because of high energy consumption and waste products presence. Probably in the future battery recycling will be oriented basically for ecological effect and observance of ecological law.

The results show that the life cycle of Li-ion battery needs about 420 kWh per each kWh of battery capacity. For a middle size battery of 40 kWh capacity, approximately 16800 kWh energy will be used during the life cycle of the BEV, if only one battery is used during the life cycle of BEV (range of 290000 km). The battery construction permits repair and change of elements. According to some authors [2] it is reasonable to make calculation for 1,5 batteries. In this case, the energy for the life cycle of battery will be one and half times more.

Taking into account the values of the energy spent for car production, battery life cycle and energy or fuel for passing a distance of 290000 km, the needed energy for the life cycle of BEV or CV can be calculated. The obtained results for primary energy used in the life cycle of BEV, produced and driven in 4 countries, are presented in Fig. 3 *a*. The energy mix and efficiency of the power station for different countries are used in the calculations.

Generally, the life cycle of the gasoline CV, produced and driven in Bulgaria, needs approximately 309750 kWh of primary energy. About 86,5% of the life cycle energy is spent on motion. This percentage depends on the energy for car production in the respective country and the last one depends on the energy mix. The energy for motion/driving changes insignificantly in different countries and basically depends on the losses in the fuel production process. Other stages of the life cycle of CV like production of vehicle and parts, their transportation etc. consume less energy – 13,5% of life cycle energy or approximately 42000 kWh.

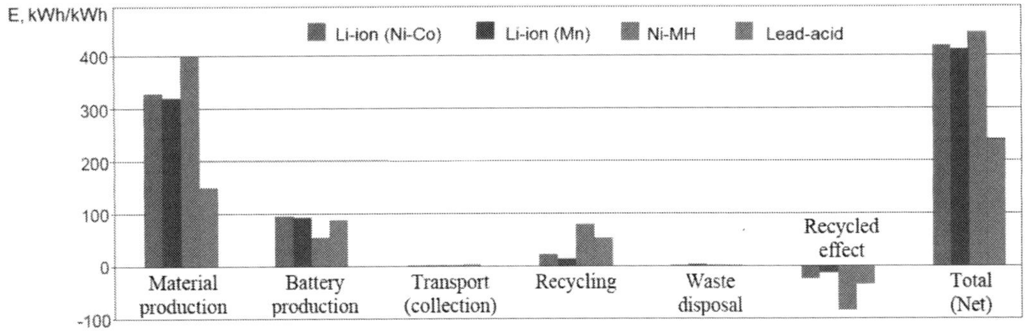

Fig. 2. Energy spent for production and recycling per 1 kWh of capacity of the different battery types.

The life cycle of BEV-1, produced and driven in Bulgaria, needs approximately 355210 kWh. In other words in these conditions electric cars need 15% more primary energy than CV (Fig. 3 *a*). The largest part of the energy goes for charging the battery – 71,5% of life cycle energy. Production of the battery has significant influence – 16,7% on life cycle energy which is equal to 59290 kWh. The energy for the production of the vehicle and parts for repair, their transportation etc. is 11,8% or 42000 kWh (value is the same like for CV).

The life cycle of the BEV-2 produced with the average energy mix of EU-28 and driven in Europe needs 16% less primary energy as compared to BEV-1.

Most effective in terms of the spent energy during its life cycle is BEV-3 produced and driven in Norway. The primary energy is two times less (49%) than the energy spent for BEV-1, 42% less than CV in Bulgaria and 38% less than CV in Norway (it is not shown in Fig. 3 *a*).

The worst scenario is the production and driving of BEV-4 in Poland – about 51% more than BEV-3 (in Norway).

The emissions of CO_2 generated during the life cycle of the vehicles in Fig. 3 *a* are calculated using data from Table 2 and are presented in Fig. 3 *b*. The BEV-1 for its life cycle generates 59940 kg CO_2 emissions. Approximately the same are the emissions of CV in Bulgaria – 59750 kg. However, priority must be given to BEV-1 because the emissions from the electric vehicle are generated where the electricity is not produced in big towns. The BEV-2 generates 31% less emissions – 40050 kg. BEV-3 generates the least – 1530 kg, which is 39 times less than BEV-1 and 26 times less than BEV-2 and 59 times less than BEV-4. In comparison with CV in Norway, the BEV-3 has 34 times less emissions.

In Table 3 the spent primary energy and generated emissions of CO_2 per 1 km distance are shown. An assumption is accepted – all vehicles consume equal secondary energy of 0,309 kWh/km. This value includes energy consumption of the BEV, and the efficiency of the charging station and battery.

BEV-4 has the highest level of energy consumption and generated emissions. It is produced and driven in Poland, where 78,6% of the electricity is from thermal power stations using coal. The most effective one is BEV-3 produced and driven in Norway, where the part of renewable energy is 98,6%. The analysis of these results shows the most effective way to increase the effectiveness and to reduce the emissions of BEV – change the energy mix of the country by using more nuclear power stations and renewable energy sources. The process will also cause change of the emissions for the life cycle of BEV. This effect is illustrated in Fig. 4 for energy mix and CO_2 emissions in Bulgaria – 0,669 kg/kWh (see Table 2).

For example, if the level of emissions decreases to 0,4 kg/kWh, for the life cycle of BEV 24126 kg of CO_2 can be saved. On the contrary, if the energy mix is changed and the part of TPS using coal increases, as a result at emission level 0,9 kg/kWh, the generated emissions for the life cycle of the BEV will exceed the respective ones for the CV by about 20717 kg.

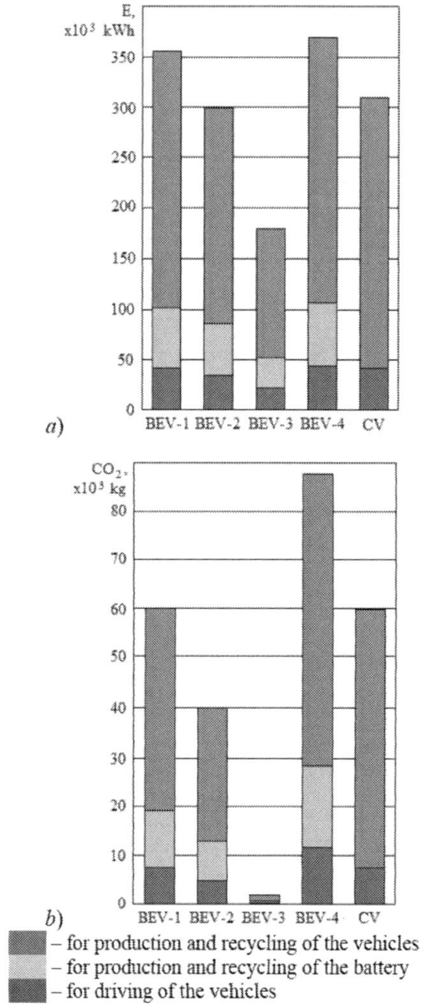

Fig. 3. Life cycle of primary energy (a) and life cycle of CO_2 emission (b).
BEV-1 – production and driving in Bulgaria;
BEV-2 – production and driving with data for EU-28;
BEV-3 – production and driving in Norway;
BEV-4 – production and driving in Poland.

Table 3. The needed primary energy and CO_2 emissions per 1 km distance.

Vehicle	Primary energy, kWh/km	CO_2, g/km
BEV-1	1,156	207
BEV-2	1,030	138
BEV-3	0,620	5,28
BEV-4	1,272	303
CV	1,068	206

The relation is well approximated with the following linear equation

$$CO_2 = -89686x + 60000\text{, kg,} \quad (1)$$

where x is level of the generated CO_2 emissions, kg/kWh.

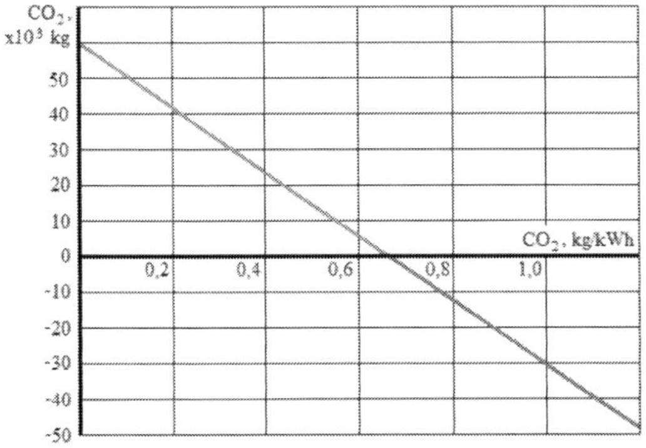

Fig. 4. Saved CO_2 emissions for the life cycle of the BEV vs emissions generated for the production of 1 kWh of electric energy.

3 Conclusion

On the basis of statistical data for the energy mix and generated CO_2 emissions for different EU countries an LCE concerning the energy consumption and CO_2 emissions of BEV and CV was done. The variants in the different countries were described. At conditions in Bulgaria (energy mix for year 2015 and level of emissions for the electricity production of 669 kg/kWh) the BEV-1 and the CV have practically equal emissions for their life cycle. The priority must be given to BEV-1 because the emissions related to the electric vehicle are generated where electricity is produced – not in big towns.

Use of RES for electricity production can reduce CO_2 emissions for the life cycle of BEV by up to 40 times in comparison to CV. A good example in this direction is Norway where 98,6% of energy is produced from RES, the life cycle energy of BEV is 40% less than the average one for EU-28 countries. The emissions are 26 times less than the average one for EU-28 countries. A negative example can be the life cycle of BEV in Poland. As a large part of electricity is produced in coal TPS the primary energy and level of the emissions are very high.

The LCE and the analysis of the results shows that the most effective way to increase the effectiveness and to reduce the emissions of BEV is changing the energy mix of the country by using more nuclear power stations and renewable energy sources.

The production technology of LI-ion battery is continuously developed, but it is still an obstacle for replacing conventional cars with battery electric vehicles. Battery recycling now is not an effective process and in the future ecological problems are possible.

References

1. D. Bakker, Battery electric vehicles. Performance, CO_2 emissions, lifecycle costs and advanced battery technology development. *Master thesis Sustainable Development, Energy and Resources*, Copernicus institute University of Utrecht, 75 (2010)
2. Aguirre K., L. Eisenhardt, C. Lim, B. Nelson, A. Norring, P. Slowik, N.Tu, Lifecycle Analysis Comparison of a Battery Electric Vehicle and a Conventional Gasoline Vehicle. California Air Resources Board, 33 (2012)
3. J. Brennan, T. Barder, Battery electric vehicles vs. internal combustion engine vehicles. A United States-Based Comprehensive Assessment, *Arthur D. Little*, 48 (2016)
4. I. Evtimov, R. Ivanov, G. Kadikyanov, A comparative analysis of the vehicles energy performance, BulTrans-2016, Sozopol (2016)
5. IS0 14040/44:2006, Environmental management – Life cycle assessment.
6. I. Palou-Rivera, J. Han, M. Wang, Updates to petroleum refining and upstream emissions, Center for Transportation Research Argonne National Laboratory, *CTR/Argonne*, 12 (2011)
7. Petroleum diesel life cycle energy demand, available at:https://hub.globalccsinstitute.com/publications/biofuels-policies-standards-and-technologies/petroleum-diesel-life-cycle-energy-demand
8. A. Moro, L. Lonza, Electricity carbon intensity in European Member States: Impacts on GHG emissions of electric vehicles. European Commission, Joint Research Centre (JRC), Italy, *Transportation Research Part D* (Elsevier, 2017).
9. K. Ishihara, N. Kihira, N. Terada, T. Iwahori, Environmental burdens of large lithium-ion batteries, Developed in a Japanese National Project, Central Research Institute of Electric Power Industry, Tokyo, Japan (2013)
10. R. Fischer, E. Elfgren, A. Toffolo, Energy supply potentials in the northern counties of Finland, Norway and Sweden towards sustainable Nordic electricity and heating sectors, Energy Engineering, Luleå University of Technology, Sweden, 31 (2018)
11. S. Skaalbones. Electricity disclosure 2015. https://www.nve.no/energy-market-and-regulation/retail-market/electricity-disclosure-2015/
12. O. Eriksson, Nuclear power and resource efficiency – A proposal for a revised primary energy factor, Department of Building, Energy and Environmental Engineering, Faculty of Engineering and Sustainable Development, University of Gävle (2017)
13. A. Scott, R. Wedmaier, The assessment and control of coal damage and loss, Project Number C3017 University of Queensland, www.acarp.com.au/abstracts.aspx?repId=C3017
14. Steam-gas power stations, Energy Review, **3** (2010)
15. Real world hydro power calculation, The Renewable Energy Website. http://www.reuk.co.uk/wordpress/hydro/calculation-of-hydro-power/
16. M. Wang, Estimation of energy efficiencies of U.S. petroleum refineries, Center for Transportation Research, Argonne National Laboratory (2008).
17. Energy production, 2005 and 2015, available at: https://ec.europa.eu/eurostat/statistics-explained/index.php?title=File:Energy_production,_2005_and_2015_(million_tonnes_of_oil_equivalent)_YB17.png

 https://doi.org/10.1051/matecconf/201823403001

A study of hydrogen fuel impact on compression ignition engine performance

Evgeni Dimitrov[1,*], *Boyko* Gigov[1], *Spas* Pantchev[1], *Philip* Michaylov[1], and *Mihail* Peychev[1]

[1]Technical University – Sofia, Department of Combustion Engines, Automobile Engineering and Transport, Bulgaria

Abstract. In this paper, a dual-fuel compression ignition engine test bench is presented. In hydrogen-diesel fuel co-combustion conditions, the engine parameters are determined – performance: effective torque, effective power and mean effective pressure; fuel economy: fuel consumption and specific fuel consumption; toxicity: carbon monoxide, carbon dioxide, nitrogen oxides, hydrocarbons, and smoke emissions (opacity). The impact of hydrogen-diesel fuel mass ratio on the performance, toxicity and economy of the engine is studied by obtaining a series of hydrogen-diesel fuel ratio variation characteristics at constant engine speed and load. Improvement of the economical parameters of the engine and reduction of carbon dioxide concentration in exhaust gases is detected under operation with hydrogen gas fuel. Significant reduction of the exhaust gases opacity is observed. It is not clear what the impact of the quantity of hydrogen, injected in the engine, on the concentration of nitrogen oxides in the exhaust gases is.

1 Introduction

Today, internal combustion engines are thermal machines which have reached a very high level of perfection – in their design and technology, as in the process of their working cycle. However, two constantly growing problems are blocking their progress: a deepening energy crisis and excessive environment pollution.

The development of the diesel engine is associated with meeting the regulations for acceptable concentration of toxic components, particle emissions and greenhouse gases in the exhaust. It is widely known that the main reason for particle emissions, mostly free carbon in soot form, of a diesel engine is the non-homogenous nature of the air-fuel mixture. In this relation, the higher level of homogenization of the air-fuel mixture in the gas-diesel cycle is a prerequisite for a fundamental improvement of diesel engine emissions.

A gas-diesel cycle could be accomplished with many gas fuels: liquefied petroleum gas – LPG (mixture of propane and butane), compressed natural gas – CNG (methane), hydrogen – H_2, etc. [1-4]. Hydrogen has a number of properties [5,6], which make it one of the most perspective alternative fuels for internal combustion engines:

- the lower heating value of hydrogen is on average by 2,8 times higher than that of gasoline and diesel fuel;
- the necessary energy for the ignition of hydrogen is a bit lower that the one necessary for hydrocarbon fuels;
- hydrogen has a very high rate of diffusion, due to which a practically homogenous hydrogen-air mixture can be achieved;
- hydrogen-air mixtures have very wide flammability boundaries- from air-fuel ratio of $\alpha = 0,2$ to air-fuel ratio of $\alpha = 10$;
- no greenhouse gases are produced during hydrogen combustion.

Two methods of achieving a gas-diesel cycle are possible. With the first one, external mixture formation, hydrogen is injected in the intake of the engine, where it mixes with the air, entering the engine. The second method consists of a direct injection of hydrogen in the cylinder – internal mixture formation [7]. Apart from the problems with hydrogen storage [8], and the lack of charge stations for it, both methods of organizing a gas-diesel cycle with hydrogen have to overcome a lot of problems, related to their practical application in vehicles.

The design of new fuel systems, hydrogen flow control systems and engine construction modification are needed for the accomplishment of a hydrogen gas-diesel cycle and internal mixture formation. Meanwhile, for the execution of a gas-diesel cycle with external mixture formation, an optimization of the mass fraction of hydrogen in the total fuel mass, entering the engine, is essential. With all parameters being constant, except for the quantity of hydrogen, the increase of hydrogen mass fraction leads to:

- improvement of engine economy [9,10];
- reduction of the exhaust gases opacity and carbon dioxide concentration [9,10];
- reduction of the engine volumetric efficiency due to the low density of hydrogen and the large volume of the intake;
- increase of the maximum in-cylinder pressure and temperature values [11-13], due to the higher of the flame front speed;
- high pressure rise rate – $dp/d\varphi$ during combustion [11,14];
- an increase of nitrogen oxide and hydrocarbon concentration in the exhaust gases [15].

* Corresponding author: etzd@tu-sofia.bg

 https://doi.org/10.1051/matecconf/201823403001

2 Purpose of the research

In relation to the aforementioned facts the purpose of the research is the quantitative assessment of the impact of the hydrogen mass fraction (in percentage) in the total fuel mass, entering a diesel engine, on its performance during a gas-diesel cycle.

3 Experimental setup

A single cylinder, four stroke engine *DV550* (Bulgaria) with air cooling is used for the purposes of the study. Several of the most important parameters of the engine are: bore D = 91,5 mm; stroke S = 85 mm; compression ratio ε = 17,5; direct injection; nominal power N_e = 8 kW at engine speed n = 3000 min^{-1}.

A schematic of the experimental setup is shown in Fig. 1. Engine 17 is connected to a direct current dynamometer "SAK 28-15", which is used to determine engine performance. Hydrogen is injected in the engine intake manifold under pressure p_{H2} = 0,65 bar, using a quick-acting gas valve. The pressure is reduced from 150 bar in the storage cylinder 18 to nominal pressure using a regulator valve 2. The hydrogen mass flow is regulated by an electronic control system 7, which is a product of cooperation between specialists of the "Department of Electrical Motion Automation Systems" and the "Department of Combustion Engines, Automobile Engineering and Transport" of the Technical University – Sofia. The system guarantees manual control (via potentiometer) of the quick-acting gas valve, i.e. of hydrogen quantity. The volume flow of the hydrogen entering the engine is measured with a gas flow meter 3, G4 type (Italy), connected between the regulator and the quick-acting gas valve. The accuracy of the hydrogen flow meter is 0,2 dm^3. A pressure gauge (with an accuracy of 5 kPa) and a type-K thermocouple connected to a digital thermometer (1 °C accuracy) are mounted in a separate chamber 4 leading to the gas flow meter. They are used to determine the temperature and pressure of the hydrogen entering the engine intake. The values of these parameters are used to calculate mass hydrogen consumption.

A volume flow meter 5, which operates by measuring the time necessary for a known quantity of liquid fuel to enter the engine, is used to determine the diesel fuel consumption. The flow meter is developed by the Department of Combustion Engines, Automobile Engineering and Transport of the Technical University – Sofia and has an accuracy of 0,2 cm^3.

The engine intake air flow is measured by a differential water pressure gauge and an orifice connected to an air receiver 12. The differential pressure gauge readings have an accuracy of 1 mm water column.

The engine exhaust gas temperature is measured with a type-K thermocouple and a digital thermometer with an accuracy of 1 °C. The thermocouple is mounted on the engine exhaust pipe, 100 mm away from the engine exhaust manifold flange.

The engine exhaust emissions are determined using a "Texa gasbox" gas analyser (Italy) 9. Apart from its other parameters, the gas analyser detects the following toxic components of the engine exhaust: carbon oxide CO (0,01% accuracy); hydrocarbons HC (1 ppm accuracy); nitrogen oxides NO_x (5 ppm accuracy); carbon dioxide CO_2 (0,1 % accuracy); oxygen O_2 (0,1% accuracy).

The exhaust gas opacity R_h is measured using the "Hartridge" method with a "Tecnotest 495/1" smoke meter (Italy) – 1 with an accuracy of 0,1%.

The engine combustion process parameters are registered with an "AVL Indiset 620" indicating system (Austria).

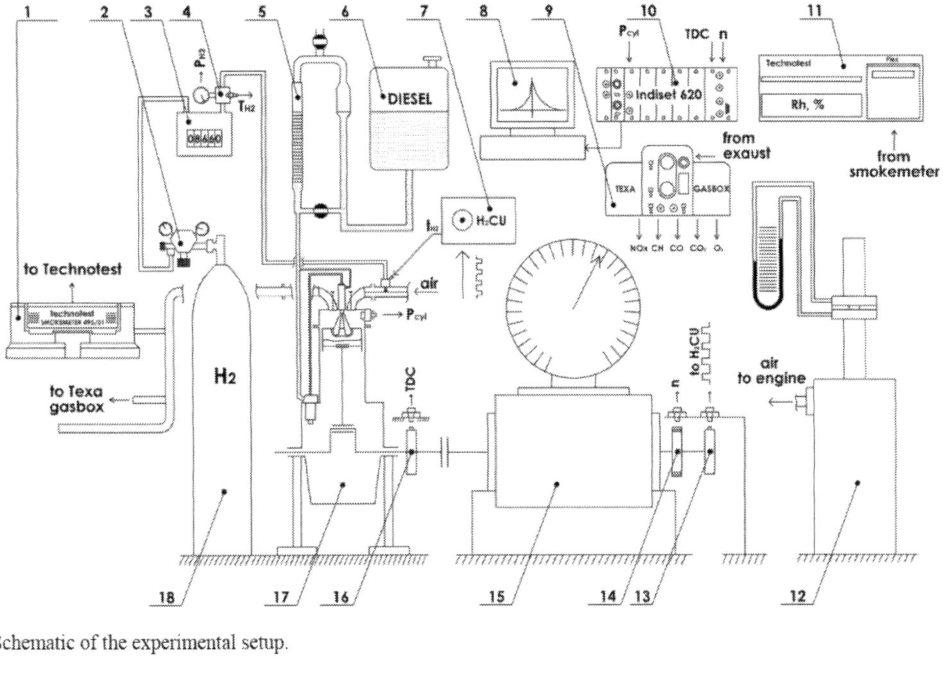

Fig.1. Schematic of the experimental setup.

4 Methodology of the study

The gas-diesel cycle requires the diesel engine to work with two types of fuel in a certain ratio. Therefore, a mass fraction coefficient for each fuel has to be defined. For hydrogen and diesel fuel, these coefficients, K_{H2} and K_D, respectively, equal:

$$K_{H_2} = \left(\frac{B_{H_2}}{B_h}\right).100, \% \qquad (1)$$

$$K_D = \left(\frac{B_D}{B_h}\right).100, \% \qquad (2)$$

where B_{H2} is the hourly hydrogen consumption, kg/h;
B_D – the hourly diesel fuel consumption, kg/h;
B_H – the total hourly fuel consumption, kg/h.

The total fuel consumption can be obtained by the equation:

$$B_h = B_D + B_{H_2}, \text{kg/h} \qquad (3)$$

The effect of hydrogen on engine performance can be most precisely evaluated by comparing the load characteristics with diesel fuel with the load characteristics with constant hydrogen mass fraction (constant K_{H2}). Practically, this could be achieved by comparing the engine characteristics at constant rpm, constant K_{H2} coefficient values and variable load.

Obtaining such characteristics directly, however, is associated with significant difficulties such as quick-acting measurement devices which can allow for recording of the corresponding parameters in real time. In this respect, the characteristics mentioned above were generated in two stages. The first stage includes obtaining regulation characteristics with variable hydrogen content. These characteristics are obtained at constant speed (n = const) and constant mean effective pressure (p_e = const, dynamometer force F_B = const) while varying the quantity of diesel fuel and hydrogen. The results are shown graphically as economical, ecological and other parameters as a function of the mass fraction of hydrogen in the total fuel mass, i.e. K_{H2}. The second stage includes determining (from the corresponding regulation characteristics) the engine parameters for a given speed and K_{H2} coefficient. With the parameters calculated in this way, the load characteristics for a certain speed and hydrogen content are obtained.

Details related to the specific execution of the so described methodology could be found in a published paper of which some of the participants in the current research are co-authors [16].

5 Results

As mentioned above, the experimental engine is a single-cylinder engine. Such engines possess some distinctive features related to their balancing, optimal counterweight and flywheel dimensions [17]. Therefore, the experimental study is conducted in engine modes typical for the engine, without high-amplitude vibrations in the mounting points.

Some of the experimental results are shown in Figures 2-12, where the following symbols are additionally used:

- b_e – brake specific fuel consumption, g/kWh;
- η_e – brake thermal efficiency;
- η_v – volumetric efficiency;
- $(dp/d\varphi)_{max}$ – maximum in-cylinder pressure rise rate, bar/deg;
- T_{EG} – exhaust gas temperature, °C.

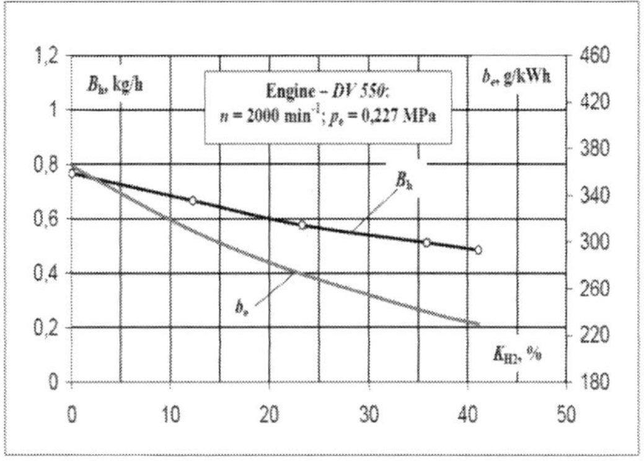

Fig.2. Variation of the economy of a *DV 550* diesel engine, working on a gas-diesel cycle as a function of the mass fraction of hydrogen.

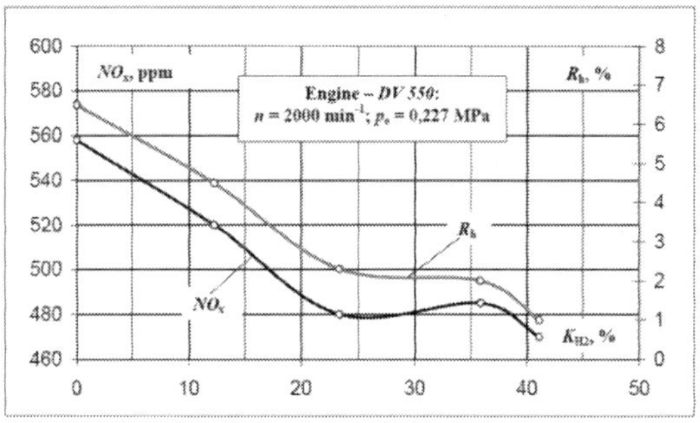

Fig.3. Variation of the nitrogen oxides concentration and the exhaust gases opacity of a *DV 550* diesel engine, working on a gas-diesel cycle in function of the mass fraction of hydrogen.

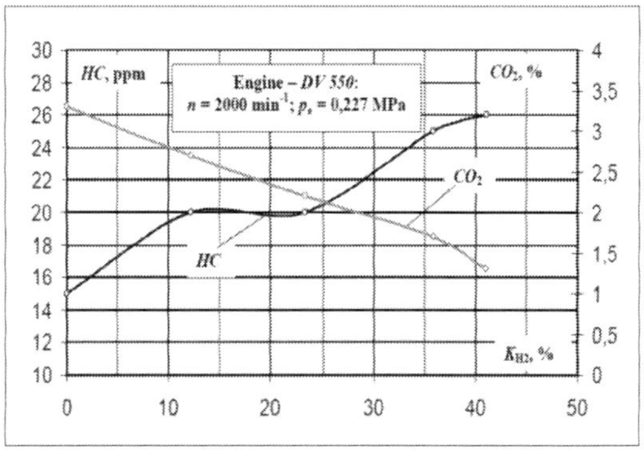

Fig.4. Variation of the hydrocarbons and carbon dioxide concentration in the exhaust gases of a *DV 550* diesel engine, working on a gas-diesel cycle as a function of the mass fraction of hydrogen.

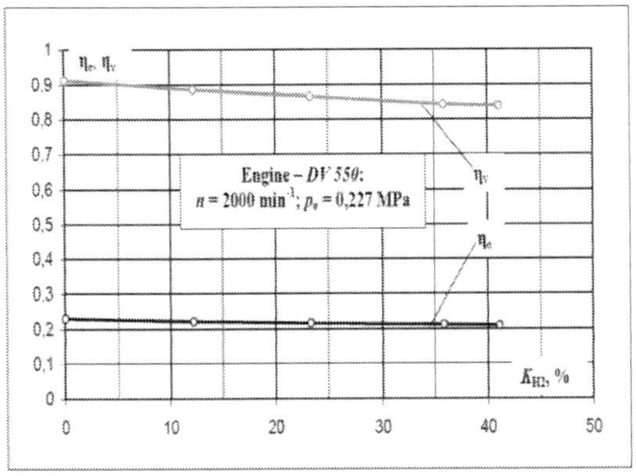

Fig.5. Variation of the brake thermal efficiency and the volumetric efficiency of a *DV 550* diesel engine, working on a gas-diesel cycle as a function of the mass fraction of hydrogen.

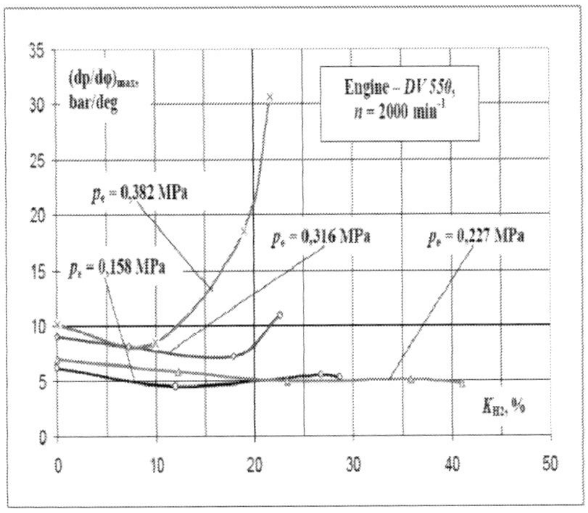

Fig.6. Variation of the maximum in-cylinder pressure rise rate in the cylinder of a *DV 550* diesel engine, working on a gas-diesel cycle as a function of the mass fraction of hydrogen.

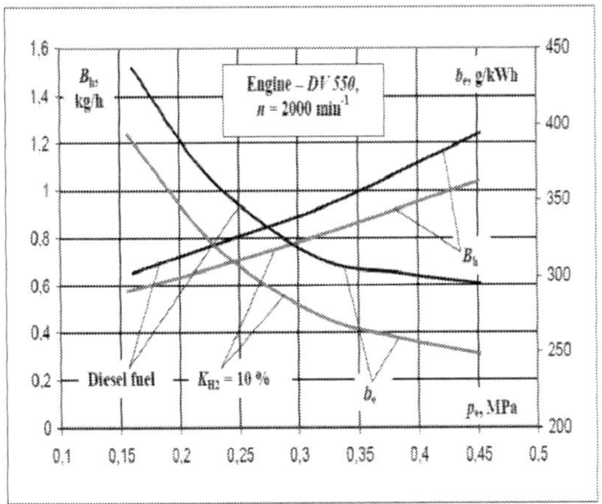

Fig.7. Comparison of the economical parameters of a *DV 550* diesel engine at an engine speed of 2000 rpm and variable load, running on: a diesel fuel and a gas-diesel cycle with hydrogen (dual fuel mode/diesel-fuel mixture) – K_{H2} = 10%.

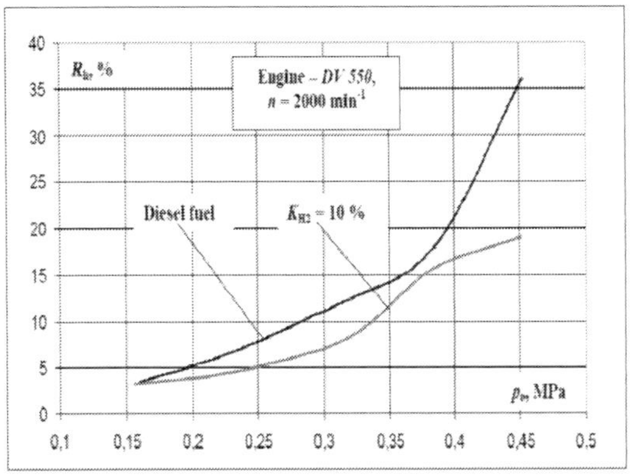

Fig.8. Comparison of the exhaust gas opacity of a *DV 550* diesel engine at engine speed of 2000 rpm and variable load, running on: a diesel fuel and a gas-diesel cycle with hydrogen (dual fuel mode/diesel-fuel mixture) – K_{H2} = 10%.

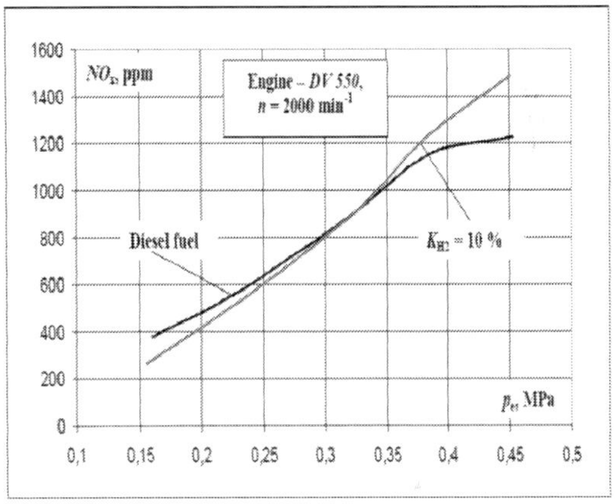

Fig.9. Comparison of the concentration of nitrogen oxides in the exhaust gases of a *DV 550* diesel engine at engine speed of 2000 rpm and variable load, running on: a diesel fuel and a gas- diesel cycle with hydrogen (dual fuel mode/diesel-fuel mixture) – K_{H2} = 10%.

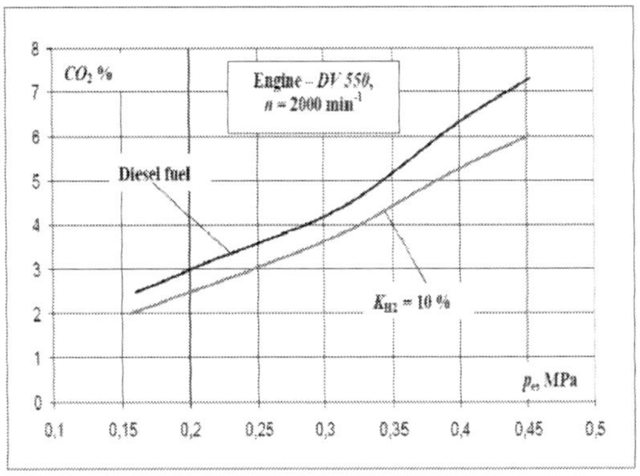

Fig.10. Comparison of the concentration of carbon dioxide in the exhaust gases of a *DV 550* diesel engine at engine speed of 2000 rpm and variable load, running on: a diesel fuel and a gas-diesel cycle with hydrogen (dual fuel mode/diesel-fuel mixture) – K_{H2} = 10%.

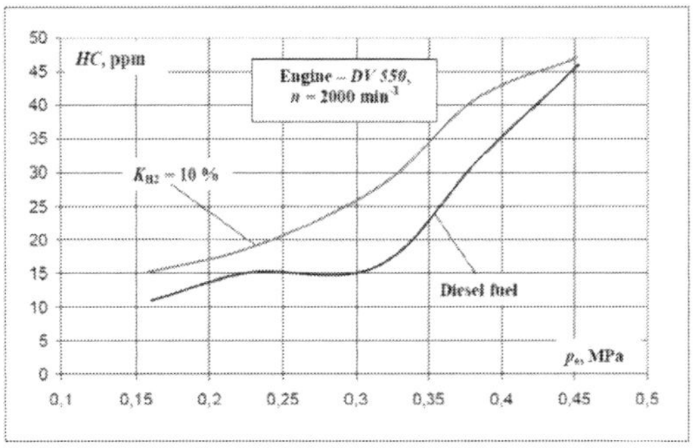

Fig.11. Comparison of the concentration of hydrocarbons in the exhaust gases of a *DV 550* diesel engine at engine speed of 2000 rpm and variable load, running on: diesel fuel and gas-diesel cycle with hydrogen (dual fuel mode/diesel-fuel mixture) – K_{H2} = 10%.

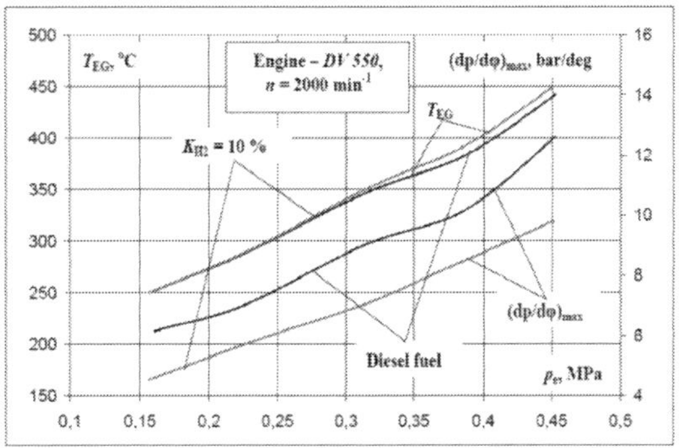

Fig.12. Comparison of the exhaust gas temperature and the maximum in-cylinder pressure rise rate of a diesel engine *DV 550* at 2000 rpm and variable load, running on: a diesel fuel and gas-diesel cycle with hydrogen (dual fuel mode/diesel-fuel mixture) – K_{H2} = 10%.

The analysis of the experimental results obtained from the examination of a *DV 550* diesel engine running on dual fuel mode showed the following:

1. A decrease in fuel consumption and brake specific fuel consumption is observed in every engine mode with the engine operating in a gas-diesel cycle as compared to the operation on standard diesel fuel. From a quantitative point of view, the percentage drop of B_h and b_e is proportional to the K_{H2} coefficient. For example, fuel consumption and brake specific fuel consumption of an engine working on a gas-diesel cycle with K_{H2} = 41,08% are 37% lower than those of an engine working on diesel fuel. The improvement in the engine economical parameters is due to the higher value of the lower heating value of hydrogen as compared to that of the diesel fuel.

2. A considerable drop in exhaust gas opacity during engine operation in a gas-diesel cycle with hydrogen is present. In some engine modes, at loads up to 30% and hydrogen mass fraction higher than 20% (K_{H2} > 20%), the exhaust gas opacity during engine operation on a gas-diesel cycle with hydrogen is 6 to 7 times lower than the opacity during engine operation on diesel fuel. For the load characteristics at 2000 rpm, the exhaust gas opacity in a gas-diesel cycle for K_{H2} = 10% is lower by 26,75% on average than in a diesel fuel cycle under the same conditions and lower by 41,42% on average for K_{H2} = 14%. The decrease of the exhaust opacity of the engine while operating in a gas-diesel cycle with hydrogen is mainly due to the high level of homogenization of the air-fuel mixture. The lower mass fraction of carbon in the total fuel mass, however, has a positive impact on opacity too.

3. The impact of the hydrogen mass fraction on the concentration of nitrogen oxides in the engine exhaust during operation on a gas-diesel cycle is not clear. With engine working in a gas-diesel cycle with hydrogen mass fraction higher than 25% (K_{H2} > 25%) and load of up to 30%, an average reduction of 23,7% in the concentration of nitrogen oxides is observed. In the range of variation of the hydrogen mass fraction from 10 to 20% (K_{H2} = 10% to K_{H2} = 20%) and engine working in a gas-diesel cycle with a load of over 50%, an increase of the nitrogen oxides concentration in the exhaust gases of 57,8% on average (as compared to the operation on diesel fuel) is observed. The impact of the hydrogen mass fraction on nitrogen oxides concentration could be explained with the variation of the engine exhaust temperature and the reduction of the total air-fuel ratio during a gas-diesel cycle operation;

4. Carbon dioxide concentration in the engine exhaust in a gas-diesel cycle with hydrogen decreases significantly. The reduction amounts to 25,75% on average in every engine mode. As mentioned above, the reduced concentration of carbon dioxide in the engine exhaust is due to the lower mass fraction of carbon in the total fuel mass;

5. An increase of hydrocarbon concentration by 28,5% on average is observed in all engine modes in gas-diesel cycle, as compared to the operation on diesel fuel. The hydrocarbon increase in the exhaust is due to the reduced air-fuel ratio in a gas-diesel cycle with hydrogen;

6. The engine exhaust gas temperature during a gas-diesel cycle operation changes similarly to the concentration of nitrogen oxides in the exhaust gases. The differences in exhaust gas temperature of the engine working on a gas-diesel cycle (for all values of the K_{H2} coefficient) and diesel fuel are insignificant for loads of up to 30%. An increase of 20°C on average in the exhaust gas temperature is observed with the engine working on a gas-diesel cycle with K_{H2} > 10% and with loads higher than 50% as compared to the operation on diesel fuel;

7. The engine volumetric efficiency η_v decreases in a gas-diesel cycle with hydrogen. The reduction is by up to 9% for loads higher than 50% and K_{H2} values higher than 15%. The decreased mass air-fuel mixture intake is an expected phenomenon due to the low density of hydrogen injected in the engine intake manifold. The impact of the hydrogen mass entering the engine on its brake thermal efficiency η_e is insignificant;

8. The diesel knock correlated to the maximum in cylinder pressure rise rate $(dp/d\varphi)_{max}$ also changes insignificantly as function of the mass of hydrogen fed into the engine. With an engine working on a gas-diesel cycle and load up to 30%, irrespective of the K_{H2} coefficient values, a reduction of $(dp/d\varphi)_{max}$ (reaching 30% for some modes) is observed. For loads higher than 50% and K_{H2} values over 15%, a gas-diesel cycle

operation is associated with dangerously high maximum in-cylinder pressure rise rate reaching 25 bar/deg. Therefore, the load characteristics is shown for K_{H2} = 10%. The high values of $(dp/d\varphi)_{max}$ when operating on a gas-diesel cycle at the abovementioned modes are undoubtedly due to the very high speeds of the hydrogen air-mixture flame front.

6 Conclusions

The analysis of the experimental study results leads to the following conclusions:

- The realization of a gas-diesel cycle with hydrogen results in an improvement of the economical parameters of the engine in its whole working range;
- The engine operation on a gas-diesel cycle guarantees significant reduction of the exhaust gas opacity and carbon dioxide concentration;
- A decrease of the nitrogen oxide concentration and decreased diesel knock is achieved during engine operation on a gas-diesel cycle with hydrogen and loads of up to 30%;
- For hydrogen mass fraction higher than 10% and with the engine working on a gas-diesel cycle at loads higher than 50%, the diesel knock and the nitrogen oxide concentration in the exhaust rise significantly;
- The diesel engine operation on a gas-diesel cycle leads to increased hydrocarbon concentration in the exhaust and reduced volumetric efficiency for external mixture formation.

In conclusion, we can state that the implementation of a gas-diesel cycle with hydrogen and external mixture formation is a promising method for improving the ecological and economical parameters of the diesel internal combustion engines. With effective and safe hydrogen storage systems, this method can be adopted for transport diesel engines as it does not require changes in their design.

References

1. E. Uludamar, Effect of hydroxy and hydrogen gas addition on diesel engine fuelled with microalgae biodiesel, International journal of hydrogen energy, (in press) (2018)
2. H. Arat, M. Baltacioglu, M. Özcanli, K. Aydin, Effect of using Hydroxy-CNG fuel mixtures in a non-modified diesel engine by substitution of diesel fuel, International journal of hydrogen energy, **41**(19), 8354-8363 (2016)
3. G. Belgiornoa, G. Blasioa, C. Beatricea, Parametric study and optimization of the main engine calibration parameters and compression ratio of a methane-diesel dual fuel engine, Fuel, **222**, 821-840 (2018)
4. B. Ashok, S. Ashok, C. Kumar, LPG diesel dual fuel engine – A critical review, Alexandria Engineering Journal, **54**(2), 105-126 (2015)
5. H. Wu, Z. Wu, Study on combustion characteristics and emissions of diesel-hydrogen mixtures by using energy-share method in a diesel engine, Applied Thermal Engineering, **42**, 154-162 (2012)
6. M. Ciniviz, H. Köse, Hydrogen use in internal combustion engine: A review, International Journal of Automotive Engineering and Technologies, **1**(1), 1-15 (2012)
7. J. Antunes, R. Mikalsen, A. Roskilly, An experimental study of a direct injection compression ignition hydrogen engine, International journal of hydrogen energy, **34**(15), 6516-6522 (2009)
8. G. Berry, A. Pasternak, G. Rambach, J. Smith, R. Schock, Hydrogen as future transportation fuel, Energy, **21**(4), 289-303 (1996)
9. H. Koten, Hydrogen effects on diesel engine performance and emissions, International journal of hydrogen energy, **43**(22), 10511-10519 (2018)
10. I. Mirica, C. Pana, N. Negurescu, A. Cernat, C. Nutu, Experimental studys of the hydrogen addition effects on diesel engine performance, *IOP Conf. Series: Materials Science and Engineering*, **147**(1) (2016)
11. N. Saravanan, G. Nagarajan, G. Sanjay, C. Dhanasekaran, K. Kalaiselvan, Combustion analysis on a DI diesel engine with hydrogen in dual fuel mode, Fuel, **87**(17-18), 3591-3599 (2008)
12. P. Dimitriou, M. Kumar, T. Tsujimura, Y. Suzuki, Combustion and emission characteristics of a hydrogen-diesel dual-fuel engine, International journal of hydrogen energy, **43**(29), 13605-13617 (2018)
13. T. Tsujimura, Y. Suzuki, The utilization of hydrogen in hydrogen/diesel dual fuel engine, International journal of hydrogen energy, **42**(19), 14019-14029 (2017)
14. I. Mirica, A. Cernat, C. Pana, N. Negurescu, C. Nutu, Performance comparison between hydrogen and diesel fuel fuelled compression ignition engine, U.P.B. Sci. Bull., Series D, **77**(4), 217-228 (2015)
15. Y. Karagöz, İ. Guler, T. Sandalcı, L. Yüksek, A. Dalkılıç, Effect of hydrogen enrichment on combustion characteristics, emissions and performance of a diesel engine, International journal of hydrogen energy, **41**(1), 656-665 (2016)
16. E. Dimitrov, A. Tashev, S. Pantchev, Methodology for Experimental Study of the Impact of Gaseous Fuels on the Performance of the Diesel Engine Operating on Gas-Diesel Cycle, *BulTrans-2015 Proceedings*, 103-107 (2015)
17. N. Pavlov, Numerical simulation on the vibration of a test bed with Engine with dual mass flywheel, International scientific journal Trans Motauto World, **3**(1), 49-51 (2018)

On the assessment of autoignition delay for Diesel fuel and Biodiesel B20

Bogdan Radu[1,*], Alexandru Racovitza[1], and Radu Chiriac[1,2]

[1]University POLITENICA of Bucharest, Faculty of Mechanical Engineering, Romania
[2]Associated Member of CMGPCE-EA7341, CNAM, 292 rue St-Martin, 75003, Paris, France

Abstract. The use of bio-fuels is a necessity nowadays, regulated by European legislation, which imposes to the EU-countries an increase in the substitution rate of classic fossil Diesel fuel. Biodiesel (B) fuel proves to be a reliable agent to fulfil this requirement, but a certain number of aspects have to be ameliorated regarding the compatibility of this kind of fuel with the existent compression ignition engines. One of these problems relies on the autoignition delay, on which the research results are still dispersed. The paper proposes an analysis of this autoignition delay when using a compression ignition (CI) engine fuelled with Diesel fuel and with blends of Diesel and Biodiesel fuels (B20 – 20% volumetric fraction of Biodiesel), starting from several correlations given by the literature, which are based on single-cycle analysis and application of the integral Livengood-Wu method. The obtained results offer an image of the in-cylinder processes complexity and of the B20 fuel behaviour related to the tested engine operation.

1 Introduction

Biofuels as an alternative to classical petroleum derived fuels mixed in different fractions have become increasingly used in the last few decades due to world-wide regulations concerning emissions. Market policies are nowadays encouraging the expanding of biofuels limits in mixtures and their use in an efficient way in internal combustion engines. Among these, combined mixtures of Diesel-Biodiesel and Diesel-Biodiesel-Alcohols play a more significant role in the whole chain of the cylinder processes, from the injection phase until the end of combustion. Therefore, evaluating and controlling the autoignition period has direct consequences on the global engine performance parameters. It is known that a larger content of alcohol (ethanol or butanol) leads to an increase of autoignition delay because of its lower cetane number (CN) value and to a decrease of the premixed air-fuel temperature due to its higher heating value (ethanol), inhibiting the NO_x formation, despite its higher molecular content of oxygen; on the contrary, the use of a predominant amount of Biodiesel leads to a decrease of this delay, ensuring a better control of the injection timing in good connection with the main variables defining the engine operation [1-3].

From the whole range of Diesel-Biodiesel fueling mixtures, B20 is a promising prospect regarding a number of evaluation criteria, such as ID (Ignition Delay – see Fig. 1, defined as the duration between the start of injection, SOI and the start of combustion, SOC [4]), BTE (Brake Thermal Efficiency), soot, CO, THC and NO_x concentrations, peak fire pressure, maximum rate of the firing pressure, according to Multi-Criteria Decision Making method (MCDM) [5]. By using Biodiesel as an additional fuel, it could also become necessary to modify the characteristics of the engine original injection system, to adjust the split-injection arrangement, the nozzle dimension, and the spray patterns, as many studies have proven [6-9].

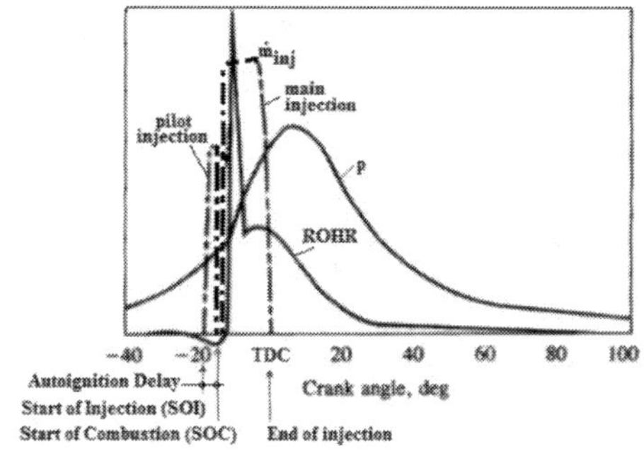

Fig. 1. Pressure p, rate of relative heat release (ROHR) and split injection characteristics in a compression ignition (CI) engine cylinder [4].

Many of the approaches used to approximate the autoignition delay are based on Arrhenius correlation and have been proposed in the case of using Diesel fuel [10], but in the case of using Biodiesel fuel there is still a lack of theoretical information. By exception, one formula [11] could be relevant, as its form was properly adjusted as following:

$$\tau_{ign}\text{, ms} = C\,p^{a}\,\Phi^{b}\,C_{n}^{c}\,\exp\left(E_{A}/\left(R_{M}T\right)\right) \qquad (1)$$

* Corresponding author: bobitaradu@yahoo.com

 https://doi.org/10.1051/matecconf/201823403002

in which C, a, b, c are constants, -, p is the in-cylinder mean pressure, atm, Φ is the equivalence ratio, -, C_n is the length of the carbon chain, -, E_A is the activation energy, kcal/mol, R_M is the gas universal constant ($1{,}987 \times 10^{-3}$ kcal/(mol·K)) and T is the in-cylinder mean temperature, K. A particular concern is the calculation of the activation energy, with different values when the fuel content varies [12].

Previous studies [10,13-15] have provided an acceptable similarity between experimental and simulation results, using the AVL-BOOST, AVL-FIRE software tools, with the pressure and the temperature picked as averaged values from a group of successive operating cycles. The present work is based on the analysis of each individual cycle from the acquired batch of consecutive cycles, when discussing pressure and temperature values, applying the Livengood-Wu method [16].

2 Hypotheses to calculate the autoignition delay

This paper uses the experimental results obtained by testing a naturally aspirated Diesel engine in tractors. Its main features, together with all the test bed systems have been discussed in a previous work [10]. Using this experimental basis, 200 individual pressure traces have been recorded, for each engine operating condition.

The autoignition delay was firstly determined experimentally and then established by considering a relationship of type (1). There is a large variety of coefficient values to form this expression, the authors propose a diversity of mathematical influences on the result. The presented paper considers the form proposed by Assanis [14], in which the pressure, p is given in bar, Φ is the equivalence ratio [-] and T is the absolute temperature, in K:

$$\tau_{ign}, \text{ms} = 2{,}4 p^{-1{,}02} \Phi^{-0{,}2} \exp(2100/(T)). \tag{2}$$

The calculation of the engine in-cylinder autoignition delay has two extremely sensitive and difficult to measure components. The first one is the start of injection (SOI), marked by the start of the injector needle lifting. The second one, even more difficult to evaluate, is that of the start of combustion (SOC). This process adds energy in cylinder leading to a pressure increase, its rate being a parameter experimentally recorded.

Paper [13] presents several methods of establishing SOC, by taking into account the 1st, the 2nd and the 3rd derivations of the pressure. Due to the sensitivity of the measurement system and since numerical derivation is an unsteady process, the obtained results were inconclusive.

Another approach consists of the relative heat release (ROHR) calculation. This is given by the sum of the piston mechanical work and the gas internal energy variation, diminished by the heat transferred to the walls during the compression process. At the end of compression both mechanical work and the heat transferred to the walls get negative values, therefore the heat release becomes also negative. It can be assumed that combustion starts corresponding to an angle for which this variable becomes positive and begins to increase.

The calculation of the autoignition delay could be achieved by calculating the pressure and the temperature average values in between a considered range period, or by calculating the relative delay on each such a range; this method is known as Livengood-Wu [16]:

$$\int_{t1}^{t2} \frac{dt}{\tau_{ign}(t)} \approx 1, \tag{3}$$

where t_1 is injection time and t_2 is the value of ignition time. In this equation t_2 is the unknown value.

These two procedures are practically equivalent, are based on the same principle, but for the second one there is a possibility to conduct a linear interpolation on the last range element to obtain a more precise result. In this paper there will be analyzed 200 individual cycles related to the autoignition delay, for an engine fuelled with Diesel and Biodiesel B20 fuels, when applying one of the proposed autoignition delay correlation and the integration calculus method. At the end of this computation a relative error analysis versus the measured value has also been completed.

3 Calculation Results

The measurements on the tested engine have been presented in [10] and listed in Table 1 (by extraction from Fig. 1):

Table 1. Injection and start of combustion angle.

Case number	1	2
Fuel/Speed	Diesel/2400 rpm	B20/2400 rpm
SOI	–5 CA	–5,5 CA
SOC	0,375 CA	–0,15 CA

where CA is the crank angle and SOC is the angle for the start of combustion.

For these cases the nominal torque was 175 Nm, corresponding to a nominal power of 44 kW.

The calculation of the autoignition delay has been developed for each single cycle and for each of them the corresponding angle where the integral (equation (3)) reaches the value 1 has been extracted and listed (see Fig. 2, Case 1).

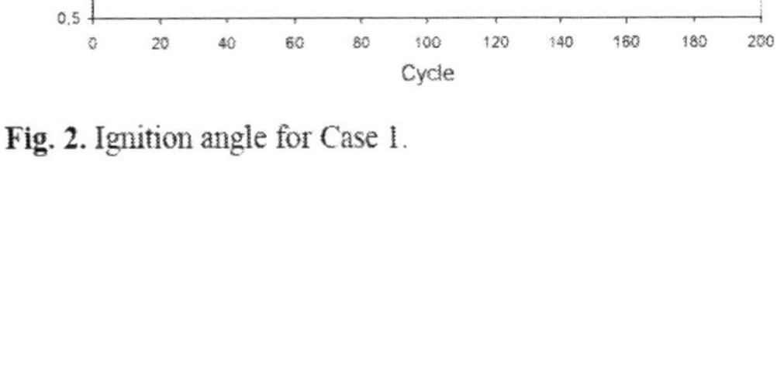

Fig. 2. Ignition angle for Case 1.

In this case, when the engine is fuelled with Diesel fuel, at full load the relative air-fuel ratio is 1.5. For these operating conditions, the angles for which the integration is 1 are included in the range 0,6 ... 0,8 deg.CA (ATDC, see Fig. 2). The average value for these data is 0,72 deg.CA as compared to the experimental average value of 0,38 deg.CA, there is a difference of 0,34 deg.CA.

Using the experimental data obtained in Case 2, when fuelling the engine with B20 fuel, the results appear more dispersed. For the ignition moment the average value is 0,3 deg. CA, very close comparing to the experimental results (0,375 CA). The experimental data for each individual cycles for start of combustion are located in domain -0,5 and 1,7 deg.CA (see Fig. 3).

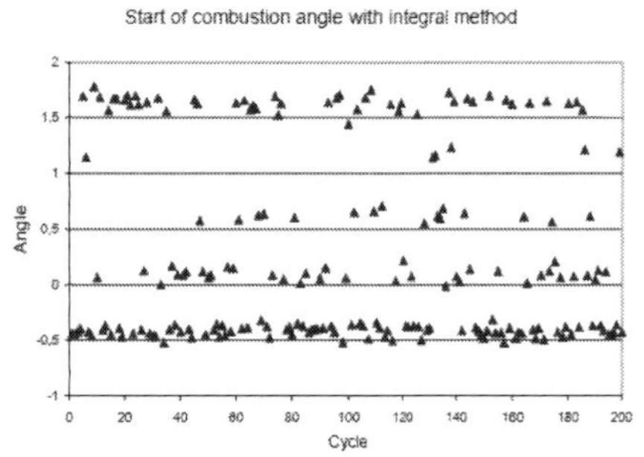

Fig. 3. Ignition angle for Case 2.

To be able to use Assanis's formula with the help of the temperature and pressure average values it is useful to expose them inside a domain of interest limited between SOI and by some points relative to top dead center (TDC). The average values for pressure and temperature from SOI to current position (where end of compression TDC = 0) are listed in Table 2.

Table 2. Pressure and temperature in cycle.

Case/CA	-5	-4	-3	-2	-1	0	1
Case 1 p	44,5	44,8	45,0	45,1	45,2	45,2	45,1
Case 1 T	1017	1016	1014	1012	1009	1007	1003
Case 2 p	46,1	46,3	46,5	46,6	46,7	46,7	46,6
Case 2T	1053	1051	1049	1046	1043	1040	1038

From this stage further the autoignition delay values for each range, starting from SOI and reaching –5 deg.CA BTDC are as follows (see Table 3, where *As*, ms is the value for Assanis formula and *dt*, ms is time elapsed from the SOI angle; in this table are presented only the values around top dead center, respectively –1, 0, 1 degree):

Table 3. Time evaluation for average values.

	–1CA		0CA		1CA	
	As	*dt*	*As*	*dt*	*As*	*dt*
Case 1	0,42	0,34	0,42	0,41	0,42	0,48
Case 2	0,38	0,34	0,38	0,41	0,38	0,48

For the analyzed engine operating conditions, applying the proposed formula, the average values define SOC in the range of –1... 0 deg.CA in the case of Biodiesel and in the range 0 ... 1 deg.CA in the case of Diesel fuel, values to be found also in most of the individual cycles.

Another analysis mode is that in which for each single cycle the adopted average values for temperature and pressure are those calculated for the period following SOI. When fuelling with Diesel (Case 1), this procedure offers values that are relatively close to the results obtained with the integral method, the average value for the integral is 0,72 CA and for the other calculation 0,69 CA (see Fig. 4).

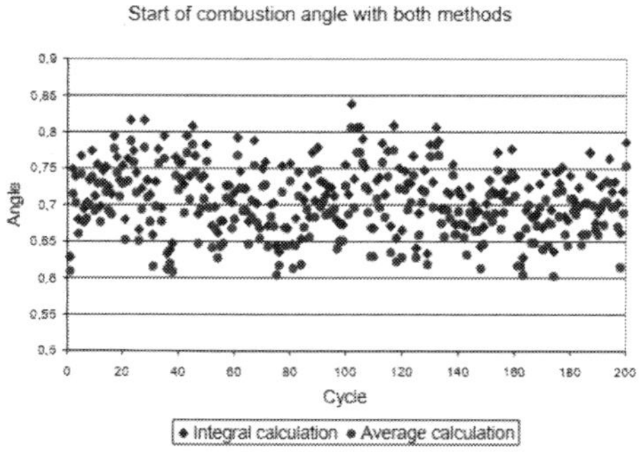

Fig. 4. Results for Case 1.

Comparing these two procedures is possible in the mean cycle obtained as an average of the 200 individual cycles. Averages for temperature and pressure are those obtained by calculating the averages over the 200 cycles and these ones resulting from the range values situated between SOI and the adopted angular limit after TDC.

Fig. 5 presents the cylinder pressure and temperature traces for the values involved in the Case 1 calculation. The mean trace represents the average values over the 200 consecutive cycles and the average trace is obtained from the values starting from the injection point.

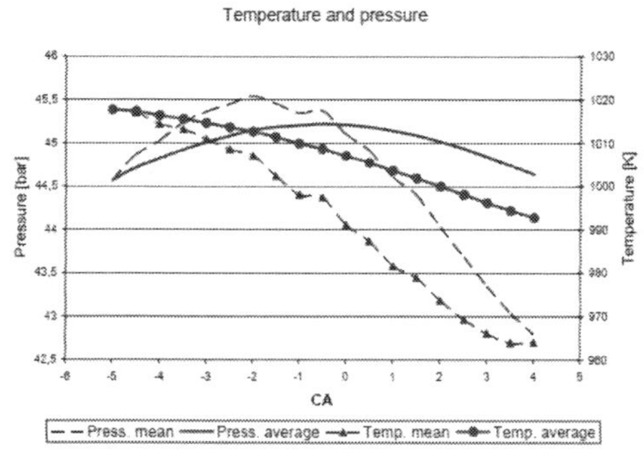

Fig. 5. Pressure and temperature traces for calculation Case 1.

Calculations were made with the following correlations (4):

$$p_{\alpha\, mean} = \frac{1}{200}\sum_{1}^{200} p_{\alpha i}, \tag{4}$$

where $p_{\alpha i}$ is the pressure measured at angle α for each individual cycle.

In order to use the Assani's formula it is necessary to define the average values for cylinder pressure and temperature in the ignition delay period. These values are calculated with the correlation (5):

$$p_{\alpha\,average} = \frac{1}{\alpha_{inj} - \alpha}\sum_{\alpha_{inj}}^{\alpha} p_{\alpha}, \tag{5}$$

where α_{inj} is the injection angle. The temperatures are calculated with the same relations.

When we plot both methods on a single diagram, we obtain insignificant differences which lead to the decision that both could be applied with same degree of certainty (see Fig. 6).

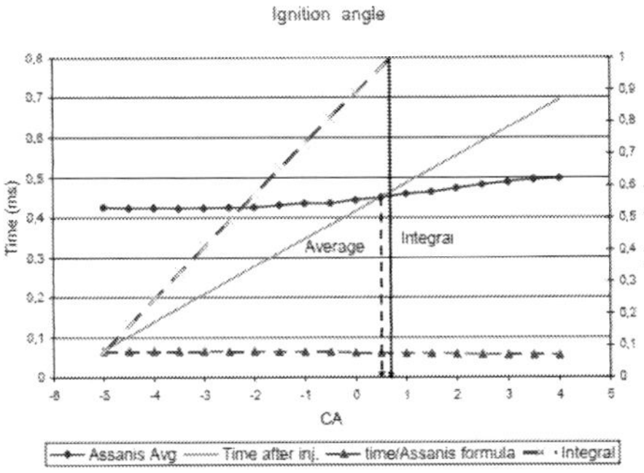

Fig. 6. Example calculation for both methods for Case 1.

When fuelling with B20 fuel, the use of the average values related to the autoignition delay reduces the dispersion of the results obtained by integration method. Using this second procedure, the range of the ignition angles becomes 0 ... 0,5 deg.CA (see Fig. 7.).

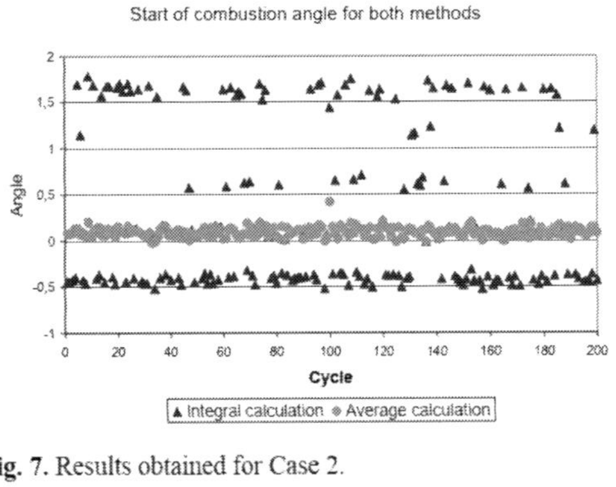

Fig. 7. Results obtained for Case 2.

In terms of the mean values obtained over 200 cycles applying both methods/procedures, the results on the SOC angle for both cases are listed in Table 4.

Table 4. Mean values of SOC angle for both cases.

	Integral calc.	Average calc.
Case 1	0,71	0,69
Case 2	0,29	0,10

4 Conclusions

The paper presents experimental data obtained when fuelling a tractor Diesel engine with Diesel and B20 Biodiesel-Diesel B20 mixture (20% vol. Biodiesel). An analysis was made to evaluate a simple calculation relationship for the autoignition delay (Assanis), depending on the in-cylinder pressure, the global uniform temperature during the autoignition delay period and the air-fuel ratio for the analyzed operating condition.

In this computation individual cycles were used, for each of them two mathematical evaluation methods have been proposed: integrating the autoignition delay relative time and evaluating the average values of temperature and pressure on a certain angular range.

The results show small differences between the two calculation modes of the autoignition delay, lower than 0,5 deg.CA.

The cyclic variability caused probably during the compression process by the variation of the inlet process combined with the different air-fuel mixture formation in case of B20 leads to an increased dispersion of SOC.

The differences calculated for both types of fuels are inside the variability domain, consolidating the conclusion that from the perspective of the autoignition delay the behaviour of these two fuels (Diesel and B20) is practically the same.

The existing deviations of the theoretical results relative to the experimental data obtained for the two different fuels range between 0,25 to 0,5 deg.CA.

This study will be continued by analyzing other engine operating conditions to extend the soundness of these conclusions.

References

1. M. Lapuerta, J. Hernandez, D. Rodriguez, A. Cova-Bonillo, Autoignition of blends of n-butanol and ethanol with Diesel or Biodiesel fuels in constant-volume combustion chamber, Energy, **118**, 613-621 (2017)
2. A. Jamrozik, W. Tutak, M. Pyrc, M. Sobiepanski, The effect of Diesel-Biodiesel-Ethanol blend on a direct injection Diesel engine, Thermal Science, **21**, 591-604 (2017)
3. D. Alviso, F. Krauch, R. Roman, M. Maldonado, R. Goncalves dos Santos, J. Rolon, N. Darabiha, Development of a Diesel-Biodiesel-Ethanol combined chemical scheme and analysis of reactions pathways, Fuel, **191**, 411-426 (2017)
4. A. Nirala, CI engine combustion (presentation), https://www.slideshare.net/akhileshroy96/ci-engine-combustion-by-akhileshwar-nirala (2014)
5. C. Sivaroja, G. Sakthivel, Compression Ignition engine performance modeling using hybrid MCDM techniques for the selection of optimum fish oil Biodiesel blend at different injection timings, Energy, **139**, 118-141 (2017)
6. H. How, H. Masjuki, M. Kalam, Y. Teoh, Influence of injection timing and split injection strategies on performance, emissions and combustion characteristics of Diesel engine fueled with Biodiesel blended fuels, Fuel, **213**, 106-114 (2018)
7. H. Koeten, A. Parlakygit, Effects of the Diesel engine parameters on the ignition delay, Fuel, **216**, 23-28 (2018)

8. M. Musthafa, T. Kumar, T. Mohanraj, R. Chandramouli, A comparative study on performance, combustion and emission characteristics of Diesel engine fueled by Biodiesel blends with and without an additive, Fuel, **225**, 343-348 (2018)
9. M. Kumaran, A. Khalid, H. Salleh, A. Razali, A. Sapit, N. Jaat, S. Norshuhaila, Effect of Algae-derived Biodiesel on ignition delay, combustion process and emission, *Intl. Eng. Research and Innovation Symposium (IRIS)*, **160**(1) (2016)
10. M. Aldhaidhawi, L. Miron, R. Chiriac, V. Badescu, Autoignition process in compression ignition engine fueled by Diesel fuel and Biodiesel with 20% rapeseed biofuel Diesel fuel, J. Energy Eng., **144**(5) (2018)
11. M. Campbell, D. Davidson, R. Hanson, Scaling relation for high temperature Bio-diesel surrogate ignition delay times, Sandia National Laboratories, Fuel, **164**, 151-159 (2016)
12. F. Soto, M. Alves, J. Valdes, O. Armas, P. Crnkovic, G. Rodrigues, A. Lacerda, L. Melo, The determination of the activation energy of Diesel and Biodiesel fuels and the analysis of engine performance and soot emissions, Fuel Processing Technology, **174**, 69-77 (2018)
13. M. Aldhaidhawi, *Study on performance, emissions and combustion characteristics of a Diesel engine fueled with Biodiesel B20*, (PhD Thesis, Politehnica Univesity Bucharest, 2017)
14. D. Assanis, Z. Filipi, S. Fiveland, M. Syrimis, A predictive ignition delay correlation under steady-state and transient operation of a direct-injection Diesel engine, ASME Transactions: J. of Eng. for Gas Turbines and Power, **125**(2), 450-457 (2003)
15. M. Aldhaidhawi, A. Racovitza, R. Chiriac, V. Badescu, Ignition delay period for compression ignition engine operating on Diesel and Biodiesel B20, UPB Sci.Bull. Series D, **79**(4), 83-94 (2017)
16. J. Livengood, P. Wu, Correlation of autoignition phenomena in internal combustion engines and rapid compression machines, *5th Symp. (Int.) on Combustion*, Reinhold Publ. Corp., 347-356 (1955)

https://doi.org/10.1051/matecconf/201823403003

HCCI combustion control using advanced gasoline direct injection techniques

Jacek Hunicz[1,*], *Maciej* Mikulski[2], and *Henryk* Komsta[1]

[1]Lublin University of Technology, Faculty of Mechanical Engineering, Nadbystrzycka 36, 20-618 Lublin, Poland
[2]TNO Automotive, Automotive Campus 30, 5708 JZ Helmond, Netherlands

Abstract. Homogeneous Charge Compression Ignition (HCCI) is a promising low temperature combustion technology for reciprocating engines that offers high fuel efficiency and extremely low exhaust emissions. However, combustion control should be improved and operating range should be widened for the technology to achieve production level. In this study an overview of different direct gasoline injection control approaches, applied to improve stability at low engine loads and to reduce pressure rise rates at high load regime, is presented. The tests are performed on a single-cylinder research engine operated in a negative valve overlap (NVO) mode for residual gasses trapping. The investigated direct injection schemes included: (i) fuel injection during the NVO period to improve mixture reactivity and take an advantage of exhaust-fuel reactions thermal effects, (ii) fuel injection during intake stroke to create homogeneous charge and (iii) late fuel injection during compression stroke to create stratified charge. The results showed that application of early NVO injection enables active control of combustion timing at nearly idle conditions. The late fuel injection, during the compression stroke, enabled mitigation of excessive pressure rise rates at high engine load regime.

1 Introduction

The recent progress in engine development towards full system flexibility, and achievements in model based engine control, greatly relaxed the constrains on system complexity. With fully flexible valve trains and variable compression ratio systems becoming standard features for next generation automotive engines [1, 2], the feasibility for advanced low temperature combustion systems like homogeneous charge compression ignition (HCCI) is being re-considered.

In HCCI, the combustion process is kinetically controlled, resulting in very fast heat release rates (HRR). This allows for optimizing the engine efficiency close to the theoretical limits of the ideal Otto cycle [3]. The premixed mixture, with uniform temperature distribution across the cylinder, supports achieving ultra-low nitrogen oxides (NO_X) and particulate matters emissions [4, 5]. On the other hand, assuring low-load combustion stability and limiting excessive pressure rise rates (PRRs) at high-loads are main challenges for the HCCI concept [6, 7].

Exhaust gas trapping via negative valve overlap (NVO) enables variety of control mechanisms that can be used to tackle the above challenges of HCCI [8, 9]. This technique introduces internal exhaust gas recreation (EGR) enabling sufficiently high intake valve closing (IVC) temperatures for gasoline auto-ignition at compression ratios low enough to avoid knock. Additionally, the EGR slows down the reaction rates and increases fuel dilution, thus contributing to further NO_X emissions reduction [10].

Note that NVO operation differs from the typical 4-stroke engine cycle (with positive valve overlap), by introducing exhaust gas recompression between exhaust and intake processes. An example of a full in-cylinder pressure trace recorded for an HCCI engine with NVO mode is shown in Fig. 1

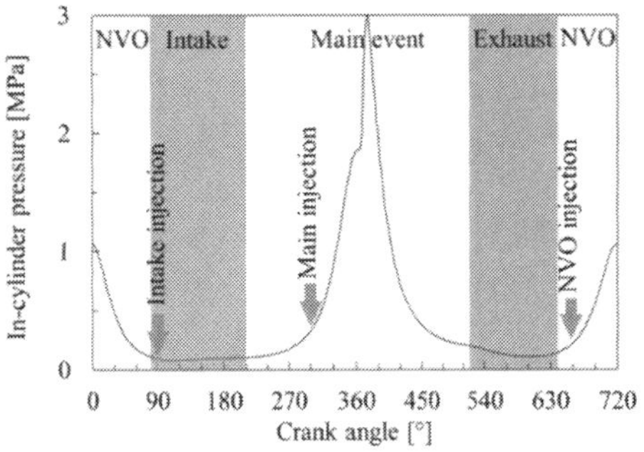

Fig. 1. Example of in-cylinder pressure and injection timings.

Further extending of the HCCI engine operating range towards low loads can be attained with direct injection of a certain portion of fuel early during the NVO period, as shown in Fig. 1. This injection strategy results in the so-called fuel reforming – a chemical process which produces highly reactive species like: ethane, ethylene, formaldehyde, methanol and acetylene [11, 12]. The mix of highly reactive reformed gasses with the air and the fuel injected during the induction process provides proper mixture ignitability in the main compression event.

* Corresponding author: j.hunicz@pollub.pl

 https://doi.org/10.1051/matecconf/201823403003

Additionally, combustion stability is improved and THC emission reduces. Furthermore, lean operation limit can be extended, positively influencing low-load engine performance and NO_X emissions [13].

Aside of the above chemical effects of fuel reforming, thermal effects of NVO injection can be used for combustion phasing control across all the engine operating conditions [14, 15]. Note however that the effects highly depend on NVO injection timing and mixture strength.

From the perspective of the expansion of engine operation area towards higher loads extensive PRRs can be avoided by introducing mixture stratification via late fuel injection during the main event. The mechanism relays on distributing the ignition of in-cylinder mixture in space and time thus elongating the combustion and making it smoother [15-17].

Bearing in mind the recent trends in engine development and the resulting, increased feasibility for advanced low temperature combustion concepts, the present work aims to provide a concise demonstration of the effects of different direct injection strategies on residually affected HCCI combustion characteristics. Namely, following techniques are showcased: (i) fuel injection during the NVO period to improve mixture reactivity and take an advantage of exhaust-fuel reactions thermal effects, (ii) fuel injection during intake stroke to create homogeneous charge and (iii) late fuel injection during compression stroke to create stratified charge. The engine experiments, supporting this demonstration, cover the whole range of engine loads attainable in the HCCI mode.

2 Experimental approach

2.1 Research engine and test stand

The discussed results are produced on a single-cylinder research engine installed on a test bed with a direct current dynamometer. The engine is equipped with a fully variable valvetrain, which enables regulation of valves lifts and timings. A detailed description of the engine and the valvetrain is presented in refs. [14, 15]. At this point only a concise decryption of the functionalities relevant to the present tests are provided.

The fully variable valvetrain allows to obtain internal EGR with the use of the NVO technique. Fuel is introduced into the cylinder with the use of a side-mounted, single-stream, swirl-type injector. An electric motor driven vane compressor and air temperature management system are installed in the intake runner to provide required boost levels. The main engine geometric parameters, along with the utilized valvetrain settings are specified in Table 1. To simplify the imaging of crank angle based figures, all timings are given in terms of degrees after top dead centre (TDC) during NVO in accordance with Fig. 1.

The engine test bench is equipped with a standard, in-cylinder pressure based combustion analysis system, by AVL. The acquisition for the in-cylinder pressure and the injector excitation current is triggered by an optical crankshaft encoder, with an angular resolution of 0,1° of crank angle. In order to provide proper input for combustion analysis, the intake mass flow rates, temperature and pressures are recorded along with the fuel consumption (measured gravimetrically). Additionally, to estimate the overall main event excess ratio (λ), a wide band lambda probe, installed in the exhaust runner, is utilized.

Table 1. Test engine data.

Parameter	Value
Swept volume, cm³	498,5
Bore, mm	84
Stroke, mm	90
Compression ratio, -	11,7
Fuel	Gasoline 95 RON
Fuel injector	Direct swirl type
Fuel pressure, MPa	10
Boost	Vane compressor
Number of valves	2
Intake valve opening (IVO), °	80
Intake valve closing (IVC), °	210
Intake valve lift, mm	3,6
Exhaust valve opening (EVO), °	520
Exhaust valve closing (EVC), °	640
Exhaust valve lift, mm	2,9

2.2 Experimental conditions and procedure

The experiments are conducted at the engine speed of 1500 rev/min with fully open throttle. The intake air temperature is set to 40 ± 1 °C. The temperature of cooling liquid at the engine outlet is maintained constant at 90 ± 1 °C. The fuel is introduced into the cylinder at strategically selected start of injection (SOI) timings, which functions were verified in our previous works [14-16]. To enable fuel reforming, the NVO injection is applied at SOI = 660°. Intake injection, applied at 90° aims at creating a homogeneous mixture without any thermal nor chemical effects which are associated to NVO injections. The role of the main injection, applied at 300°, is to create a stratified mixture in order to slow-down the combustion rates at elevated loads.

The experimental matrix includes three operating points covering the whole attainable HCCI load range. The operating parameters, along with the applied injection strategies are shown in Table 2. The set points of λ and boost pressure are adjusted to optimal conditions for a given load case. Those settings were determined in previous research endeavours as the best trade-offs between efficiency and NOx emissions [14, 15]. Note that the valve settings used to achieve NVO in the present study are not per-case optimized. This is maintained the same in all operating points, in order to provide a common benchmark for comparing the effects of NVO injection.

The internal EGR ratio definition used in the present work (Table 2) relays on compering the calculated mass of trapped residuals at exhaust valve closing (EVC) with the same quantity at the IVC event. The mass indexes are computed using the AVL Boost software, basing on the measured in-cylinder and manifold conditions. The same software is further used for the HRR analysis. The net

 https://doi.org/10.1051/matecconf/201823403003

HRR traces presented in the results section are calculated according to the first law of thermodynamics, with consideration of the effects of temperature and composition on a the instantaneous specific heats ratio.

The Equivalence ratio (Φ) is calculated using the following formula:

$$\Phi = \frac{m_F(1+L_t)}{m_{Air} + m_{Exh}}, \quad (1)$$

where m_F, m_{Air} and m_{Exh} are masses of fuel, fresh air and residuals respectively, and L_t is the theoretical air-fuel ratio for gasoline. One should note that the above definition represents the total fuel dilution by air and exhaust gasses. The values of Φ, presented in Table 2 show how extensive is the dilution in each operating point despite nearly stoichiometric excess air. It should be further noted , that some values shown in Table 2 are averaged for different injection strategies for the sake of clarity of the presentation. In fact, these values were changing in some extent due to effects of fuel injection on mass of aspirated air, temperature, etc.

Table 2. Detailed engine settings for analysed cases.

Parameter		Case 1	Case 2	Case 3
IMEP, MPa		0,19	0,31	0,61
λ, -		1,25	1,0	1,07
EGR, %		56	47	29
Φ [-]		0,354	0,507	0,693
p_{int}, kPa		101	101	152
SOI, °	A	660	660	660
	B	660; 90	660; 90	660; 90; 300
	C	90	90	660; 300
m_F, mg	A	7,79	11,05	20,65
	B	2,6; 5,03	2,6; 8,06	2,6; 14,9; 2,6
	C	7,95	9,86	2,6; 19,59
η_{therm}	A	0,28	0,32	0,35
	B	0,31	0,33	0,37
	C	0,29	0,33	0,34

3 Results and discussion

At the lowest investigated engine load (case 1 in Table 2) the fuel is injected in a single dose early during NVO (injection strategy A), in a single dose during the intake stroke (strategy C) and with split fuel injection (strategy B). At the latter case, only a small portion of the fuel (so-called pilot fuel), is injected during the NVO phase. In order to better illustrate the effects of variable injection strategies on in-cylinder processes, the pressure-volume diagrams in logarithmic scale are shown in Fig. 2. Careful analysis of the p-V diagram, for the injection strategy A, shows that the system performs positive work around NVO TDC. This is the result of partial fuel oxidation during NVO. Since the mixture is lean at this operating point, the trapped exhaust gases contain some amounts of oxygen, which allows for exothermic reactions to occur. While the in-cylinder volume increases, the work becomes negative, with values comparable to those obtained for the single intake injection strategy. For the split injection case, a positive loop near TDC can observed, similarly to strategy A. Note however, that the negative work at higher cylinder volumes is much smaller than in the remaining cases. This observation allows to state that the reduction of the NVO injected fuel mitigates the re-compression loop losses in comparison to strategy A. It should be also noted from Fig. 2 that the intake injection strategy results in much higher expansion pressure. It is an effect of improved cylinder aspiration and the resulting higher energy introduced with the fuel at constant λ.

Furthermore, the main compression-expansion loops in Fig. 2 reveal the effect of injection strategies on combustion timing. To provide more insight into combustion evolution, Fig. 3 shows HRR curves for the analysed case 1. It should be noted that NVO fuel injection results in very early combustion. Nearly all fuel is oxidized before TDC, which increases compression work and deteriorates overall cycle efficiency. Shifting the injection from NVO period to the intake stroke enables delaying the start of combustion by approximately 10 °CA. Additionally, the reaction rates are substantially reduced, which is manifested by lower peak HRR value and longer combustion duration. Note that the split fuel injection (strategy B), provides moderate peak HRR values and optimal combustion timing. The properly aligned combustion timing and discussed above reduced NVO losses provide improvement of overall thermal efficiency of the cycle by 10%, comparing to all fuel injection during NVO (see Table 2).

At mid-load (case 2 in Table 2) the engine was operated at stoichiometric mixture. At given valve timings, this was the maximum attainable load under naturally aspirated conditions. The applied injection strategies were the same as in the previously discussed case 1.

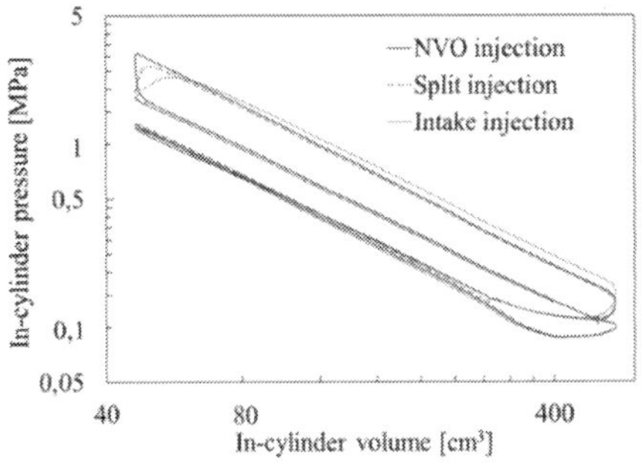

Fig. 2. Pressure-volume diagrams for case 1.

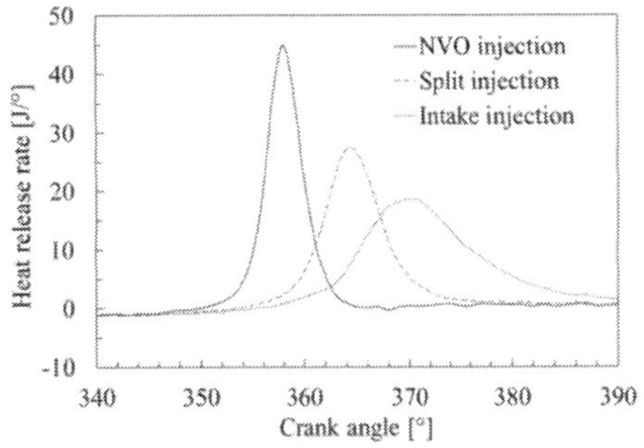

Fig. 3. Heat release rates for case 1.

 https://doi.org/10.1051/matecconf/201823403003

Pressure-volume diagrams in Fig. 4 show that the effects of fuel injection on the performed work are substantially different than for lean mixture realized in case 1. The strategy A, realizing a single NVO injection, results in high recompression losses. It is evident from Fig. 4, that the pressure rise during the NVO compression is diminished by fuel vaporisation. It is not compensated by partial fuel oxidation due to the lack of oxygen in the exhaust, while operating close to the stoichiometric conditions. Without additional oxidation the evaporative cooling deteriorates cycle efficiency, as shown in Table 2. Note however, that at the same time, the strategy reduces temperature and pressure at IVO which improves fresh charge aspiration. This effect in some extent compensates for efficiency drop, as engine load (amount of fuel injected) itself increases.

Following the same reasoning, with all the fuel injected during the intake stroke (strategy C), the recompression loop is much thinner, while, adding a small pilot injection (strategy B) increases the pressure delta between NVO compression and expansion.

Fig. 5 shows peculiar behaviour it terms of the auto-ignition timing. Despite in Fig. 4, the thermal effect associated to different fuel injection strategies are clearly visible, the ignition timing is nearly the same no matter the SOI. It is plausible that the NVO fuel reforming produces some auto-ignition promoting species, however, changes in mixture reactivity are balanced-out by the thermal effects of fuel evaporation. Nevertheless, the combustion timing for the discussed mid-load, stoichiometric case is perfectly aligned from the cycle efficiency point of view. The reaction rates are substantially affected by the NVO injection, similarly to the lean mixture operation in case 1. This finally suggests that the mixture reactivity improves as a result of injection strategy A, even at stoichiometric conditions. For more details on quantification of reactive species production as a result of NVO fuel injection the reader is referred to the following works [11, 18].

In order to further increase the engine load, precluded in case 2 by insufficient fresh air aspiration, it was necessary to elevate the intake pressure beyond the ambient level. For the case 3, this is set to the absolute value of 152 kPa. As it was earlier demonstrated in [15], HCCI combustion under loads excessing 0,5 MPa in terms of IMEP produces unacceptable PRRs. Thus, in this case investigated injection strategies are altered in comparison to the lower load cases, where exceeding the PRR limit does not pose an issue. At high load, late injection at SOI = 300° is incorporated to take the advantage of introducing in-cylinder mixture stratification during the main event. Two stratification patterns were realized via different amounts of fuel injected during the compression stroke. The injection strategies for case 3 and fuel division between injections are provided in Table 2. To avoid excessive emissions of CO and unburned hydrocarbons from locally fuel-rich zones which might arise as a result of stratification, the engine is operated under overall slightly leaner mixture compared to case 2. It should be also noted that during boosted operation, the engine was unable to realize autonomous HCCI operation without the support of the NVO pilot injection. The reason for this behaviour is the substantial reduction of IVC temperatures, caused by trapped residuals being replaced by the colder intake air. To compensate for that drop in temperature, the mixture reactivity has to be improved by incorporating NVO fuel reforming effects.

Fig. 6 shows how impactful are the effects of applied fuel injection schemes at high load. Similarly to the medium load case, strategy A (NVO injection) produces high recompression losses. These losses can be minimized by reduction of the NVO fuel amount, as evident when analysing the Fig. 6 for strategies B and C (split injection with different levels of stratification). However, for two cases with the same pilot fuel injection recompression loops are different. It is a result of different amounts of trapped residuals and different exhaust temperatures. Namely, the share of the thermal losses for large stratification was higher, which deteriorated the cycle efficiency, as shown in Table 2.

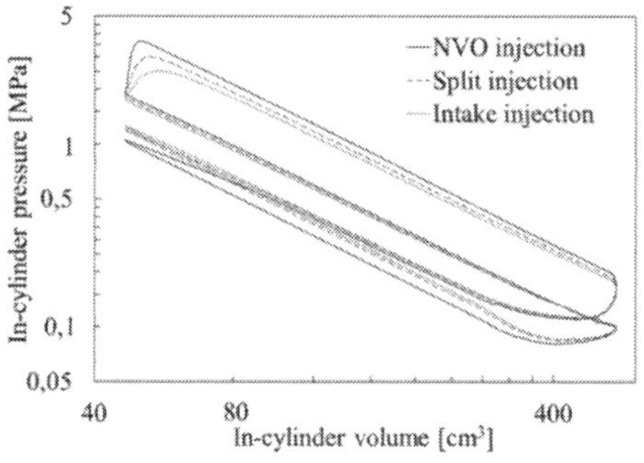

Fig. 4. Pressure-volume diagrams for case 2.

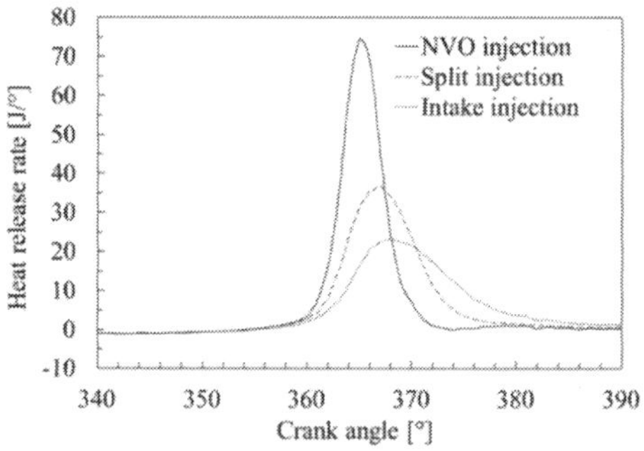

Fig. 5. Heat release rates for case 2.

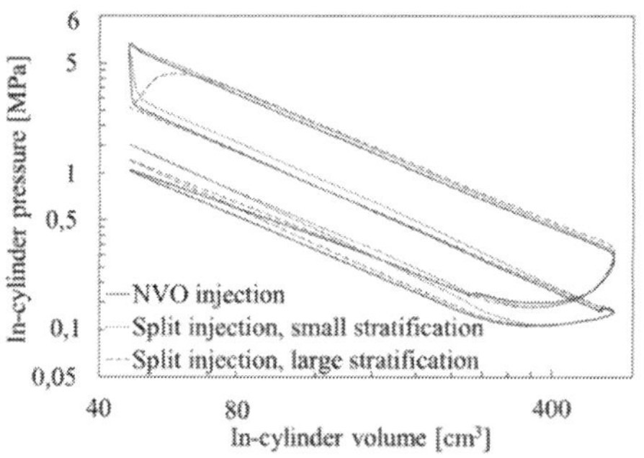

Fig. 6. Pressure-volume diagrams for case 3.

https://doi.org/10.1051/matecconf/201823403003

Fig. 7 shows that for the primary, NVO injection strategy A the maximum HRR exceeds 200 J/°. Additionally, this point appears at TDC, which causes very high peak pressure and excessive PRR at a level of 13 GPa/s, where commonly acceptable limit is approximately 5 GPa/s [19]. Interestingly, the small stratification achieved with strategy B indeed reduces combustion rates, but advances auto-ignition. In contrast, large stratification reduces the combustion rates and delays the auto-ignition at the same time. This peculiar effect can be explained as follows. With relatively short injection during the compression phase, the fuel has sufficient time to evaporate and largely premix before the auto-ignition starts. Still, the evaporating fuel creates thermal stratification with locally hot zones acting as ignition sources. Nevertheless, the following combustion process is slower because cold zones ignite later. This thesis was proven on the basis of modelling and experimental studies [20-22]. At extensively stratified mixture (strategy C) the fuel does not premix well before the auto-ignition. Thus, the mixture is stratified not only thermally, but also compositionally. As a result, in the fuel-rich zones, that are more prone to auto-ignition temperature is heavily impacted by the fuel vaporization. In contrast in the regions of high temperature, the mixture is too lean to initiate the combustion. This introduces ignition delay time required for the mixture to reach optimal auto-ignition properties (so-called physical ignition delay related to mixture preparation). The combustion rates are reduced on the basis of the same mechanisms as for mild stratification. In this case, however, the effect is reinforced by the combustion being retarded.

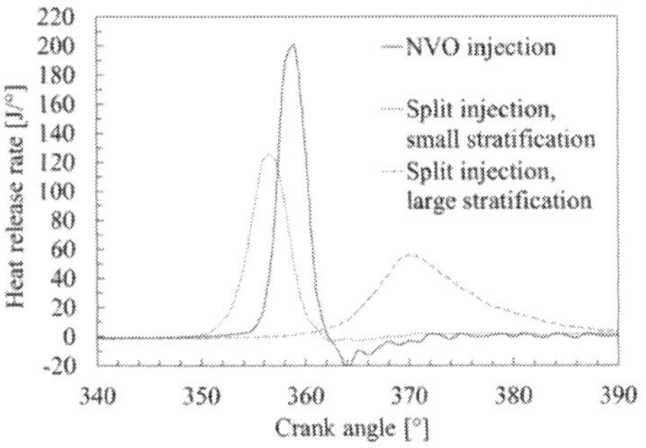

Fig. 7. Heat release rates for case 3.

Summary

The present work provides a concise summary of the authors experiences with residually affected HCCI concept development related to combustion controllability and load-range extension. The results show that application of early NVO injection enables active control of combustion timing at nearly idle conditions. The late fuel injection during the compression stroke enables mitigation of excessive pressure rise rates at high engine load regime. Furthermore, applying injection split between NVO and intake phases can be an effective control measure to attain optimal combustion phasing with fast response times not attainable with air-path related measures.

The research was funded by the National Science Centre, Poland under grant No. 2015/17/B/ST8/03279. The authors wish to thank AVL List GmbH for making simulation software available within a framework of AVL University Partnership Program.

References

1. H. Hong, G. Parvate-Patil, B. Gordon, Review and analysis of variable valve timing strategies-eight ways to approach, *Proceedings of the Institution of Mechanical Engineers, Part D: Journal of Automobile Engineering*, **218**, 1179-1200 (2004)
2. M. Mikulski, P. Balakrishnan, E. Doosje, C. Bekdemir, Variable valve actuation strategies for better efficiency load range and thermal management in an RCCI engine, *SAE Technical Paper* 2018-01-0254 (2018)
3. M. Yao, Z. Zheng, H. Liu, Progress and recent trends in homogeneous charge compression ignition (HCCI) engines, *Progress in Energy and Combustion Science*, **35**, 398-437 (2009).
4. T. Sharma, G. Amba, P. Rao, K. Murthy, Effective reduction of NOx emissions of a HCCI (Homogeneous charge compression ignition) engine by enhanced rate of heat transfer under varying conditions of operation, Energy, **93**, 2102-2115 (2015)
5. B. Çeper, M. Yıldız, S. Akansu, N. Kahraman, Performance and emission characteristics of an IC engine under SI, SI-CAI and CAI combustion modes, Energy, **136**, 72-79 (2017)
6. K. Lee, S. Cho, N. Kim, K. Min, A study on combustion control and operating range expansion of gasoline HCCI, Energy, **91**, 1038-1048 (2015)
7. J. Valero-Marco, B. Lehrheuer, J. López, S. Pischinger, Potential of water direct injection in a CAI/HCCI gasoline engine to extend the operating range towards higher loads, Fuel, **231**, 317–327 (2018)
8. J. Lavy, J. Dabadie, C. Angelberger, P. Duret, J. Willand, A. Juretzka, J. Schäflein, T. Ma, Y. Lendresse, A. Satre, C. Schulz, H. Krämer, H. Zhao, L. Damiano, Innovative ultra-low NO_X controlled auto-ignition combustion process for gasoline engines: the 4-SPACE project, *SAE Technical Paper* 2000-01-1837 (2000)
9. G. Kontarakis, N. Collings, T. Ma, Demonstration of HCCI using a single cylinder four-stroke SI engine with modified valve timing, *SAE Technical Paper* 2000-01-2870 (2000)
10. J. Dec, Y. Yang, Boosted HCCI for High Power without Engine Knock and with Ultra-Low NOx Emissions – using Conventional Gasoline, SAE International Journal of Engines, **3**, 750-767 (2010)
11. J. Hunicz, An experimental study into the chemical effects of direct gasoline injection into retained

https://doi.org/10.1051/matecconf/201823403003

residuals in a homogeneous charge compression ignition engine, International Journal of Engine Research, **17**, 1031-1044 (2016)

12. I. Ekoto, B. Wolk, W. Northrop, N. Hansen, K. Moshammer, Tailoring charge reactivity using in-cylinder generated reformate for gasoline compression ignition strategies, Journal of Engineering for Gas Turbines and Power, **139**, 122801 (2017)
13. T. Urushihara, K. Hiraya, A. Kakuhou, T. Itoh, Expansion of HCCI operating region by the combination of direct fuel injection, negative valve overlap and internal fuel reformation, *SAE Technical Paper* 2003-01-0749 (2003)
14. J. Hunicz, M. Mikulski, Investigation of the thermal effects of fuel injection into retained residuals in HCCI engine, Applied Energy, **228**, 1966-1984 (2018)
15. J. Hunicz, M. Geca, P. Kordos, H. Komsta, An experimental study on a boosted gasoline HCCI engine under different direct fuel injection strategies, Experimental Thermal and Fluid Science, **62**, 151-163 (2015)
16. J. Hunicz, A. Tmar, P. Krzaczek, Effects of mixture stratification on combustion and emissions of boosted controlled auto-ignition engines, Energies, **10**, 2172 (2017)
17. A. Turkcan, M. Altinkurt, G. Coskun, M. Canakci, Numerical and experimental investigations of the effects of the second injection timing and alcohol-gasoline fuel blends on combustion and emissions of an HCCI-DI engine, Fuel, **219**, 50-61 (2018)
18. J. Hunicz, P. Kordos, P. Ignaciuk, Experimental investigation into thermal and chemical effects of negative valve overlap injection in a gasoline HCCI engine, *SAE Technical Paper* 2014-01-2660 (2014)
19. D. Yang, Z. Wang, J. Wang, S. Shuai, Experimental study of fuel stratification for HCCI high load extension, Applied Energy, **88**, 2949-2954 (2011)
20. M. Sjöberg, J. Dec, N. Cernansky, Potential of thermal stratification and combustion retard for reducing pressure-rise rates in HCCI engines, Based on multi-zone modeling and experiments, *SAE Technical Paper* 2005-01-0113 (2005)
21. A. Turkcan, A. Ozsezen, M. Canakci, G. Coskun, H. Soyhan, U. Demir, An experimental and modeling study to investigate effects of two-stage direct injection variations on HCCI combustion, Combustion Science and Technology, **187**, 642-658 (2015)
22. M. Mikulski, C. Bekdemir, Understanding the role of low reactivity fuel stratification in a dual fuel RCCI engine – A simulation study, Applied Energy, **191**, 689-708 (2017)

https://doi.org/10.1051/matecconf/201823403004

The impact of selected biofuels on the performance parameters of the Common Rail power system in the utility engine

Paweł Krzaczek[1,*], *Arkadiusz* Rybak[2], and *Andrzej* Bochniak[3]

[1]University of Life Sciences, Department of Power Engineering and Transportation, Głęboka 28, 20-612 Lublin, Poland
[2]Lublin University of Technology, Faculty of Mechanical Engineering, Nadbystrzycka 36, 20-618 Lublin, Poland
[3]University of Life Sciences, Department of Applied Mathematics and Computer Science, Głęboka 28, 20-612 Lublin, Poland

Abstract. The aim of the research was to determine the impact of biofuels from waste materials of plant and animal origin on the parameters of the common rail power supply system in the utility engine. The tests included identification of power system operation parameters in the whole load range of the tested engine, taking into account the limit and diagnostic parameters of the injectors operation. Then, for certain parameters, the engine injectors were tested on the test bench: injection pressure in the range of 25-135 MPa, injection time in the range of 200-1600 µs. In the tests, as reference fuel for testing injectors were used diesel fuel and three types of methyl esters of higher fatty acids: vegetable, animal and WCO origin. The measurements for individual fuels were made in the operating temperature range 30-60°C. The tests have shown significant changes in the volume of the fuel injection rates depending on the fuel used. Particularly, visible changes concerned the power system operation parameters for high engine speeds and the maximum working pressure of the common rail system in the engine.

1 Introduction

Today, it is estimated that transport generates about 20% of global energy demand. Due to high energy efficiency, high torque, high energy density of fuel in the different means of transport, self-ignition internal combustion engines play a dominant role [1, 2]. This particularly applies to the use of internal combustion engines in heavy duty vehicles. In addition, the use of diesel engines makes it possible to use a wide range of fuels to supplement or replace diesel [3].

The growing, restrictive requirements for the emission of toxic exhaust components and the increase in requirements for improving the efficiency of diesel engines, also in the context of reducing CO_2 emissions, necessitate the introduction of new design solutions (e.g. diesel particulate filters, various types of catalysts, use of high-pressure fuel systems) [4, 5] or the use of new combustion systems [6]. This causes complications in the construction of motors and significantly reduces the reliability of drive units, increases maintenance activities (e.g. for units with SCR system) [7], and consequently, leads to potential interruption of transport processes. In addition, the introduction of new constructional solutions entails a delay of several years in the exploitation practice of the vehicle and in some cases the need to adapt the transport infrastructure to them.

Regarding vehicles and means of transport in operation, it seems preferable, instead of implementing new construction solutions, to introduce fuels that have a significantly smaller impact on the environment than diesel [1, 8, 9]. The most commonly used fuels are methyl esters of higher fatty acids of various origins [10]. The most popular ones are first generation biofuels based on oils of edible plants (rapeseed [11], sunflower, soybean, or palm) [12], the use of which does not require significant modifications in usable or traction engines [13].

On the one hand, biofuels allow for the reduction in the emission of the toxic components of exhaust fumes, carbon monoxide, hydrocarbons or particulates (due to the oxygen content), and, on the other hand, to fulfill the indicative targets of individual countries regarding the use of renewable energy sources [2]. However, the level of energy demand makes it necessary to look for new sources of raw materials for the production of biofuels. The use of raw materials in the form of non-edible plant oils or waste oils (animal, frying) allows for the increase in the potential of liquid biofuels. However, fuels made from them often do not meet the requirements for fuels used in modern diesel engines, especially in common rail systems. In addition, the production of biofuels from a variety of raw materials causes a variable chemical composition of the obtained fuel and also involves changes in physical properties, especially low-temperature ones. In the case of biofuels, an important parameter is their viscosity from the point of view of controlling the injection process. On the other hand, the chemical composition, especially oxygen content and energy value of fuel have significant impact on the fuel combustion process [14, 15, 16].

The reasons mentioned for the introduction of biofuels of various origin resulted in undertaking research aimed

* Corresponding author: pawel.krzaczek@up.lublin.pl

 https://doi.org/10.1051/matecconf/201823403004

at assessing the impact of pure biofuels on the performance parameters of the common rail power system in the utility engine. The tests were based on an internal combustion engine, which is used in heavy road vehicles, farm tractors or industrial engines.

2 Research focus

The aim of the research was to determine the impact of biofuels from waste materials of plant and animal origin on the parameters of the common rail power supply system in the utility engine.

3 Object and methods of research

The tests included identification of the operating parameters of the John Deere JD 4045HF285 fuel system. The basic parameters of the tested engine are shown in Table 1.

Table 1. Specification of the tested engine.

Parameter	Engine
Engine type/model	automotive variable speed 4-stroke engine/4045HF285
Application	diesel truck, tractors, industrial motors
No. of cylinders	4
Compression ratio	19:1
Rated speed, rpm	2400
Rated power, kW	74 @ 2400 rpm
Rated torque, Nm	353 @ 1600 rpm
Displacement, cc	4530
Aspiration system	turbocharged without Intercooler
Fuel delivery system	common rail
Fuel Injection Pump	Denso HP3
Maximum fuel pressure, MPa	105
Minimum fuel pressure at idle speed, MPa	40
Injector type/model	CR/Denso 095000-6310

The engine is equipped with a common-mode Denso power supply system and an electronic control system that allows for the reading and recording of fuel injection pressures in the system. The engine was mounted on a dynamometric test bench coupled with an electro-spin brake and Automex AMX data control and archiving system. In order to determine the fuel pressure level, the engine was loaded with a torque *T* from 0 Nm to full load, every 25 Nm in the speed range *n* of 1200 to 2400 rpm, every 100 rpm.

In the next step, the Denso CR injectors type 095000-6310 were removed from the engine and mounted on the test stand for characteristics of injectors and common rail injection pumps. A detailed description of the test stand has been included in the works [17, 18].

The tests were conducted for reference fuel (Calibrol), commercial diesel oil and three biofuels in the form of methyl esters respectively: rapeseed oil – BIO 1, animal fats – BIO 2, plant waste cooking oil – BIO 3. The chemical composition and basic properties of the tested fuels are presented in Table 2 and 3.

Table 2. The chemical composition of the tested biofuels.

Common name of faty acide methyl ester	BIO 1	BIO 2	BIO 3
Myristic C14:0	-	-	0,12
Palmitic C16:0	0,32	20,8	6,51
Palmitoleic C16:1	4,1	-	0,35
Stearic C18:0	1,58	6,3	3,7
Oleic C18:1	62,04	44,6	63,7
Linoleic C18:2	19,81	15,7	21,27
Linolenic C18:3	8,97	0,85	1,2
Arachidic C20:0	0,77	-	0,27
Eicosenoic C20:1	1,16	-	0,65
Others	1,25	11,75	2,23

Table 3. Property of the tested fuels.

Fuel property	BIO 1	BIO 2	BIO 3	Diesel	Calibrol
Density @ 30°C, kg/m^3	870,8	862,8	875,0	832,3	815,0
Density @ 40°C, kg/m^3	865,2	856,0	868,4	825,1	807,6
Density @ 50°C, kg/m^3	858,4	849,3	861,8	817,8	800,2
Density @ 60°C, kg/m^3	851,7	842,6	855,2	810,6	792,9
Viscosity @ 30°C, mm^2/s	5,54	7,09	6,07	3,65	3,10
Viscosity @ 40°C, mm^2/s	4,43	5,45	4,80	2,97	2,55
Viscosity @ 50°C, mm^2/s	3,63	4,49	3,94	2,5	2,14
Viscosity @ 60°C, mm^2/s	3,05	3,73	3,30	2,15	1,81

The density of fuel was determined by the hydrometer in accordance with PN-EN ISO 3675 at 30, 40, 50, 60°C. The kinematic viscosity of the studied fuels was determined by capillary according to PN-EN ISO 3104 at temperatures of 30, 40, 50 and 60°C (Table 3). The composition of the various methyl esters was determined in accordance with PN-EN 14103 (Table 2), due to the significant impact of different types of esters on the physicochemical property of fuel.

All fuels were tested at temperatures from 30 to 60°C, at fuel pressures ranging from 25 to 135 MPa. Injectors opening time varied from 200 to 1600 μs. Therefore, the extreme values of injection parameters were also checked, which were determined by the manufacturer in points P1 (injection pressure 25 MPa, injector opening time 600 μs) and P2 (injection pressure 135 MPa, injector opening time 850 μs). On the basis of the obtained fuel injection quantities for individual parameters (pressure and injection time), the active flow cross section was calculated. For this purpose, the effective injection time was determined as the ratio of the amount of fuel injected at the test stand to the time of injection. The effective time was calculated as the difference between the injector coil feeding time and the time determined by extrapolation to the ordinates of the curve formed on the basis of the injection values at a given point at constant pressure. The active flow cross section identified is directly proportional to the volumetric flow rate of the injected fuel and inversely proportional to the square root of the ratio between the value of twice the product of the differential pressure before and after the injector and the density value of the tested fuel. A detailed definition of the active flow cross section is presented in [17].

 https://doi.org/10.1051/matecconf/201823403004

4 Test results

4.1 Injection pressure of the fuel supply system

The obtained values of fuel injection pressure at different rotational speeds allowed us to distinguish between three characteristic curves of pressure curves as a function of load, which were presented in Figures 1 to 3.

In Fig. 1 for speeds of 1200-1400 rpm it can be seen that the fuel pressure rises from about 70 to 90 MPa at 150 Nm, and then remains at this level. Changes in the course of the pressure curve can be observed at a rotational speed of 1500 rpm. For 50 Nm load the pressure increases slightly, then decreases by 10 MPa at 200 Nm, after which it gradually increases to the level of 90 MPa.

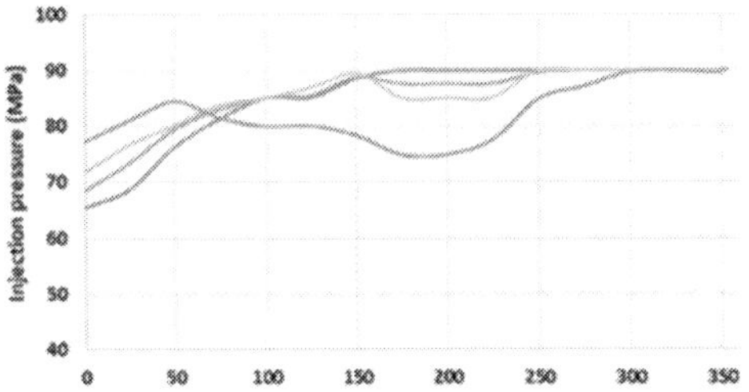

Fig. 1. Fuel injection pressure in the fuel supply system at 1200-1500 rpm.

The next group of curves is shown in Fig. 2 and refers to the rotational speed from 1600 to 1900 rpm, for which the curves have a similar character. Initially, the pressure drops from 85-88 MPa to the local minimum at a load of 150 Nm, which ranges from 50 MPa to 70 MPa. Then the pressure increases to 75-88 MPa at maximum load. It should be noted that the tested engine acquires maximum torque at a speed of 1600 rpm, for which the lowest fuel injection pressure level has been recorded.

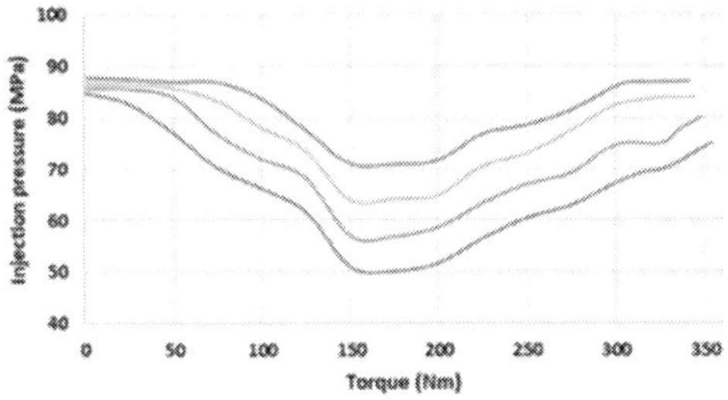

Fig. 2. Fuel injection pressure in the fuel supply system at 1600-1900 rpm.

The third group comprises the pressure curves for speeds from 2000 to 2400 rpm (Fig. 3), which have a relatively flat course. The initial injection pressure is around 90 MPa. The lowest pressure values were at 2000 rpm, where the pressure drops to 78 MPa (150 Nm), after which it increases to 90 MPa. The highest fuel pressure values correspond to the rated speed of the engine. In this case, the pressure course is a variable characteristic, it increases to 100 MPa, then drops to 95 MPa and at full load again increases to a maximum pressure of 105 MPa.

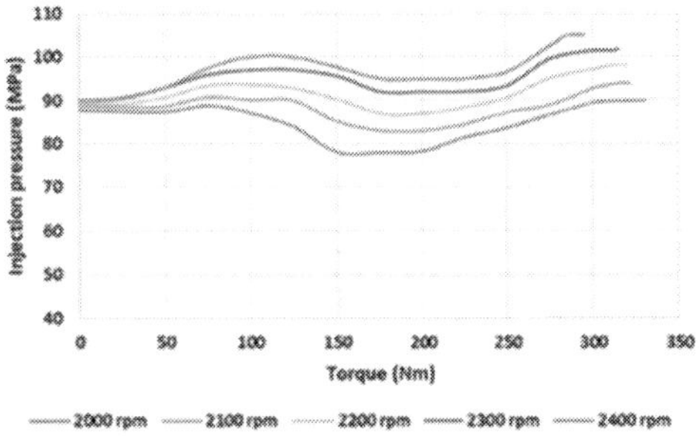

Fig. 3. Fuel injection pressure in the fuel supply system at 2000-2400 rpm.

Fig. 4 shows the dependence of the injection pressure on the rotational speed at zero load and for full engine load. In the absence of a load, the injection pressure increases with the increase of the rotational speed, the fastest being between 1300 and 1600 rpm. At full load, the injection pressure does not change to 1500 rpm and is about 90 MPa, then it decreases to a minimum of 74 MPa at 1600 rpm, after which it increases linearly to 105 MPa at 2400 rpm.

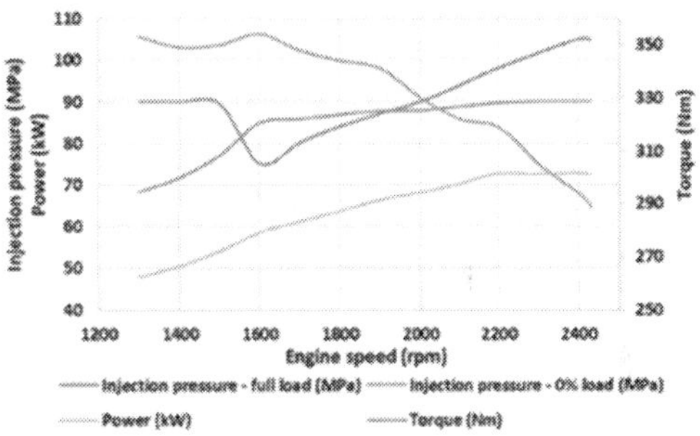

Fig. 4. Injection pressure, power output and torque characteristics in the function of engine revolutions at maximum load and idle speed.

Analyzing the entire working area of the engine, it can be concluded that the injection pressure varies with changes in the engine load in the range of 50 to 105 MPa. It should be pointed out that the minimum injection pressure for a given speed corresponded to the load, at which the air pressure generated by the turbocharger increases rapidly. For the minimum idling speed (800 rpm) the injection pressure was 40 MPa, therefore the injection pressure in the range of 40 to 100 MPa was taken into account when analyzing the discharge coefficient.

4.2 The active flow cross section

The tested fuels differed significantly in chemical composition and physical properties (Tables 2 and 3). The fuels used in the research represent the most popular types of raw materials for the production of biofuels, i.e.

 https://doi.org/10.1051/matecconf/201823403004

1 – generation biofuel vegetable oils – BIO 1, technical animal fats – BIO 2, waste cooking oil – BIO 3. Certain values of density, viscosity or heat of combustion are close to those described in the literature [16, 19]. The lowest density and the highest viscosity was noted for BIO 2 biofuel. In contrast, the highest density and calorific value was found for BIO 3 frying oil. The obtained physical parameters result directly from the chemical composition of the fuel [16, 20]. In addition, during tests it was found out that animal fuels have a high pour point, in the case of BIO 2 the blocking temperature of the cold filter was 11°C. Analogous results for biofuels of animal origin were obtained in [16, 18, 19, 20]. This parameter is the main obstacle to the use of clean biofuels during the actual operation of internal combustion engines.

Fig. 5 shows the size of the fuel dose at the diagnostic points specified by the manufacturer for the tested injectors. The parameters of point P1 correspond to the moment of opening the injector, while point P2 determines the maximum injection pressure provided for he tested injector type. The differences in the amount of reference fuel, diesel oil and biofuels tested by the injector are clearly visible, especially in P2 point. However, all injection volume values are within the range specified by the manufacturer, i.e. for P1 points they are 3.2 ± 1 mm^3 and for P2 – 60 ± 4 mm^3. Taking into account the viscosity of fuels (Table 3), there was a clear dependence of the fuel doses (Fig. 5) on this parameter.

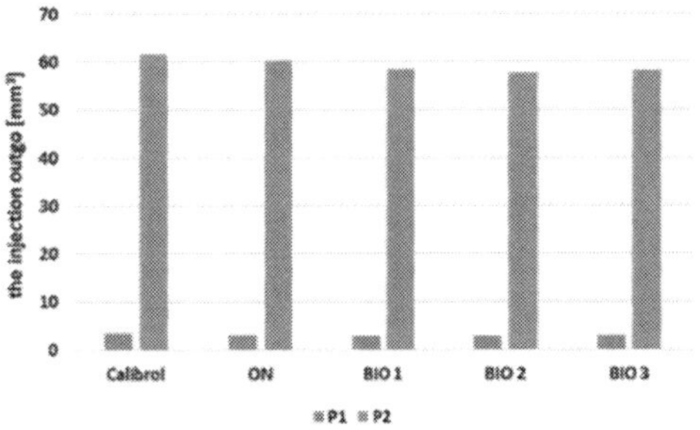

Fig. 5. Amount of fuel injected at diagnostic points.

In Fig. 6, the calculated active flow cross section at 30°C increases from 40 to 80 MPa for all fuels, in which case the value of the coefficient is slightly higher for diesel fuel. For diesel and BIO 1, the factor reaches the maximum at 80 MPa, after which it is lower, especially for diesel. The maximum values for BIO 2 and BIO 3 are shifted by 20 MPa. This is especially evident for fuel of animal origin.

Due to the low temperature CFPP for BIO2 no measurements below 30°C were taken, it can be assumed that it corresponds to the engine heating phase and any correction of the injection parameters should be introduced for points with pressures above 80 MPa, i.e. for low and high engine loads.

The comparison of the active flow cross section curves depending on the injection pressure (Fig. 7) for the fuel temperature of 40°C was found to be similar to those in Fig. 6. The local maximum of the coefficient does not change, but the differences in the coefficient value increase above 60 MPa. The lowest values were recorded for biofuel BIO 2, while for pressure 100 MPa for diesel oil. Since the fuel temperature of 40°C corresponds to low and medium loads, e.g. during transport work, correction of injection maps should be introduced for pressures above 60 MPa, and its variable size associated with the increase of pressure, especially at high engine load.

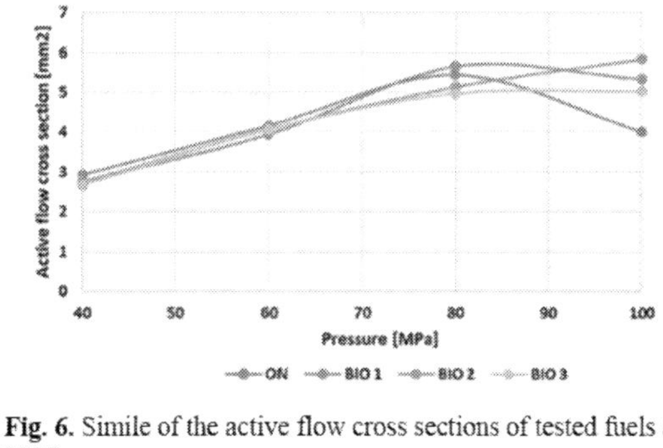

Fig. 6. Simile of the active flow cross sections of tested fuels at 30°C.

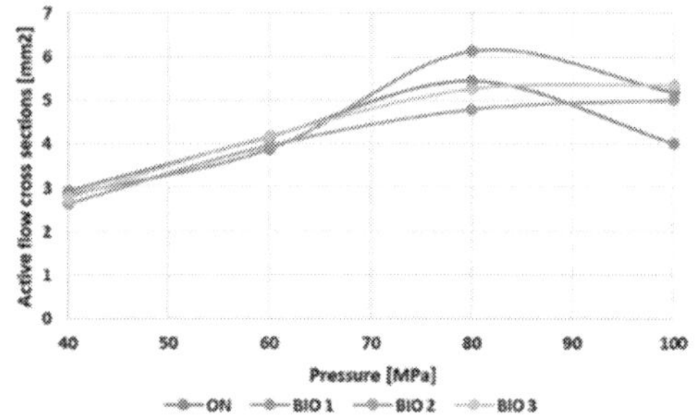

Fig. 7. Simile of the active flow cross sections of tested fuels at 40°C.

An increase in fuel temperature of up to 50°C corresponds to medium and long-term engine operation conditions. In this case, the active flow cross section (Fig. 8) in the range of 60 to 80 MPa is clearly higher for diesel and its maximum is shifted between 60 and 80 MPa. Moreover, above this point the coefficient for other fuels is higher, especially for BIO 1 fuel, which reaches its maximum again at 80 MPa. The coefficients for BIO 2 and BIO 3 are similar and reach a maximum at 100 MPa. At this temperature, the correction of the injection parameters should cover the entire tested pressure range, but their changes should be of a different nature.

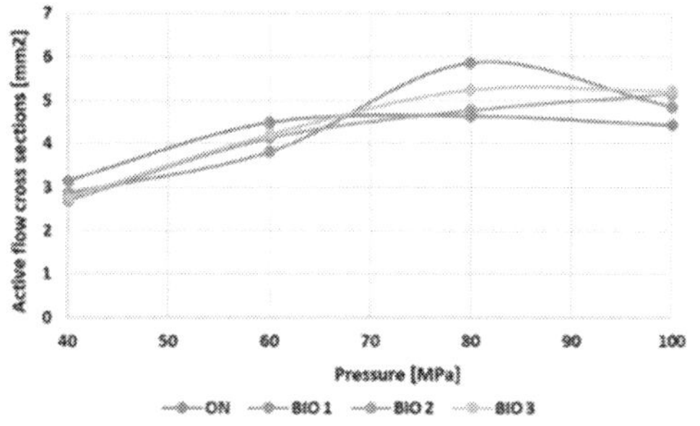

Fig. 8. Simile of the active flow cross sections of tested fuels at 50°C.

https://doi.org/10.1051/matecconf/201823403004

In the research at 60°C, there were clear differences in the active flow cross section for diesel and all biofuels in the pressure range of 40 to 80 MPa (Fig. 9). It should be pointed out that this is an area where there is an increase in turbocharger efficiency for medium rotational speeds and medium loads. In addition, in this pressure range, the entire load characteristic for the maximum torque speed is included. Therefore, when the motor is used for a long time with high thermal loads (high ambient temperature, long-lasting operation at almost full load) it requires adjustment of the controller settings, especially at pressures of 60 MPa and 100 MPa.

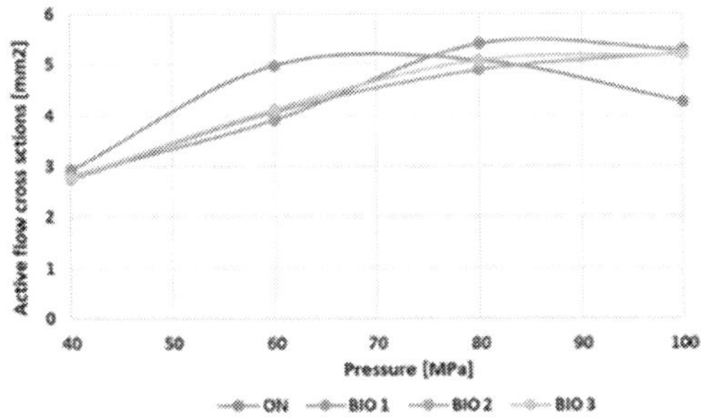

Fig. 9. Simile of the active flow cross sections of tested fuels at 60°C.

5 Conclusion

The engine tests carried out on the dynamometric bench allowed us to determine the scope of the fuel supply system depending on the load and rotational speed. It was found out that fuel operating pressures range from 40 to 105 MPa, the value of which is significantly lower than the pressure levels tested in the literature [21, 22]. It has been shown that for medium loads and high level of engine recharging, fuel pressure is reduced even to 50 MPa for the maximum torque speed.

Analyzing the calculated active flow cross section it was concluded that the differences in its value for diesel oil and individual biofuels increase with the temperature of the examined fuel. For temperatures of 30 and 40°C, clear differences in the value of the coefficient occur above 80 MPa, which is related to the need to adjust the injection parameters for biofuels. Further increase of fuel temperature to 50 or 60°C deepens the differences in the coefficient for diesel oil and biofuels within the whole range of the power system operation, which is from 40 to 100 MPa.

In addition, it was found out that for BIO 1 fuel above 80 MPa, the active flow cross section has a higher value than for diesel fuel. However, for BIO 2 and BIO 3 fuels, this is the case only at 100 MPa.

References

1. K. Duda, S. Wierzbicki, M. Śmieja, M. Mikulski, Comparison of performance and emissions of a CRDI diesel engine fueled with biodiesel of different origin, Fuel, **212**, 202-222 (2018)
2. S. Mishra, K. Anand, S. Santhosh, P. Mehta, Comparison of biodiesel fuel behavior in a heavy duty turbocharged and a light duty naturally aspirated engine, Applied Energy, **202**, 459-470 (2017)
3. B. Sajjadi, A. Raman, H. Arandiyan, A comprehensive review on properties of edible and non-edible vegetable oil-based biodiesel: Composition, specifications and prediction models, Renewable and Sustainable Energy Reviews, **63**, 62-92 (2016)
4. H. Howa, H. Masjuki, M. Kalam, H. Teoh, Influence of injection timing and split injection strategies on performance, emissions, and combustion characteristics of diesel engine fueled with biodiesel blended fuels, Fuel, **213**, 106-114 (2018)
5. S. Ren, B. Wang, J. Zhang, Z. Wang, J. Wang, Application of dual-fuel combustion over the full operating map in a heavy duty multi-cylinder engine with reduced compression ratio and diesel oxidation catalyst, Energy Conversion and Management, **166**, 1-12 (2018)
6. J. Hunicz, A. Tmar, P. Krzaczek, Effects of mixture stratification on combustion and emissions of boosted controlled auto-ignition engines, Energies, **10**, 2172 (2017)
7. W. Urzędowska, Z. Stępień, Prediction of threats caused by high FAME diesel fuel blend stability for engine injector operation, Fuel Processing Technology, **142**, 403-410 (2016)
8. G. Edara, Y. Murthy, P. Srinivas, J. Nayar, M. Ramesh, Effect of cooled EGR on modified light duty diesel engine for combustion, performance and emissions under high pressure split injection strategies, Case Studies in Thermal Engineering, **12**, 188-202 (2018)
9. L. Lešnik, I. Biluš, The effect of rapeseed oil biodiesel fuel on combustion, performance, and the emission formation process within a heavy-duty DI diesel engine, Energy Conversion and Management, **109**, 140-152 (2014)
10. C. Hayes, D. Burgess, J. Manion, Combustion pathways of biofuel model compounds: A review of recent research and current challenges pertaining to first-, second-, and third-generation biofuels, Advances in Physical Organic Chemistry, **49**, 103-187 (2015)
11. G. Zając, W. Piekarski, P. Krzaczek, Comparison of an effect of FAME and FAEE addition to diesel fuel on energetic parameters of an engine, TEKA Kom. Mot. i Energ. Rol. – OL PAN, **8a**, 217-223 (2008)
12. M. Bhuiya, M. Rasul, M. Khan, N. Ashwath, A. Azad, M. Hazrat, Prospects of 2nd generation biodiesel as a sustainable fuel – Part2: Properties, performance and emission characteristics, Renewable and Sustainable Energy Reviews, **55**, 1129-1146 (2016)
13. V. Goel, N. Kumar, P. Singh, Impact of modified parameters on diesel engine characteristics using biodiesel: A review. Renewable and Sustainable Energy Reviews, **82**, 2716-2729 (2018)

14. S. Slavinskas, G. Labeckas, T. Mickevicius, Experimental study on injection characteristics of diesel and biodiesel fuel blends with common-rail injection system, *Engineering for rural development*, Jelgava, 2134-2140 (2018)
15. M. Das, M. Sarkar, A. Datta, A. Santra, Study on viscosity and surface tension properties of biodiesel-diesel blends and their effects on spray parameters for CI engines, Fuel, **220**, 769-779 (2018)
16. E. Giakoumis, A statistical investigation of biodiesel physical and chemical properties, and their correlation with the degree of unsaturation. Renewable Energy, **50**, 858-878 (2013)
17. A. Rybak, M. Gęca, P. Krzaczek, A. Mazanek, Determination of common rail injector flow characteristics with the use of diesel and biodiesel fuels, Journal of KONES, **23**(4), 443-450 (2016)
18. A. Rybak, J. Hunicz, P. Krzaczek, W. Golimowski, D. Marcinkowski, Effect of different biofuels on common rail injector flow rate, Combustion Engines, **171**(4), 39-43 (2017) DOI: 10.19206/CE-2017-407.
19. D. Marcinkowski, B. Rukowicz, W. Golimowski, M. Czechlowski, P. Krzaczek, W. Piekarski, Effect of selected depressants on cold filter plugging point for methyl esters obtained from transesterification of waste vegetable and animal fats, Przem. Chem, **96**(9), 1927-1930 (2017), DOI: 10.15199/62.2017.9.22
20. N. Kumar, S. Varun, R. Chauhan, Performance and emission characteristics of biodiesel from different origins: A review, Renewable and Sustainable Energy Reviews, **21**, 633-658 (2013)
21. L. Duan, S. Yuan, L. Hu, W. Yang, J. Yu, X. Xia, Injection performance and cavitation analysis of an advanced 250 MPa common rail diesel injector, International Journal of Heat and Mass Transfer, **93**, 388-397 (2015)
22. J. Bergthorson, M. Thomson, A review of the combustion and emissions properties of advanced Transportation biofuels and their impact on existing and future engines, Renewable and Sustainable Energy Reviews, **42**, 1393-1417 (2015)

https://doi.org/10.1051/matecconf/201823403005

Detailed speciation of emissions from a diesel engine fuelled with canola methyl ester

Grzegorz Koszalka[1,*], and *Jacek* Hunicz[1]

[1]Lublin University of Technology, Faculty of Mechanical Engineering, Nadbystrzycka 36, 20-618 Lublin, Poland

Abstract. Although the effects of biodiesel combustion on emissions of regulated toxic components have been widely studied, the issue of specific hydrocarbon species production has not yet been comprehensively understood. This study compares detailed exhaust emissions from a compression ignition engine fuelled with canola methyl ester and mineral diesel, as a reference fuel. Additionally, blends of these two fuels were examined. The experiments were performed on a 4-cylinder diesel engine, where fuel was injected in a single dose. The experimental matrix included two engine load sweeps at rotational speeds corresponding to maximum torque and maximum power. Detailed exhaust composition was measured with the use of a Fourier transform infra-red analytical system. To enable unbiased evaluation of the effect of different fuels on hydrocarbons emissions fuel carbon conversion into species carbon was considered. The results showed that there is not a monotonic effect of the content of biodiesel fuel on particular hydrocarbon species. In the case of some hydrocarbon species, the lowest emissions were recorded for mixture of the two fuels.

1 Introduction

Global trends in energy sources diversification heavily rely on the increasing share of renewable transportation biofuels. Additionally, the use of biofuels has significant pro-environmental aspect. The biofuel carbon footprint is nearly zero, making these energy carriers completely neutral in terms of CO2 emissions. Utilization of biofuels in diesel engines does not require any large scale hardware modifications. Thus, there is no additional energy and materials consumption associated with introduction of biofuels.

In the relevant literature, a great number of studies have been dedicated to research into the effects of biofuels on combustion, emissions of toxic compounds and energy consumption [1-14]. Evaluation of the performance and durability of engines is also of the great importance [2, 3]. From the point of view of the environmental impact, there is a common opinion that the use of bio-diesel decreases engines performance, but reduces all toxic exhaust gas compounds, except nitrogen oxides [2-5]. Duda et al. [6] demonstrated that both, plant and animal-originated biofuels combustion reduce gaseous components and smoke emissions in comparison to mineral diesel fuel. Nevertheless, despite the origin of the biofuel, reduction in brake specific efficiency by approximately 2% was noted. When analysing the results of studies concerning the effects of the use of renewable components on exhaust gas emissions attention is drawn to the small number of studies on the emissions of specific hydrocarbon components. In other words, the authors usually compare the so-called regulated toxic compounds. However, there are dozens of different hydrocarbon species having different effects on living organisms. For example Claxton [7] indicated that 1,3-butadiene deserves special consideration because of its carcinogenic potential. It is considered to be one of the most harmful compounds among all toxic compounds emitted by motor vehicles. For gasoline fuel Zervas et al. [8] studied the effects of fuel composition on particular hydrocarbon species emissions from spark ignition engine. The authors determined the pathways of creation of specific hydrocarbons depending on combustion conditions. Hunicz and Krzaczek [9] demonstrated the effects of pre-combustion reactions on exhaust hydrocarbon compositions in low-temperature combustion system. Gupta and Agarval [1] utilized FTIR exhaust gas analytical system to study the effects of non-edible vegetable oil on emissions of particular hydrocarbons. They concluded that addition of biofuel to mineral diesel reduces the content of the most harmful hydrocarbons in exhaust gases. It should be noted that some hydrocarbons pose a challenge in terms of catalytic oxidation, because at low loads diesel exhaust temperature is far below their light-off threshold [10]. Hunicz and Medina [11] demonstrated the effects of exhaust hydrocarbon compositions on oxidation catalyst efficiency. Despite a confirmation of slow oxidation properties of methane they demonstrated inhibiting effects of acetylene on oxidation of other hydrocarbons.

In this paper the results of measurements of the concentration of regulated and non-regulated exhaust components from a diesel engine fuelled with mineral diesel fuel, bio-diesel and their blend are presented. The effect of fuel and engine operation conditions on shares of particular hydrocarbons to total hydrocarbon is discussed.

* Corresponding author: g.koszalka@pollub.pl

2 Measurements

The research was conducted on an engine test stand equipped with eddy-current dynamometer (AVL-Zollner Alpha 240), fuel consumption meter (AVL Dynamic Fuel Balance) and other instrumentation necessary for proper engine operation and control its basic parameters. The tests were performed on a 4-cylinder diesel engine with swept volume of 2.4 dm^3, maximum torque of 146 Nm at 2500 rpm and rated power of 51.5 kW at 4200 rpm, according to the engine specification. The fuel was injected in one dose by an in-line injection pump.

AVL SESAM FTIR (Fourier Transform Infrared) multi-component gas analytical system was used to measure the concentrations of 23 compounds in exhaust gases. The system allowed for simultaneous measurement of the content of specific hydrocarbons, nitrogen oxides and other chemical compounds. The measured compounds are listed in Table 1. Exhaust gases flew from the engine to the FTIR analyser through a heated line. The sampling rate of the analyser was around 1 s. The time-averaged value of the 30 s measurement period was taken as a result of a single reading.

Table 1. Exhaust compounds measured.

Name of compound	Symbol
Methane	CH4
Acetylene	C2H2
Ethylene	C2H4
Ethane	C2H6
Propylene	C3H6
Propane	C3H8
1,3 butadiene	C4H6
N-octane	NC8
Aromatic hydrocarbons (toluene)	AHC
Formaldehyde	HCHO
Acetaldehyde	CH3CHO
Formic acid	HCOOH
Isocyanic acid	HNCO
Hydrogen cyanide	HCN
Ammonia	NH3
Nitrogen monoxide	NO
Nitrogen dioxide	NO2
Nitrous oxide	N2O
Carbon monoxide	CO
Carbon dioxide	CO2
Sulphur dioxide	SO2
Carbonyl sulphide	COS
Water	H2O

Concentrations of particular exhaust components were measured for engine fuelled with regular diesel oil (D 100%), bio-diesel based on canola methyl ester (CME 100%) and their blend with 50% mass fractions. The basic parameters of the studied fuels are presented in Table 2.

The measurements were done at two engine rotational speeds: 2500 rpm (speed at which the engine achieved the maximum torque) and 3800 rpm (speed at which the engine actually achieved maximum power). At each selected rotational speed an engine load was changed from 0 to 100% of torque attainable at the given speed. Before each single measurement the engine worked at steady-state conditions for minimum 2 minutes to assure also steady thermal conditions. During measurements relative air-fuel ratio calculated from carbon balance equation varied from 7.2 at idling to 1.3 and 1.4 at full load at 2500 and 3800 rpm, respectively.

Table 2. Parameters of the studied fuels.

Parameter	Unit	CME 100%	D 100%
Content of fatty acid methyl esters	% (m/m)	97,3	
Density at 15°C	kg/m^3	883	839
Kinematic viscosity at 40°C	mm^2/s	4,47	2,74
Sulphur content	mg/kg	4,7	6,3
Water content	mg/kg	117	76
Total impurities	mg/kg	17	11
Cetane number	–	51	51,6
Flash-point	°C	130	65
Temperature of cold fuel filter blockage	°C	-23	-32

3 Results and discussion

Figures 1-9 show the raw species concentration measurement results for all studied conditions. The concentration of carbon monoxide in the exhaust gases of engine fuelled with CME based fuel was smaller by 25% in comparison to that in engine fuelled with mineral diesel. For the blend of CME and mineral fuel that concentration was smaller by 13%. The content of CO decreased with the increase in engine load in the range of small and medium loads, despite the fuel used. However, for the highest loads CO production increased. This increase was especially rapid at the rotational speed of the maximum torque, as shown in Fig. 1. It should be noted that the increase of biofuel content reduced CO emissions considerably because of the oxygen content of the fuel. Thus, the global excess air ratio for biofuel was higher and combustion more complete.

A similar effect of the fuel was observed in the case of sulphur dioxide. On average, the content of SO2 in exhausts was by 27% smaller for CME fuel, and by 15% for the blend in comparison to mineral diesel (Fig. 1). These differences were comparable to the differences in the sulphur content in the fuels (Table 2). The concentrations of SO2 strongly increased with the increase in engine load, which was related to increase in fuel content in the combustion mixture.

In the case of nitrogen oxide the results were not that clear, although on average the concentration for CME fuel was 3% higher, and for blend 5% lower than for mineral diesel. For all tested fuels the content of nitrogen oxide increased with the increase in the engine load (Fig. 2).

The content of nitrogen dioxide in exhaust of engine fuelled with CME fuel was much higher than in the engine fuelled with mineral fuel – by 32% on average (Fig. 2). For the blend of fuels emission it was also slightly higher – by 4%. Contrary to NO, the concentrations of NO2 decreased with the increase in the engine load, especially for 2500 rpm, where the content at full load was about

https://doi.org/10.1051/matecconf/201823403005

10 times higher than at idling. Although the concentration of NO2 to NOx is several times smaller than the concentration of NO, that big difference caused the concentration of NOx for CME fuel to be on average by 6% bigger. For the blend it was on average by 4% smaller than for mineral diesel. The changes in the content of NOx as function of load were similar to the content of NO – they increased together with the increase of load (Fig. 3). A total concentration of NOx was determined by NO, NO2 species, although the N2O compound was also considered. Nevertheless, the N2O contributed to the total NOx emissions solely in a small extent.

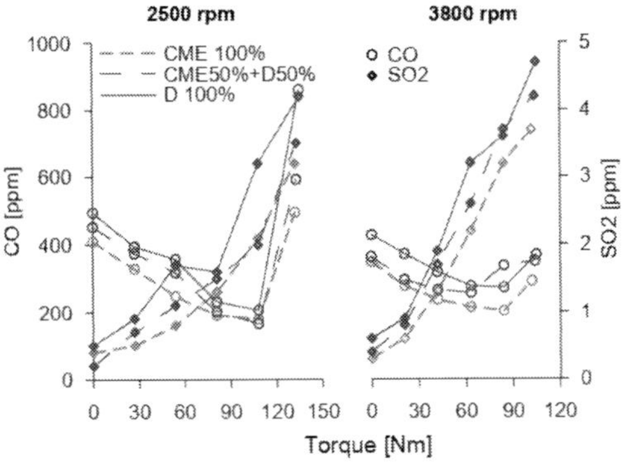

Fig. 1. Concentrations of carbon monoxide and sulphur dioxide.

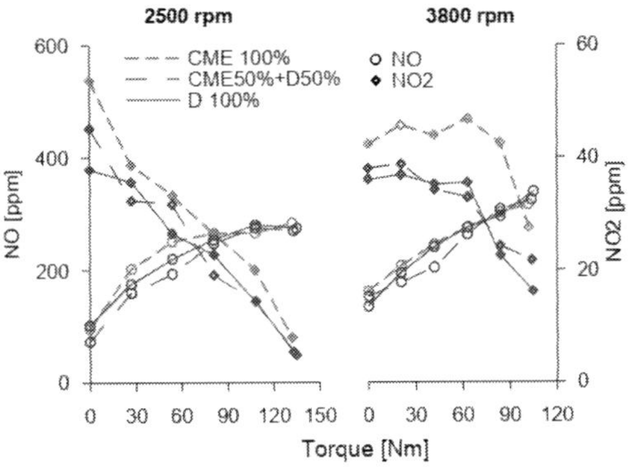

Fig. 2. Concentrations of nitrogen monoxide and nitrogen dioxide.

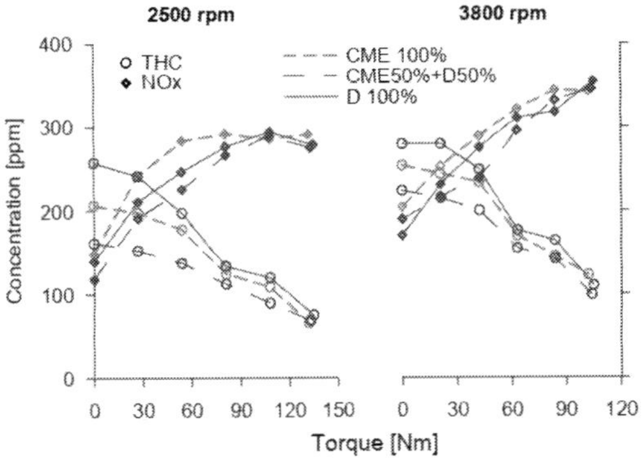

Fig. 3. Concentrations of total hydrocarbons and nitrogen oxides.

Concentrations of both determined aldehydes were increased by 11% for CME based fuel. However, the smallest concentrations were for the blend: 15% lower for formaldehyde and 11% lower for acetaldehyde in comparison to mineral diesel. The content of both formaldehydes decreased with the increase in engine load (Fig. 4).

Also concentrations of isocyanic acid were by 20% higher for CME fuel and by 14% smaller for the blend in comparison to mineral fuel (Fig. 5). Concentrations of strongly toxic hydrogen cyanide were very low and comparable for all tested fuels (Fig. 5).

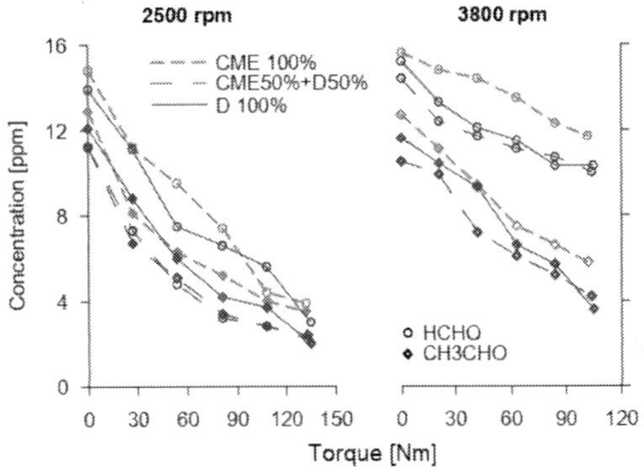

Fig. 4. Concentrations of formaldehyde and acetaldehyde.

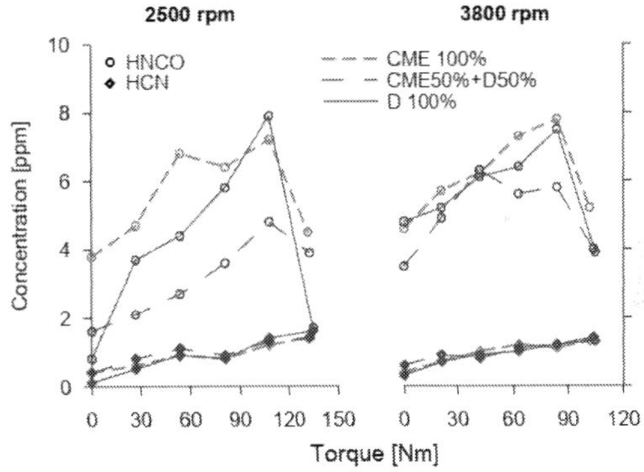

Fig. 5. Concentrations of isocyanic acid and hydrogen cyanide.

A concentration of total hydrocarbon related to one carbon atom (THC) was calculated as a sum of the concentration of particular hydrocarbons (the first 9 compounds in Table 1) taking into account the number of atoms of carbon in a specific hydrocarbon. The content of THC in exhausts of engine fuelled with CME fuel was lower, on average, by 8% than in the engine fuelled with mineral diesel. However, the lowest THC content was for the engine fuelled with the blend (13% lower than for diesel). These differences were particularly noticeable at small engine loads (Fig. 3). Concentrations of THC decreased with the increase in engine load for all fuels.

Concentrations of specific hydrocarbons in the exhaust gases of the engine operated on different fuels are presented in Figs. 6-9. Considering the mineral diesel as a baseline, the following average decreases of concentrations for CME fuel were recorded for particular

https://doi.org/10.1051/matecconf/201823403005

hydrocarbon species: C_2H_6 by 50%, AHC by 37%, C_3H_8 by 19%, C_4H_6 by 17%, and NC8 by 9%. In contrast, average concentrations were higher for: CH_4 by 12% and C_2H_4 by 11%. The lowest concentrations of almost all determined hydrocarbons were manifested by the blend of the two basic fuels. For the mixture, especially significant drop was noted in the case of N-octane (by 24% in comparison to mineral fuel), which had the largest concentration to THC. In the case of C_2H_2 and C_3H_6 no distinction could be made between the fuels.

Regarding the effect of load on hydrocarbons content in the exhaust gases, the results showed that concentrations of CH_4, C_3H_8, NC8 and AHC decreased together with the increase in engine load. In the case of C_2H_6, C_3H_6 and C_2H_4 moderate increases were observed, and finally, in the case of C_4H_6, the concentration was not affected by the load.

Fig. 10 shows the cumulative content of hydrocarbon species expressed as C1 for all studied conditions. In other words, the graphs show the distribution of the conversion of fuel carbon into different hydrocarbon species. Without any detailed analysis it can be noted that despite different fuel compositions the trends in fuel conversion into unburned hydrocarbon species were similar. N-octane had the biggest concentration to THC. In the range of small loads this concentration even exceeded 60% and dropped with the increase in engine load to 20% at maximum load. The concentration of propane also slightly decreased with the load. On the other hand, the concentrations of ethylene, ethane, propylene, 1,3-butadiene and acetylene increased with the load.

Considering the effect of fuel it can be stated that despite the different fuels the composition of hydrocarbons in the exhaust gases did not changed significantly. Only concentrations of ethane, 1,3 butadiene, which is the most toxic hydrocarbon compound emitted by motor vehicles [7], and aromatic hydrocarbons were slightly smaller for CME fuel. In the case of other hydrocarbon species there was no clear influence of the fuel on specific hydrocarbons concentration into the total hydrocarbon.

4 Summary

In this work the concentrations of regulated and non-regulated components in the exhaust gases of a compression ignition engine fuelled with canola methyl ester, mineral diesel, as a reference fuel, and a blend of these two fuels were examined. The experiments were performed in an engine test bench on the engine operated in steady-state conditions at two rotational speeds corresponding to maximum torque and maximum power, and loads changed in a full range: from idling to maximum torque attainable at the given rotational speed. Detailed exhaust composition was measured with the use of a Fourier transform infra-red analytical system.

Combustion of biofuel reduced concentrations of CO and SO_2, and increased concentrations of NOx for all studied conditions. Both trends are related to the molecular structure of the fuels. Namely, biofuels contain less sulphur and more oxygen than mineral fuels. For the

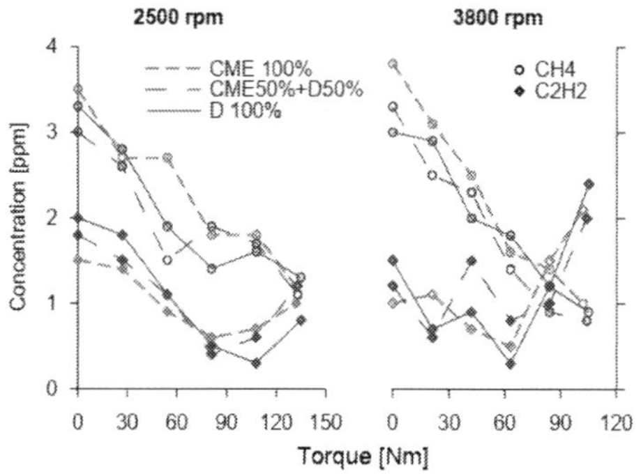

Fig. 6. Concentrations of methane and acetylene in exhausts of the engine fuelled with studied fuels.

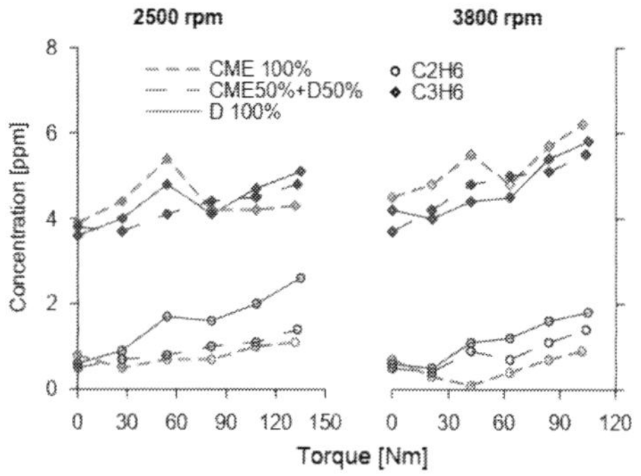

Fig. 7. Concentrations of ethane and propylene in exhausts of the engine fuelled with studied fuels.

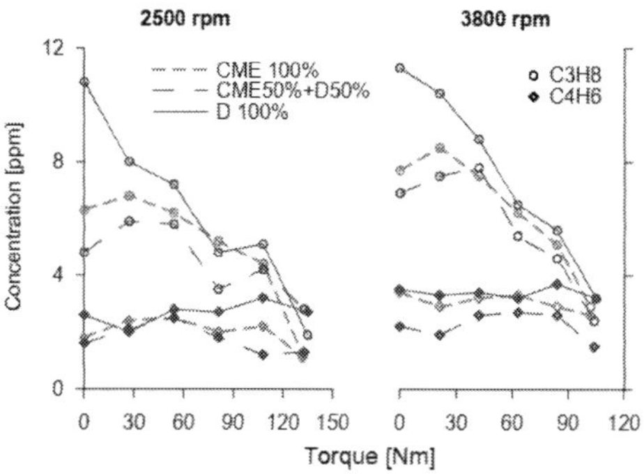

Fig. 8. Concentrations of propane and 1.3-butadiene in exhausts of the engine fuelled with studied fuels.

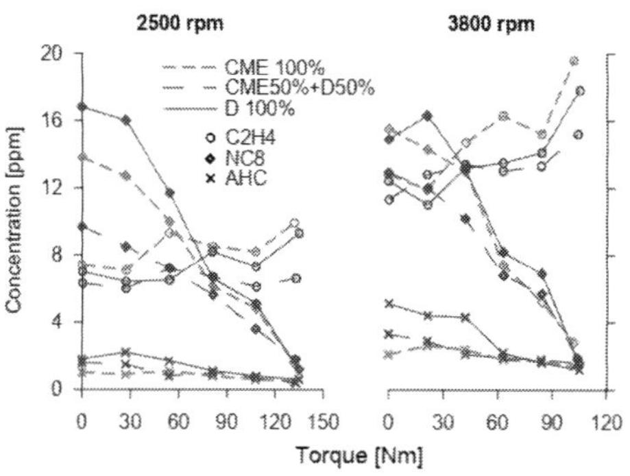

Fig. 9. Concentrations of ethylene, N-octane and aromatic hydrocarbons in exhausts of the engine fuelled with studied fuels.

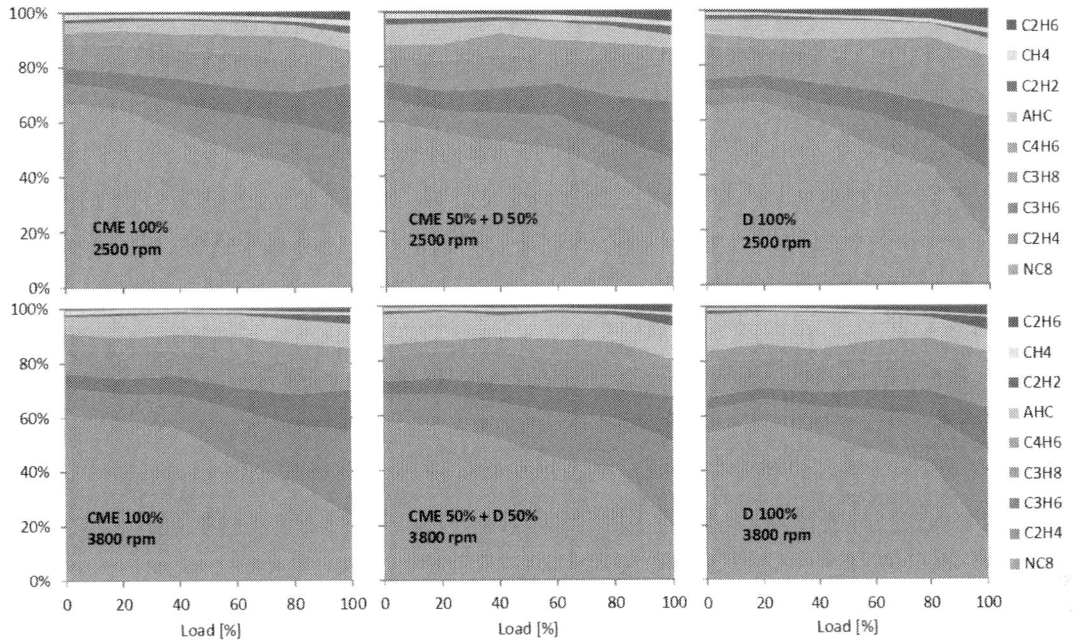

Fig. 10. Cumulative content of particular hydrocarbon species in total hydrocarbon expressed as C1 for all studied fuels and conditions of engine operation.

same reason the combustion of biofuels increased the concentrations of aldehydes. For all studied conditions biofuel combustion reduced the concentrations of total hydrocarbon (THC). However, it is worth noting that this trend was not monotonic. The lowest THC production was recorded for the blend of fuels. This peculiar effect was already observed at high load regime at whole rotational speed range of the engine [15]. It can be attributed to the balance between auto-ignition and diffusion properties of the tested fuels [16].

Regardless of the fuel used, the increase in the engine load resulted in reduction of the long-chained saturated alkanes. A plausible reason of this reduction is more intense cracking of heavy fuel hydrocarbon species at higher combustion temperatures appearing at the higher engine loads. This effect is most evident for diesel fuel at the highest load. 1,3 butadiene (C4H6) also deserves comment, as it is the most toxic hydrocarbon compound emitted by motor vehicles. Combustion of fuel based on canola methyl ester reduced the production of this compound. Combustion of this fuel also slightly reduced the production of ethane and aromatic hydrocarbons. In the case of other hydrocarbon species there was no clear influence of the fuel on specific hydrocarbons emissions.

References

1. J. Gupta, A. Agarwal, Unregulated and regulated emissions from biodiesel fuelled CRDI SUV engine, *SAE Technical Paper* 2015-01-0889 (2015)
2. J. Krahl, A. Munack, M. Bahadir, L. Schumacher, N. Elser, Review: Utilization of rapeseed oil, rapeseed oil methyl ester or diesel fuel: Exhaust gas emissions and estimation of environmental effects, *SAE Technical Paper* 962096 (1996)
3. F. Staat, P. Gateau, The effects of rapeseed oil methyl ester on diesel engine performance, exhaust emissions and long-term behavior – A summary of three years of experimentation, *SAE Technical Paper* 950053 (1995)
4. M. Lapuerta, O. Armas, J. Rodrıguez-Fernandez, Effect of biodiesel fuels on diesel engine emissions, Progress in energy and combustion science, **34**, 198-223 (2008)
5. F. Wu, J. Wang, W. Chen, S. Shuai, A study on emission performance of a diesel engine fueled with five typical methyl ester biodiesels, Atmospheric Environment, **43**, 1481-1485 (2009)
6. K. Duda, S. Wierzbicki, M. Smieja, M. Mikulski, Comparison of performance and emissions of a CRDI diesel engine fuelled with biodiesel of different origin, Fuel, **212**, 202-222 (2018)
7. L. Claxton, The history, genotoxicity, and carcinogenicity of carbon-based fuels and their emissions. Part 3: Diesel and gasoline. Mutation Research/Reviews in Mutation Research, **763**, 30-85 (2015)
8. E. Zervas, X. Montagne, J. Lahaye, Influence of fuel and air/fuel equivalence ratio on the emission of hydrocarbons from a SI engine, 1 Experimental findings, Fuel, **83**, 2301-2311 (2004)
9. J. Hunicz, P. Krzaczek, Detailed speciation of emissions from low-temperature combustion in a gasoline HCCI engine, Polish Journal of Environmental Studies, **25**, 137-145 (2016)

10. C. Dubien, D. Schweich, G. Mabilon, B. Martin, M. Prigent, Three-way catalytic converter modelling: fast- and slow-oxidizing hydrocarbons, inhibiting species, and steam-reforming reaction, Chemical Engineering Science, **53**, 471-481 (1998)
11. J. Hunicz, A. Medina, Experimental study on detailed emissions speciation of an HCCI engine equipped with a three-way catalytic converter, Energy, **117**, 388-397 (2016)
12. J. Hunicz, An experimental study into the chemical effects of direct gasoline injection into retained residuals in a homogeneous charge compression ignition engine, International Journal of Engine Research, **17**, 1031-1044 (2016)
13. G. Koszalka, J. Hunicz, A. Niewczas, A comparison of performance and emissions of an engine fuelled with diesel and biodiesel, SAE International Journal of Fuels and Lubricants, **3**, 77-84 (2010)
14. R. Pešić, A. Davinić, D. Taranović, D. Miloradović, S. Petković, Experimental determination of double vibe function parameters in diesel engines with biodiesel, Thermal Science, **14**, 197-208 (2010)
15. G. Koszalka, J. Hunicz, A. Niewczas, Comparative analysis of emission from engine fuelled with diesel and bio-diesel, Journal of KONES, **16**, 165-171 (2009)
16. O. Kuti, J. Zhu, K. Nishida, X. Wangb, Z. Huang, Characterization of spray and combustion processes of biodiesel fuel injected by diesel engine common rail system, Fuel, **104**, 838-846 (2013)

Automotive compressor: effect of an electric throttle in the upstream circuit on the surge limit

Pierre Podevin[1,*], *Amélie* Danlos[1], *Michaël* Deligant[2], *Plamen* Punov[3], *Adrian* Clenci[4], and *Guillaume* De La Bourdonnaye[5]

[1]Laboratoire de Chimie Moléculaire, Génie des procédés Chimiques et Energétique, EA7341, Conservatoire National des Arts et Métiers, HESAM Université, 75003 Paris, France
[2]Laboratoire DynFluid, EA 92, Arts et Métiers ParisTech, 75013 Paris, France
[3]Université technique de Sofia, 8 rue Kliment Ohridski, 1000, Sofia, Bulgarie
[4]Department Automobiles and Transport, University of Pitesti, 110040 Pitesti, Romania
[5]Valeo, 95892 Cergy Pontoise, France

Abstract. On vehicles equipped with a turbocharged engine, there is a risk of compressor surge. This surge generates instabilities that lead to driving inconvenience, or even mechanical failure of the supercharging system. In general, the surge appears rather in transient operation: sudden closing of the throttle valve on gasoline engine, regulation of the EGR on diesel engine linked also to turbine regulation (VNT device or Waste Gate). On a turbocharger test stand, we set up the surge line in a "conventional way": stationary experiments. Then we set up this line in transient conditions for different positions of an electric throttle placed upstream the compressor. It appears that: the surge limit is pushing back to lower flow rates when it is determined in transient; the surge limit is pushing back to lower flow rates when closing the throttle valve. The tests were carried on by the transient analysis of the surge during a quick closing-opening of the electric throttle valve.

1 Introduction

Most automotive engines are turbocharged to reduce CO_2 emission and fuel consumption. On an automotive engine, the manufacturer ensures that the engine operates out of the surge phenomenon zone of the compressor in stationary regime. To achieve this condition, the manufacturer keeps a margin from surge limit defined by the turbocharger supplier. This margin can be more or less important according to the manufacturer's skill and knowledge.

In the case of vehicles, if the automotive manufacturer has correctly set the surge margin for steady operation, the appearance of surge will concern only transient engine regimes, or when control systems are activated (throttle vane, EGR, bypass of fixed geometry turbines, nozzle vane of variable ones...).

The surge limit has been researched in numerous studies seeking to determine a criterion to detect the occurrence of surge [1, 2] but also on the validity of this limit on engine operation. These studies were mainly carried out for stationary operating regimes.

Phenomena leading to surge instabilities complicate the study of surge occurrence which may be associated to the compressor stall and may achieve a mild surge depending on the circuit acoustic resonance. For a high rotational speed of the compressor, the limit is difficult to characterize, and it is possible to switch abruptly to a deep surge [3, 4].

In order to use turbochargers in a larger operating range, it is necessary to determine precisely the surge limit and to identify mild surge and deep surge [5-8]. Sinusoidal signals of pressure ratio or mass flow rate characterize the deep surge and may be a better criterion for identifying the surge limit [9,10]. Other methods as the analysis of the standard deviation of temperature signals [11] or frequency domain analysis [12, 13] have also been applied on automotive turbocharger performances studies. Real-time quantification of the surge margin associated with an appropriate control would achieve the best performance of the system while maintaining stable operation, but it is still a challenge [3, 4].

After presenting the test bench, procedures for determining surge limits in stationary and transient tests are explained. The surge limits are compared.

Transient surge limits are also compared for different openings of an electric throttle placed upstream the turbocharger.

Finally, appearance of surge is studied for a quick closing of the throttle valve.

2 Experimental set-up

2.1 Test bench description

A prototype compressor from Valeo has been used for the purposes of this experimental study.

* Corresponding author: pierre.podevin@lecnam.net

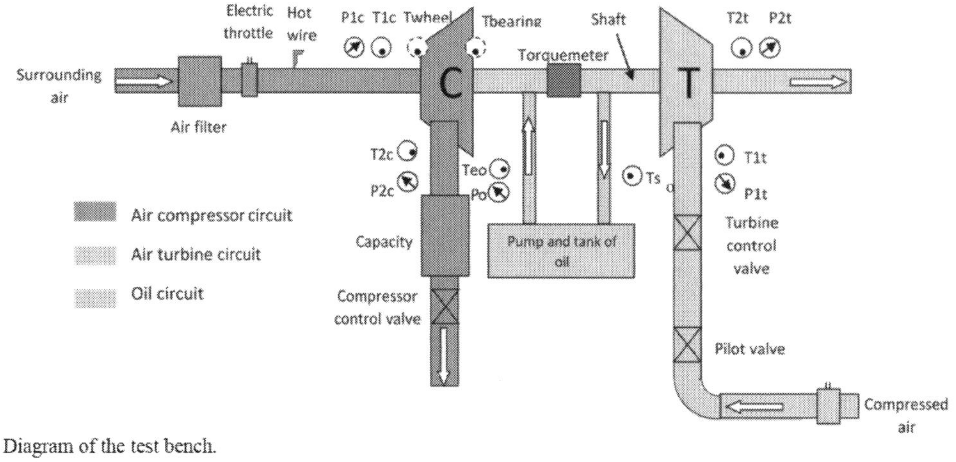

Fig. 1. Diagram of the test bench.

The characteristic curves for this compressor have been measured. In order to evaluate precisely the power at the output of the wheel, a dedicated torquemeter has been installed between the compressor and the compressor drive turbine.

On the downstream circuit, a 5L capacity is integrated to amplify the surge phenomenon and to make it easier to detect [14]. A diagram of the test bench is presented in Fig. 1. In this study, only the downstream circuit has been modified.

Experimental parameters have been measured using classical sensors (piezoresistive pressure sensors, PT100 wires probes). For measurements occurring in deep surge, a temperature sensor, Twheel, is added just upstream the compressor wheel using a K thermocouple with a diameter of 0.5 mm. The flow rate is measured with a hot wire probe upstream the compressor. The hot wire mass flow sensor was calibrated using the information given by a thermal mass flowmeter [14].

The valve placed upstream the compressor is an electric throttle usually used on turbocharged Diesel engines (Fig.2). This component is powered with 12V using the PWM (Pulse Width Modulation) method by the ECU device (Engine Control Unit).

Fig. 2. Electric throttle.

The opening of this valve is controlled by a 0-5V signal. A Hall effect position sensor powered with 5V allows for the control of its position. For the transient measurements, a Producer-Consumer type loop is developed with the Labview software in order to be able to simultaneously control the valve and acquire signals.

An example of control signal (red line) and of the position signal (blue line) is presented in Fig. 3. The valve is fully opened for 4.5V and fully closed for 0V.

The control is generated by a PID module in Labview. In the experimental measurements presented in this paper, the voltage values of the control are given in the form of a table, see Fig. 3. The position follows the control for 0.2s delays. Below this value, a time lag is created.

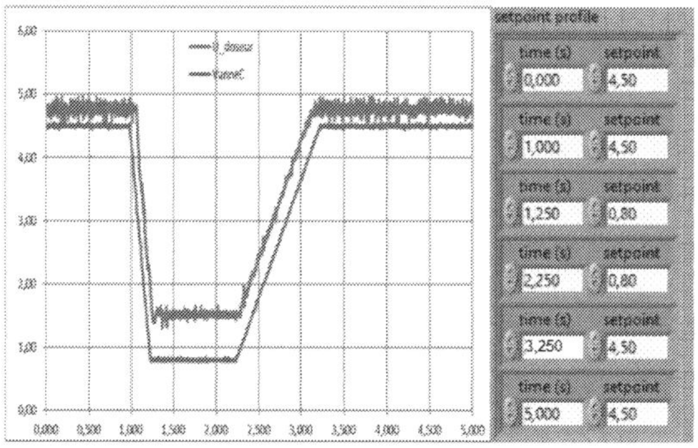

Fig. 3. Control signal and position signal of the electric throttle.

2.2 Definition of the surge limit in stationary test

The surge limit is determined by the operator from the noise generated once it occurs or by the signal analysis (for example, pressure, temperature, flow rate...). This limit depends on upstream and downstream conditions in the compressor circuit. In our experimental set-up, the geometry is the same for all measurements. The signal used in this paper is measured with the hot wire anemometer as it seems the most sensitive sensor to surge. We have defined two surge levels: the appearance of the surge, called mild surge and the deep surge. The mild surge is identified with a Fast Fourier Transform (FFT) analysis of the signal when a frequency peak occurrence (Fig. 4 (b)). The raw signal presents some regular fluctuations (Fig. 4 (a)). The deep surge corresponds to a very high peak of frequency detected by the Fourier analysis (Fig. 4 (d)). In this case, the raw signal has a sinusoidal shape as it is shown in Fig. 4 (c).

In order to draw this surge limit, we adjust the rotation speed for a first point at low regime. Next, we close gradually the valve at the compressor outlet until mild surge appears. If the rotation speed is too high, we adjust the working point by reducing the turbine velocity acting on the turbine control valve. The point where the surge phenomenon appears is then acquired. The compressor valve is closed again until the deep surge and the working point are recorded. Finally, the same process is applied for another rotation regime.

The appearance of the surge and the deep surge have been determined for a full opened position of the electric throttle valve. The determination of these limits follows these steps: the valve placed downstream the compressor is closed to operate in proximity to the surge limit. This position is evaluated from previous results when measurements were conducted in iso-speed conditions.

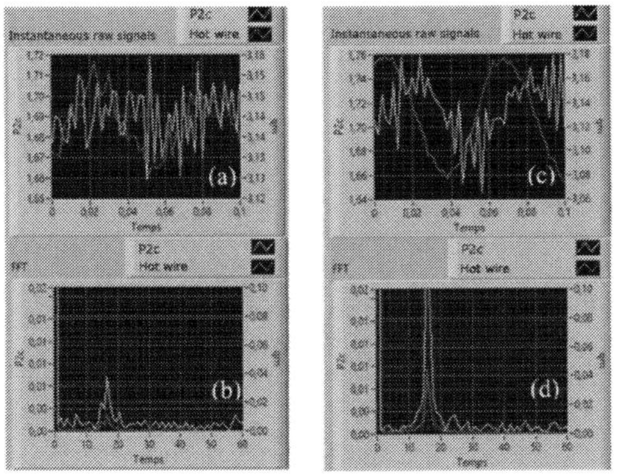

Fig. 4. Raw signals of instantaneous pressure in outlet of the compressor (red line) and flowrate (white line) for (a) the mild surge and (c) the deep surge and corresponding FFT signals calculated on same signals respectively in case of (b) mild surge and (d) deep surge.

Measurements have then been repeated for other rotation speeds to build surge limits presented in Fig. 5.

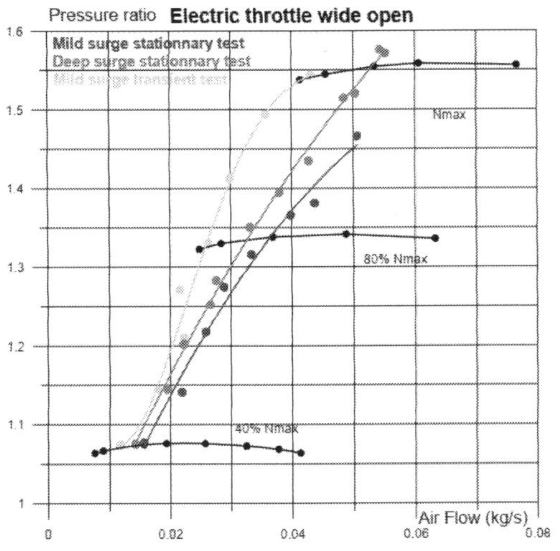

Fig. 5. Comparison of surge lines obtained in stationary and transient tests.

2.3 Definition of the surge limit in transient test

The setup of this line is carried out as follows: for a selected rotation speed, the valves after the compressor and before the turbine are adjusted to set an operating point close to mild surge. Then the compressor downstream valve is progressively closed until deep surge is achieved. Data are recorded for a duration of 5 s at 1 kHz acquisition rate.

Fig. 6 presents the evolution of air flow and the pressure downstream the compressor P2c for an initial speed close to 80% Nmax.

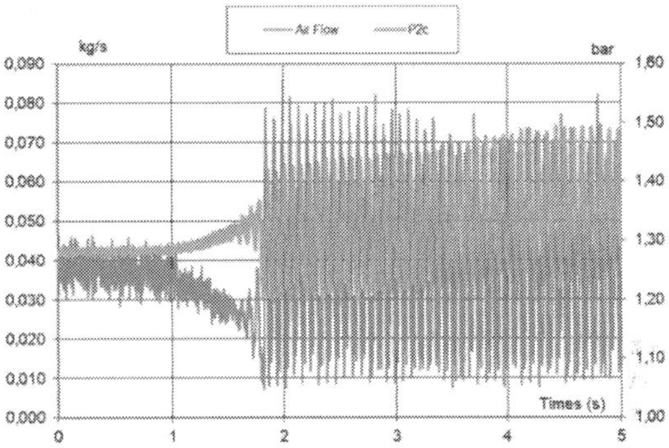

Fig. 6. Evolution of Air Flow and relative pressure downstream the compressor P2c versus time.

The beginning of the surge is determined from the RMS (Root Mean Square) or SD (Standard Deviation) calculation of the air flow signal. The SD shows a greater sensitivity [14] as shown in Fig. 7.

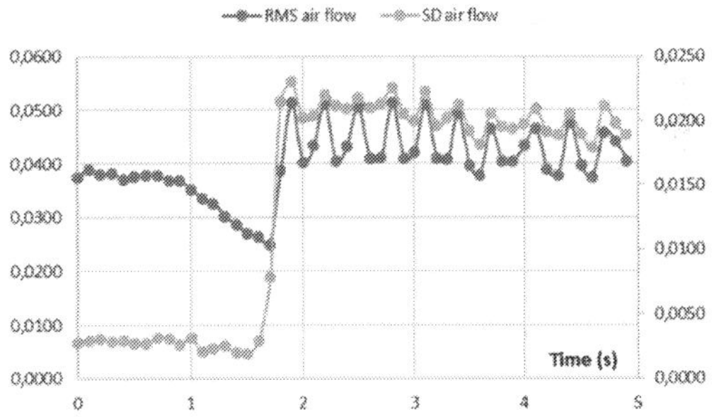

Fig. 7. RMS and SD of Air Flow versus time.

The tests are then repeated for other speeds and openings of the electric throttle thus making it possible to construct the surge limits of Fig. 8.

3 Surge limits comparison

Fig.5 shows the surge lines obtained for a wide-open position of the electric throttle. The surge lines obtained in the stationary test for mild surge and deep surge are logically slightly offset. On the other hand, a clear difference exists between the lines of mild surge obtained for stationary and transient tests. A margin on air flow of about 30% is earned at 80% of Nmax. This difference may be related to the aerodynamic inertia of the fluid column.

https://doi.org/10.1051/matecconf/201823403006

Nevertheless, this limit seems more representative of what actually happens during the transient operation of an automotive engine. This would allow the engine to operate briefly beyond the stationary surge limit without any inconvenience for the driving, thus increasing the low speed torque.

Fig. 8 shows the results obtained for different positions of the valve of the electric throttle.

The results of these tests show that the closer the electric throttle valve, the lower the surge limit. This confirms previous results obtained during tests on the effect of an upstream valve [14]. A margin of more than 60 % is obtained between the open position and the position $U = 1.0$ V.

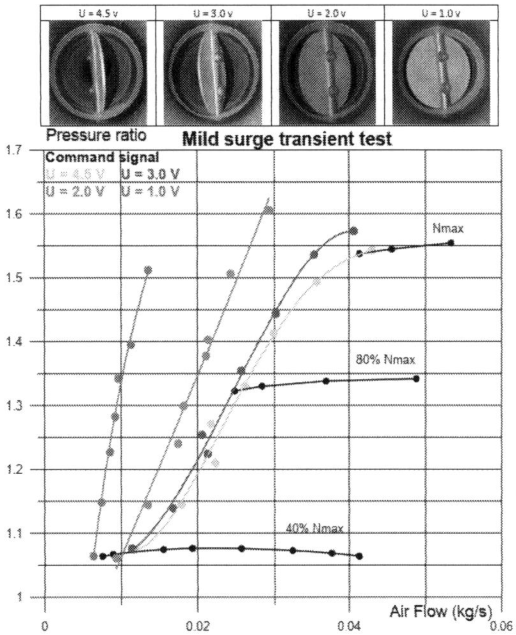

Fig. 8. Comparison of surge lines obtained in transient tests for different openings of the electric throttle.

4 Surge limit evolution closing-opening

Some experiments have been carried out for closing and opening of the electric throttle. Different sets of profiles have been performed from a wide open initial position and a compressor speed close to 80% of Nmax. For this speed, the opening of the valve downstream the compressor is set to obtain a point close to the mild stationary surge line.

Test results presented are for setpoint profile according to Fig. 3. The measurements are recorded for 5 s at a data acquisition rate of 1 kHz. The turbine pilot valve being fixed, compressor speed increases when closing and decreases when opening, as shown in Fig. 9. Special care is needed during experiments to avoid overshoot of Nmax.

The results of the measurements are presented in Fig. 10. It is possible from this figure to extract the surge loop with a time step of 0.025 s. In Fig. 11 some surge loops are presented for different initial time.

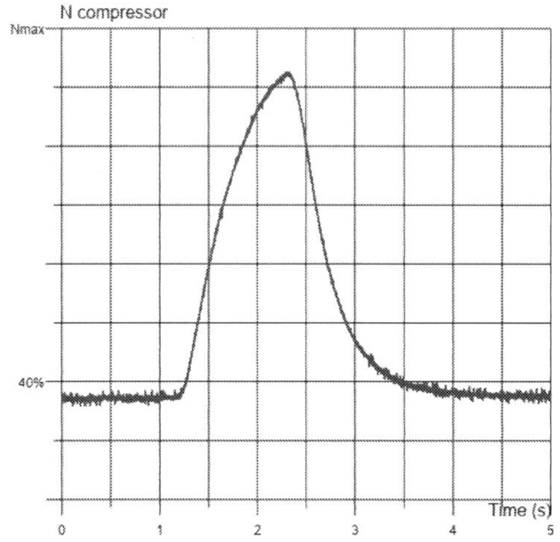

Fig. 9. Speed evolution during closing-opening.

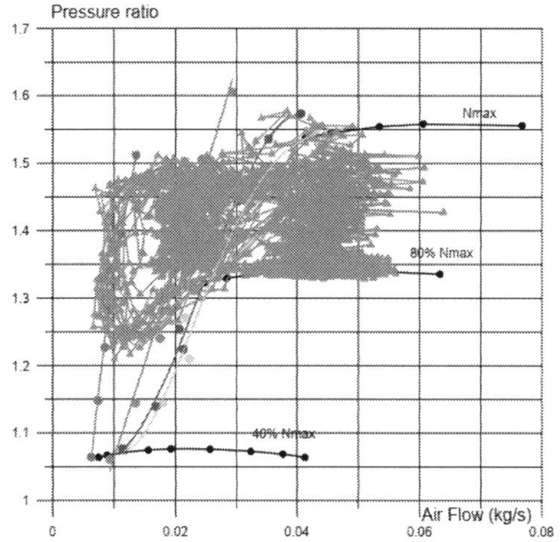

Fig. 10. Air Flow and Pressure evolutions during closing-opening.

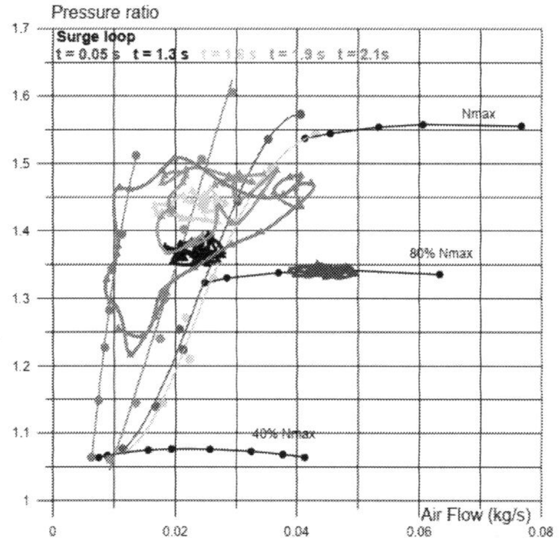

Fig 11. Surge loops evolutions during closing-opening for different initial time.

5 Conclusion

Experiments show that:

- the surge limit appears for lower flow rates when it is determined in transient;
- the surge limit appears for lower flow rates when closing the throttle valve.

These results are very interesting as a larger margin could be expected from turbocharger manufacturer surge line curve in the case of a valve closing at an inlet of the compressor.

Previous experiments showed that closing a valve at inlet of a compressor increases the margin line but this one could be decreased for very important closing [14]. So, additional experiments have to be done to confirming the present behaviour for more important closing.

Also, the complete or almost complete quick closure of the electric throttle valve must be experienced. At present, the pilot valve before the inlet of the turbine can be controlled electrically but it takes a long time. So we can have during these experiments turbocharger overspeed and risk of damage. So, new methods of controlling the turbine speed should be considered for future experiments

Measurement with a hot wire does not give the direction of air flow which could be negative for deep surge. In this case, we have to correct the sign of the air flow. A special procedure has been already applied [8]. At present, we are using an ultrasonic air flow meter. The first results are promising.

Tests are also foreseen to establish the effect of closure at the outlet of the compressor.

References

1. B. Semlitsch, M. Mihaescu, Flow phenomena leading to surge in a centrifugal compressor, *Energy,* **103,** 572 (2016)
2. E. Guillou, M. Gancedo, R. DiMicco, E. Gutmark, F. Hellström, L. Fuchs, A. Mohamed, Surge characteristics in a ported shroud compressor using PIV measurements and large eddy simulation, 9th International conference on turbochargers and turbocharging-institution of mechanical engineers, combustion engines and fuels group, **161** (2010)
3. J. Galindo, A. Tiseira, R. Navarro, D. Tari, C. Meano, Effect of the inlet geometry on performance, surge margin and noise emission of an automotive turbocharger compressor, *Applied Thermal Engineering,* **110**, 875 (2017)
4. X. Zheng, Z. Sun, T. Kawakubo, H. Tamaki, Experimental investigation of surge and stall in a turbocharger centrifugal compressor with a vaned diffuser, *Experimental Thermal and Fluid Science*, **82**, 493 (2017)
5. E. Greitzer, F. Moore, A theory of post-stall transients in axial compression systems: part II: application, *ASME Paper,* **GT-1985-172** (1985)
6. T. Camp, I. Day, A study of spike and modal stall phenomena in a low-speed axial compressor, ASME Journal of Turbomachinery, **120**, 393 (1998)
7. V. Garnier, A. Epstein, E. Greitzer, Rotating waves as a stall inception indication in axial compressors, ASME Journal of Turbomachinery, **113**, 290 (1991)
8. M. Deligant, A. Danlos, P. Podevin, A. Clenci, S. Guilain, Surge detection on an automotive turbocharger during transient phases, *IOP Conf. Ser. Mat. Sci. Eng.*, **252** (012082) (2017)
9. R. Bontempo, M. Cardone, M. Manna, G. Vorraro, A statistical approach to the analysis of the surge phenomenon, Energy, **124**, 502 (2017)
10. J. Andersen, F. Lindström, F. Westin, Surge definitions for radial compressors in automotive turbochargers, SAE Int J. Engines, **1**, 218 (2008)
11. A. Liu, X. Zheng, Methods of surge point judgment for compressor experiments, Experimental Thermal and Fluid Science, **51**, 204 (2013)
12. M. Toussaint, A. Danlos, G. Descombes, Analyse du fonctionnement en pompage d'un compresseur centrifuge refoulant à travers une vanne rotative, *COFRET'14*, Paris, France (2014)
13. J. Galindo, J. Serrano, C. Guardiola, C. Cervello, Surge limit definition in a specific test bench for the characterization of automotive turbochargers, Experimental Thermal and Fluid Science, **30**, 449 (2006)
14. P. Podevin, M. Toussaint, A. Danlos, M. Deligant, G. De La Bourdonnaye, Effets du vannage amont sur la marge au pompage des turbocompresseurs de suralimentation, *COFRET'18*, Strasbourg, France (2018)

Experimental study of multiple pilot injection strategy in an automotive direct injection diesel engine

Plamen Punov[1,*], *Tsvetomir* Gechev[1], *Svetoslav* Mihalkov[1], *Pierre* Podevin[2], and *Dalibor* Barta[3]

[1] Technical University – Sofia, Department of combustion engines, automobiles and transport, 1000 Sofia, Bulgaria
[2] Laboratoire de Chimie Moléculaire, Génie des procédés Chimiques et Energétique, EA7341, Conservatoire National des Arts et Métiers, HESAM Université, 75003 Paris, France
[3] University of Zilina, Faculty of Mechanical Engineering, Univerzitna 8215/1, 010 26, Žilina, Slovakia

Abstract. The pilot injection strategy is a widely used approach for reducing the noise of the combustion process in direct injection diesel engines. In the last generation of automotive diesel engines up to several pilot injections could occur to better control the rate of heat release (ROHR) in the cylinder as well as the pollutant formation. However, determination of the timing and duration for each pilot injection needs to be precisely optimised. In this paper an experimental study of the pilot injection strategy was conducted on a direct injection diesel engine. Single and double pilot injection strategy was studied. The engine rated power is 100 kW at 4000 rpm while the rated torque is 320 Nm at 2000 rpm. An engine operating point determined by the rotation speed of 1400 rpm and torque of 100 Nm was chosen. The pilot and pre-injection timing was widely varied in order to study the influence on the combustion process as well as on the fuel consumption.

1 Introduction

Recently, a more demanding emissions homologation procedure for light duty vehicle was imposed by the EU. It includes a new test cycle called WLTC as well as a real drive emission test [1]. In order to validate the results a conformity factor was proposed which would be decreased in the following years. Moreover, in 2020 the reference CO_2 emissions will be decreased to 95 g/km which represents a reduction by 27% in comparison with 2015. Although many companies have already announced that diesel engines will not be used in their new passenger cars any more, the main advantage of diesel engines, i.e. higher thermal efficiency, can still be important in order to reduce CO_2 emissions.

A number of studies are now focused on post-treatment in diesel engines. However, the emissions, noise and fuel consumption could be improved by split injection strategy usually provided by the last generation common-rail fuel injection systems [2, 3]. The split injection strategy can also be used for control of nonconventional combustion processes such as HCCI, RCCI, LTC, etc. [4].

Pilot injections are mainly used to reduce the ignition delay period, which leads to low combustion noise. Finesso and Spessa in [5] studied ignition delay in the case of multiple injection for a light-duty DI diesel engine. On the basis of an experimental and numerical approach they proposed controlled-orientated correlations for prediction of the ignition delays for both the pilot and the main injection. According to [6], the ROHR in the case of a pilot injection could be interpolated by double Vibe function. A numerical study of the influence of pre-injection strategy on ignition and combustion was presented by Punov and Evtimov in [7]. In the case of single pre-injection, the maximum ROHR can be reduced by 25% at 2000 rpm and BMEP of 8 bars. In [8] the pilot-main injection strategy was studied by means of endoscopic visualisation. The results revealed that small pilot injection and late start of the pilot injection reduced the ignition delay period, so that the rate of heat release peak decreased. Additionally, it leads to lower soot emission due to better oxidation of the soot.

Moreover, pilot injections affect the formation of NO_x and PM due to the influence on the premixed and mixed combustion phase. Park et al. [9] studied the combustion and emission characteristics of a single cylinder diesel engine. Three different strategies of split injection were proposed. The results revealed that the increase of the injection interval between the first and second injection caused an increase of NO_x and HC while PM and CO were reduced. Late SOI leads to lower NO_x, CO and HC. At the same time the PM was increased. Xu et al. [10] reported a significant impact of the start of the pilot injection when a diesel engine is fuelled with natural gas. Pre-injection timing higher than 70° BTDC leads to lower maximum pressure and lowered NO_x emission by 35%. The split injection strategy was studied by How et al. in [11] for a diesel engine using biodiesel blends. The experimental test was conducted on a light-duty diesel engine as single, double and triple injections were applied for different fuel blends. It was found out that considerably lower level of NO_x emission could be achieved by retarding the SOI in case of triple injection for B20 and B50 fuel blends. Li et al. studied the effect of multiple-injection strategies on the

* Corresponding author: plamen_punov@tu-sofia.bg

 https://doi.org/10.1051/matecconf/201823403007

particle size and structure in [12]. The pilot-injection strategy in comparison with the single injection leads to increased number and mass of the particles with diameter above 0.1μm. However, due to later oxidation in combustion, the particle number and diameter are reduced.

The swirl in combination with split injection could provide a positive effect on the combustion [13]. The higher the swirl ratio, the shorter the combustion duration. The combustion duration can be reduced by 19.5-25.7% depending on the pilot quantity. At the same time the BSFC and Soot emission are reduced.

The optimization of the multiple injection strategy is not easy to be done due to the impact of many injection parameters as well as the control system limitation. Millo et al. [14] used a genetic algorithm in order to optimize control parameters such as SOI, rail pressure, injection quantity, boost pressure, etc. in the case of a split injection strategy by 5 separate injections.

On the basis of the literature survey a lot of split injection strategies were studied in DI diesel engines. However, not a lot of research was reported on the influence of dual pilot injection on the combustion characteristics. In view of that, the aim of this paper is to study the impact of single and double pilot injection strategies applied to a passenger car DI diesel engine.

2 Experimental equipment

2.1 Engine under study

It is a DI diesel engine with displacement of 2 l, which complies with EURO IV emission standard. The engine is equipped with a Common-Rail injection system, VG turbocharger, water cooled exhaust gas recirculation (EGR), catalytic converter, etc. The maximum engine power is 100 kW at 4000 min^{-1} and the maximum torque is 320 Nm at 2000 min^{-1}. The boost pressure is limited to 1.3 bar. The main parameters of the engine are listed in Table 1.

Table 1. Engine specifications.

Engine type	HDI
Number of cylinders	4
Displacement	2 l
Bore	85 mm
Stroke	88 mm
Compression ratio	17.6
Maximum Rail Pressure	1600 bar

2.2 Engine test bed

The engine was mounted on an engine test bed as it is shown in Fig. 1. An eddy-current dynamometer was used with a maximum power of 160 kW and maximum speed of 10000 rpm. The brake is electronically controlled by means of Schenck control unit. In order to study the in-cylinder combustion the first cylinder is equipped with piezoelectric pressure sensor. The sensor replaces the glow plug of the cylinder. Signal from the sensor is transmitted to a Kistler Type 5011 charge amplifier, then to the AVL measurement unit. The crankshaft angular position is registered by means of optical encoder. The fuel consumption is measured by means of a volumetric flow meter Rotronics RCC101.

Table 2. Test equipment specification.

Eddy-current brake	160 kW/10000 rpm
Brake control	Schenck
Pressure sensor AVL GH13P	250 bars
Signal amplifier Kistler	Type 5011
Angle encoder AVL 364C	0.1 CAD
Optical signal converter	AVL 364
Measurement system	AVL Indiscop 647
Fuel flow measurement	Rotronics RCC 101

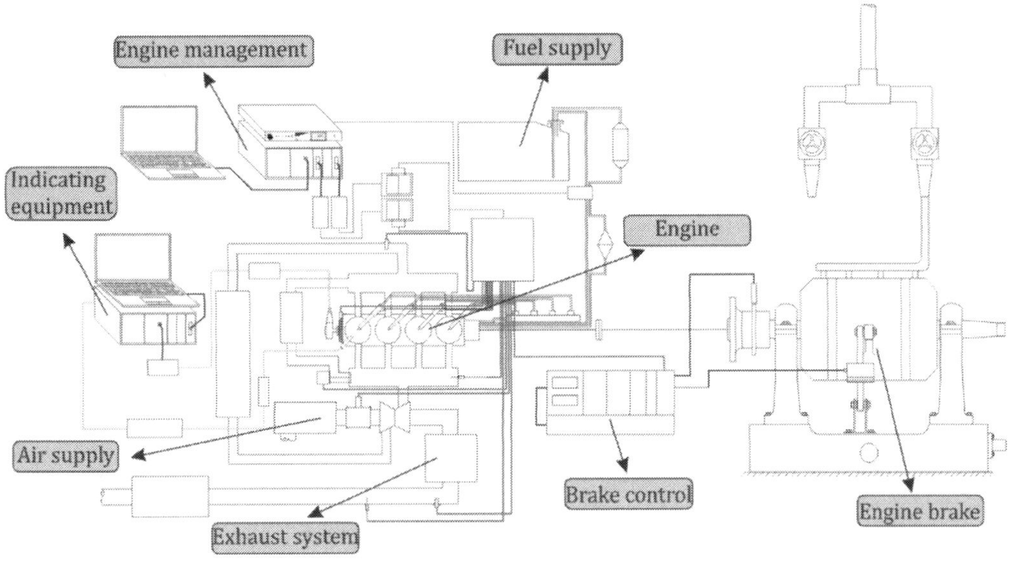

Fig. 1. Schematic diagram of the engine test bed.

2.3 Engine management and data logging

The engine management system is based on a National Instruments real-time controller. The hardware of the system consists of PXI chassis – NI PXI-1031, Real-Time Embedded Controller – NI PXI-8106 RT, FPGA module – NI PXI-7813R, two R Series Expansion Chassis – NI CRIO 9151 as well as the modules developed by Drivven. In our project we used four modules: DI Driver – 2 pcs., Low Side, AD Combo and O_2 sensor module. All of these modules are based on NI C series interface.

The entire system provides a large functionality for real-time control and monitoring of the engine control parameters such as: injection control by up to five injections per cycle, injection pressure control, boost pressure control, exhaust gas recirculation (EGR) control, closed-loop control of injection by means of wide band oxygen sensor in exhaust gases, etc.

A LabVIEW project was developed in order to determine the control logic of the management system. The real-time data were visualized by means of the software called CalView developed by Drivven. This software was also used to set the operating parameters in real-time.

3 Experimental results

The experimental test was conducted at 1400 rpm and engine torque of 100 Nm. This operating point is a representative point in both the WLTC and RDE tests for passenger cars. The EGR rate is zero. Two pilot injection strategies were studied: a single pilot injection and a double pilot injection. In order to study the influence on the combustion characteristics as well as on the fuel consumption, the engine output torque was constant. The torque was controlled through the duration of the main injection.

3.1 Single pilot-injection

Firstly, a single pilot injection strategy was studied. In this case the pilot injection was provided with a constant duration (constant energizing time). In order to study the influence of the pilot injection, the start of the pilot injection was varied within the range of 0.8 ms to 2ms before the start of main injection. The start of the main injection was constant 2° BTDC. The torque of the engine was controlled by variation of the main injection duration. It was conducted automatically by means of the engine management system. The variation of the main engine control parameters is listed in Table 3.

Table 3. Injection parameters of single pilot injection.

Engine speed	1400 rpm
Engine output torque	100 Nm
Rail pressure	620 bars
Start of main injection	2° BTDC
Duration of main injection	variable
Start of pilot injection	0.8 … 2 ms
Duration of pilot injection	0.35 ms

The start of the pilot injection was varied by a step of 0.2 ms. Estimated in CAD the variation is in the range of 7° BTDC to 18.8° BTDC. The results concerning the indicated diagram are presented in Fig. 2.

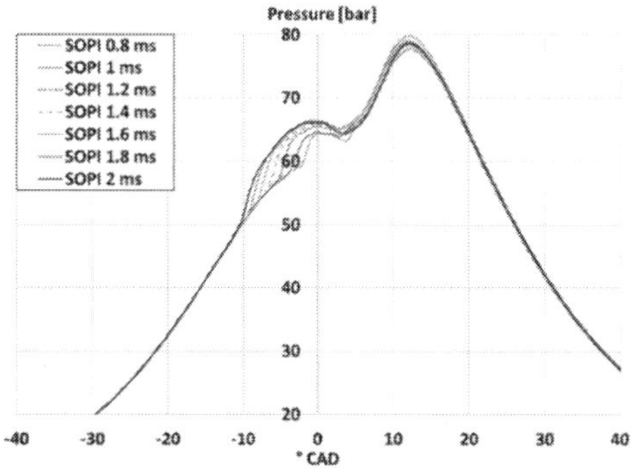

Fig. 2. In-cylinder pressure as a function of start of pilot injection.

The start of the pilot injection has a significant impact on the in-cylinder pressure. The early injection leads to an early start of combustion and to higher pressure before TDC. No big impact on the ignition delay period of pilot injection or on the ignition delay of the main injection was not found. The maximum of the pressure peak was observed to be at the same CAD for all studied cases. The highest in-cylinder pressure was observed in the case when the pilot injection started 1.8 ms before the start of the main injection – it amounted to 79.8 bar. The lowest in-cylinder pressure peak was observed when the pilot injection start was 0.8 ms before the start of the main injection. In this case the maximum in-cylinder pressure was observed to be 77.8 bars. The fuel consumption varied by 1.44 % and the lowest consumption was observed at the latter pilot injection.

3.2 Double pilot-injection

The second study was conducted as one more injection, so called pre-injection, was added between the pilot-injection and the main injection. The start of the pilot injection (2 ms BTDC), as well as its duration, were non-variable during this study. The start of the main injection was also set to 2° BTDC while its duration was used to control the engine torque. In this case the impact of the start of the pre-injection was studied. The pre-injection duration was fixed to 0.25 ms (energizing time). The start of the injection was varied within the range of 0.5 ms to 1.3 ms before the start of main injection. These are shown in Fig. 3.

The pre-injection leads to higher pressure before the start of combustion of the main injected fuel while the maximal pressure in the cylinder has very similar value to the one in the first study. The highest pressure peak was observed when the pre-injection starts 1 ms before the main injection while the lowest pressure peak was observed at the pre-injection starting 1.3 ms before the main injection. In this study a small impact on the engine fuel consumption was observed within the range of 0.75 %.

 https://doi.org/10.1051/matecconf/201823403007

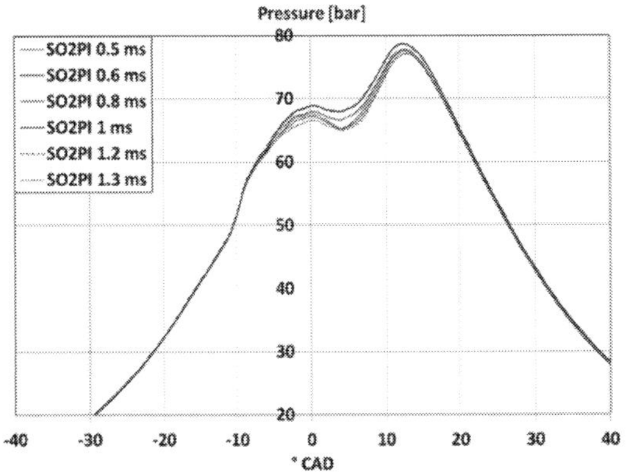

Fig. 3. In-cylinder pressure as a function of start of pre-injection.

4 Analysis of the combustion characteristics and discussion

In order to better understand the impact of the multiple pilot injection strategy on the combustion process, the ROHR was compared for each studied case. The ROHR is related to the combustion noise as well as to the pollutant formation. The ROHR was estimated by means of simulating programme in the range of 30° BTDC to 90° ATDC. The results are shown in Figures 4 and 5 as well as in Table 4.

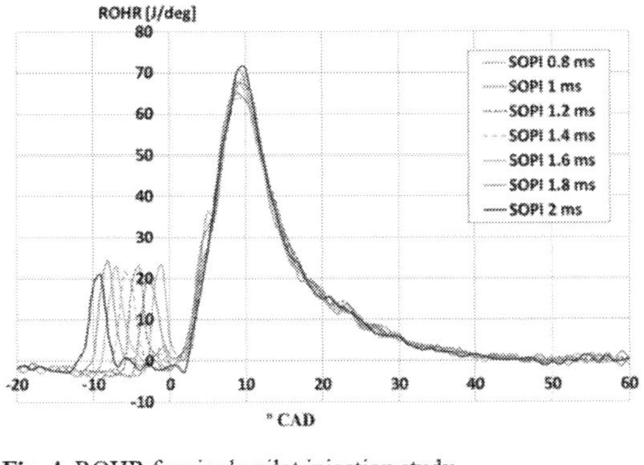

Fig. 4. ROHR for single pilot injection study.

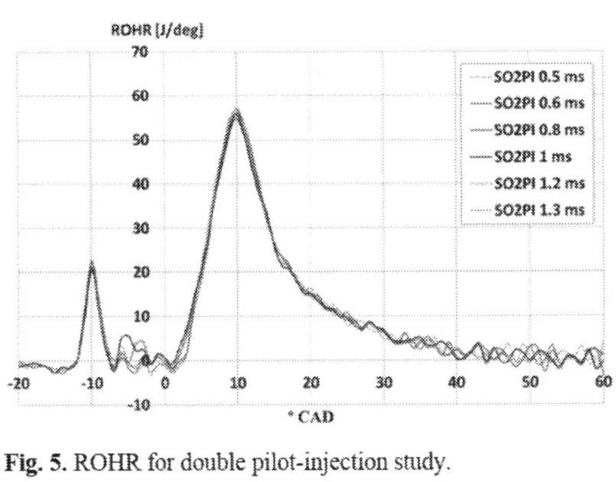

Fig. 5. ROHR for double pilot-injection study.

The start of the pilot injection in the case of single pilot-injection strategy has a slight impact on the peak of ROHR. Early pilot injection leads to a higher peak while the closer the start of the pilot injection to the main injection, the lower the ROHR peak. In the first case, the maximum ROHR was estimated to be 71.2 J/deg while the maximum decreases to 63.5 J/deg in retarded pilot injection.

Obviously, double pilot-injection strategy has a positive impact on the ROHR peak. Although, the pre-injection in our study was quite small as quantity, the peak was reduced to 54 J/deg. The start of the pre-injection does not lead to significant variation in the ROHR curve. However, we could expect lower noise levels as well as less NO_x emissions.

Table 4. Maximum in-cylinder pressure and ROHR.

Single pilot-injection strategy			
Start of pilot injection, ° BTDC	Start of pre-injection, ° BTDC	Maximum pressure, bar	Maximum ROHR, J/deg
8.7	-	77.8	63.5
10.4	-	78.9	65
12.0	-	78.5	67.1
13.8	-	77.6	67.9
15.4	-	79.2	67.8
17.1	-	79.8	70.5
18.8	-	78.7	71.2
Double pilot-injection strategy			
Start of pilot injection, ° BTDC	Start of pre-injection, ° BTDC	Maximum pressure, bar	Maximum ROHR, J/deg
18.8	6.2	77.7	54.8
18.8	7.0	76.8	56.9
18.8	8.7	77.4	54.8
18.8	10.4	78.6	55.8
18.8	12.0	77	57.4
18.8	12.9	77.1	56.3

5 Conclusion

Multiple pilot-injections strategy was experimentally studied in a modern diesel engine intended for passenger cars. Two strategies were compared: single pilot-injection and double pilot-injection. In both cases, the varied parameter was the start of the pilot or pre-injection. For each of the injections the in-cylinder pressure values were recorded and the ROHR was then estimated.

On the basis of our experimental results, it could be stated that the double pilot injections strategy has a significant potential to reduce the maximum of the ROHR in the combustion chamber by up to 13.7 % in comparison to the single pilot injection. The start of the pilot injection in the case of a single pilot injection strategy and the start of pre-injection in the case of the double pilot injection offer a slight impact on the maximum of the cylinder pressure. Early start of the pilot injection leads to higher negative piston work. Therefore, it increases fuel consumption by 1.44 % in comparison to the latter injection. In double pilot injection strategy, the pre-injection timing has lower impact to the to the fuel consumption up to 0.75 %. However, in case of double pilot-injection the fuel consumption increases due to the second injection before the main injection.

In order to fully optimise the fuel injection in DI diesel engines more injection parameters, such as injection pressure, injection duration and post-injection, need to varied. This will be the subject of another research that will take into consideration pollutant emissions as well.

Special acknowledgements to Mr. Pierre Podevin from laboratory CMGPCE of Cnam, Paris, who provided to the Department of Combustion engines, automobiles and transport of TU-Sofia an eddy-current brake and an electronic control device. Also, the authors would like to thank Mr. Theo Bosaert and Mr. Cyril Fresque who during their internship helped us to install the engine test equipment in our laboratory.

Nomenclature

CO_2	carbon dioxide
CO	carbon monoxide
NO_x	nitrogen oxides
HC	hydrocarbons

Abbreviations

ATDC	after top dead center
BMEP	brake mean effective pressure
BSFC	brake specific fuel consumption
BTDC	before top dead center
CAD	crank angle degree
DI	direct injection
EGR	exhaust gas recirculation
HCCI	homogenous charge compression ignition
LTC	low-temperature combustion
PM	particulate matter
RCCI	reactivity controlled combustion ignition
RDE	real drive emission
ROHR	rate of heat release
SOI	start of injection
VG	variable geometry
WLTC Cycles	Worldwide harmonized Light vehicles Test

References

1. A. Clenci, Real Driving Emissions (RDE): enjeux et défis, *Conférences CNAM-SIA-LMP-ENSPM* (2018)
2. P. Punov, Research the fuel injection characteristics of a Common Rail solenoid injector, *Proceedings of BulTrans-2015*, 153-158 (2015)
3. G. Dober, S. Tullis, G. Greeves, N. Milovanovic, M. Hardy, S. Zuelch, The impact of injection strategies on emissions reduction and power output of future diesel engines, *SAE Conference Proceedings*, 2008-01-0941 (2008)
4. K. Mathivanan, J. Mallikarjuna, A. Ramesh, Influence of multiple fuel injection strategies on performance and combustion characteristics of a diesel fuelled HCCI engine – An experimental investigation, Experimental Thermal and Fluid Science, **77**, 337-346 (2016)
5. R. Finesso, E. Spessa, Ignition delay prediction of multiple injections in diesel engines, Fuel, **119**, 170-190 (2014)
6. P. Punov, N. Milkov, C. Perilhon, P. Podevin, T. Evtimov, Study on the combustion process in a modern diesel engine controlled by pre-injection strategy, *IOP Conference Series: Materials Science and Engineering*, **252**(1), 23, Article number 012090, (2017)
7. P. Punov, T. Evtimov, Combustion optimization in a modern diesel engine by means of pre-injection strategy, International Journal MTM, **4**, 41-44 (2015)
8. X. Li, W. Yang, L. Zhao, F. Liu, The influence of pilot-main injection matching on DI diesel engine combustion using an endoscopic visualization system, Fuel, **188**, 575-585 (2017)
9. S. Park, H. Kim, D. Shin, J. Lee, Effects of various split injection strategies on combustion and emissions characteristics in a single-cylinder diesel engine, Applied Thermal Engineering, **140**, 422-431 (2018)
10. M. Xu, W. Cheng, Z. Li, H. Zhang, T. An, Z. Meng, Pre-injection strategy for pilot diesel compression ignition natural gas engine, Applied Energy, **179**, 1185-1193 (2016)
11. H. How, H. Masjuki, M. Kalam, Y. Teoh, Influence of injection timing and split injection strategies on performance, emissions, and combustion characteristics of diesel engine fueled with biodiesel blended fuels, Fuel, **213**, 106-114 (2018)
12. X. Li, C. Guan, Y. Luo, Z. Huang, Effect of multiple-injection strategies on diesel engine exhaust particle size and nanostructure, Journal of Aerosol Science, **89**, 69-76 (2015)
13. X. Li, H. Zhou, L. Zhao, L. Su, H. Xu, F. Liu, Effect of split injections coupled with swirl on combustion performance in DI diesel engines, Energy Conversion and Management, **129**, 180-188 (2016)
14. F. Millo, P. Arya, F. Mallamo, Optimization of automotive diesel engine calibration using genetic algorithm techniques, Energy, **158**, 807-819 (2018)

Influence of plastic deformation of S235JR steel rods on their mechanical properties and corrosion behavior in NaCl solution

Veselin Tsonev[1,*], *Nikolay* Nikolov[1], and *Yordanka* Marcheva[2]

[1]Technical University – Sofia, "Strength of materials" Department, Bulgaria
[2]Technical University – Sofia, "Chemistry" Department, Bulgaria

Abstract. Studies show that although commonly used steels are standardized, some of their mechanical properties (such as tensile strength) vary widely and knowledge of their corrosion behavior is insufficient. Additional treatments, such as plastic deformation, alter the structure of carbon steels and affect their properties. This article explores one of the most widely used materials in mechanical and civil engineering – steel S235. Two types of rods, 6 mm in diameter, from hot-rolled non-alloy structural steel (S235JR, BDS EN 10025-2: 2005) and bright cold drawn steel (S235JRC, BDS EN 10277-2: 2008) have been tested. Tensile tests have been carried out, stress-strain curves are constructed and compared, the main mechanical properties such as yield strength, tensile strength and modulus of elasticity are determined. The typical consequences of plastic deformation such as increased yield strength have been identified. The assessment of corrosion behavior was done by means of the weight loss method in a 3,5% NaCl water solution for 5 weeks. It was found out that in the studied period the two types of rod exhibit close corrosion resistance, with the tendency for the cold drawn steel to have a higher uniform corrosion rate over a longer period.

1 Introduction

Low-cost carbon steels are used as the preferred construction material across industries and are considered to be a more economical option than the costly corrosion-resistant alloys [1]. Carbon steels typically contain less than 1,5% of carbon (C). Depending on the carbon content they are divided into 3 groups – low carbon steels (<0,25% C), medium carbon steels (0,25-0,70% C) and high carbon steels (0,70-1,05% C). The different carbon content leads to different mechanical properties of steel such as strength, ductility, hardness, etc., which determines their application. Thus, for example, steel with a content of about 0,08 wt% C is good to form and has good ductility, steel with 0,18 wt% C is useful for general application and good for welding, etc.

Carbon steels are used in a wide range of applications, such as structural components, industrial pipes, pipelines that supply gas and oil, construction and others. In general, they are susceptible to corrosion under the conditions of industrial operations.

An important consequence of steel corrosion is the deterioration of its mechanical properties. Change in steel behavior can lead to unexpected problems and even to the undesired brittle failure due to a reduced cross section [2-4]. That is why it is of great importance to study the impact of corrosion on the mechanical properties in order to be able to make a correct assessment of the corroded steel structures. This allows for the building of models for adequate forecasting and evaluation of the behavior and safety of these constructions.

It has been found out that mechanical treatment impacts the corrosion resistance of steel. Deformation conditions, such as hot rolling and cold rolling parameters affect texture development in steel. Rolling schedule, rolling temperature, reheating time and temperature, etc., are some of the important parameters to consider for texture development, which can cause significant changes to material properties. Surface texture (also referred to as preferred orientations), is one of the important parameters related to corrosion. Other parameters such as microstructure, chemical composition, defects and surface energy of crystal planes, also affect the corrosion properties of steel [1].

Hot plastic deformation is carried out in temperature-rate conditions, which lead to full recrystallization during the process, and does not cause work strengthening of metals and alloys. Cold plastic deformation takes place in temperature-rate conditions, which do not allow for recrystallization to occur. This leads to significant work strengthening of metals and alloys [5].

The purpose of this paper is to study the mechanical characteristics and corrosion behaviour of two varieties of one of the most widely used non-alloy structural steel – S235. The steel is formed into rods with a diameter of 6 mm. In one variety the rods are hot rolled, and the material has the standard designation S235JR, BDS EN 10025-2: 2005. In the other variety the rods are cold drawn, and the material has the standard designation S235JRC, BDS EN 10277-2: 2008. The mechanical characteristics are determined before the beginning of the corrosion process. Corrosion behaviour is studied in a model corrosion

* Corresponding author: tzonev@tu-sofia.bg

solution – 3,5% water solution of NaCl. The results presented here belong to the initial stage of a comprehensive study of the impact of corrosion on the mechanical properties of steel S235JR and S0235JRC rods.

2 Specific features of hot rolling and cold drawing

Hot rolling is a process, which removes defects from previous technological operations in manufacturing the workpiece and achieves the required form and quality of the surface. In hot screw rolling (Fig. 1*a*) the workpiece passes through crossed rolls, which rotate in the same direction and give it a rotary and progressive motion. The process of deformation leaves a spiral trace on the workpiece as can be seen in the picture of the rod in Fig.1*b*.

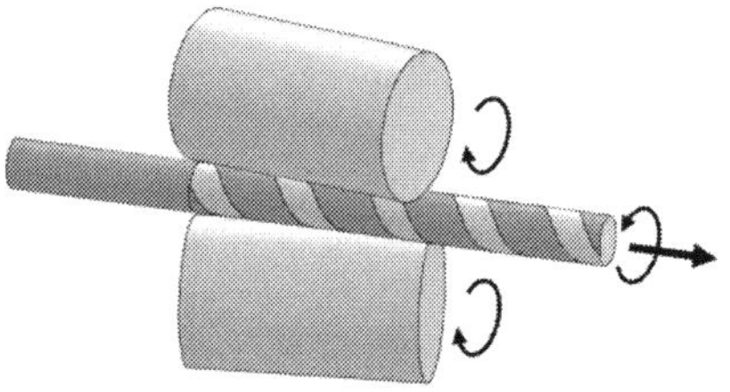

a)

b)

Fig. 1. Hot screw rolling. *a*) Basic schematic view of the process; *b*) Picture of the surface of the rod, from which the tested test pieces have been made.

Drawing through a nozzle is a technological process of metal working using plastic deformation, where workpieces are deformed by compulsory passing them through an opening under tensile force [5], Fig. 2*a*. An exact form and size of the cross section of the workpieces is achieved as well as clean and smooth surface. Fig. 2*b* shows a picture of the surface of cold drawn steel.

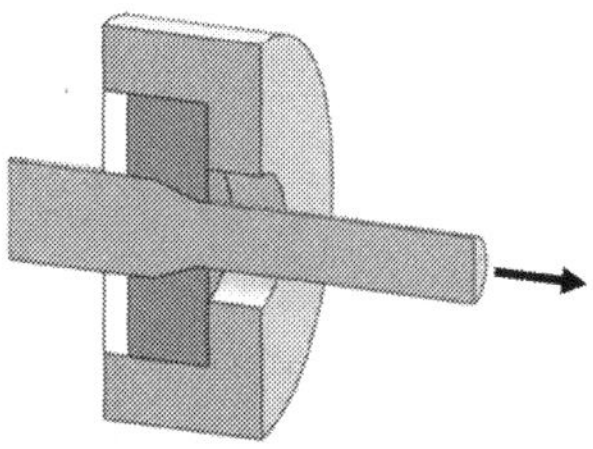

a)

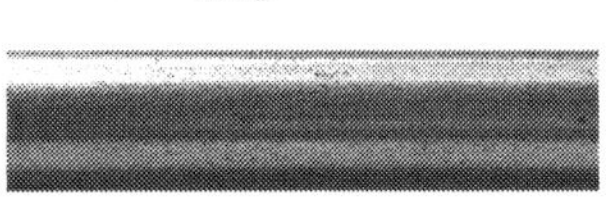

b)

Fig. 2. Cold drawing. *a*) Schematic view of the process of drawing through a nozzle; *b*) Picture of the surface of the rod, from which the tested test pieces are made.

In order for the process of drawing to take place the stress created by the drawing force in the already drawn part of the workpiece must not reach the yield limit. That condition can be satisfied only if plastic deformation goes together with work hardening. This means that the process of drawing through a nozzle cannot take place in the conditions of hot plastic deformation. Plastic deformation and work hardening in cold drawing considerably change the stress-strain curve of material and its strength-deformation characteristics. The changed structure and surface impact its corrosion behaviour.

3 Chemical composition

The chemical composition of the tested test pieces is given in Table 1. The table compares them with the standard values too. As can be seen, the chemical composition as a whole corresponds to the stipulations in the standard. There are differences between the two steels, however. In cold drawn steel S235JRC, the content of C and Mn is considerably lower than the limit value, but on the other hand small quantities of Si and Cr have been added. The composition of the test pieces of S235JRC is better than that of the hot rolled steel S235JR in its content of sulphur and phosphorus – apart from the fact that the content of these elements is lower in the cold drawn material, it is considerably lower than the limit value set in the standard. Obviously, the chemical composition has been changed in order to improve deformability, which is very important for cold drawing. Variations in the chemical composition, which are permitted in the respective standards, however, inevitably affect the mechanical properties and the corrosion behaviour. Table 1 shows that with the commonly used types of steel, even when they fully meet the standards, it is necessary to test each batch if their properties are to be defined exactly.

Table 1. Chemical composition of the test pieces and standard values.

Element	Hot rolled steel S235JR		Cold drawn steel S235JRC	
	Test pieces	EN10025-2: 2005, max	Test pieces	EN10277-2: 2008, max
C, %	0,17	0,17	0,08	0,17
Si, %	-	-	0,2	-
Mn, %	1,4	1,4	0,08	1,4
P, %	0,035	0,035	0,027	0,04
S, %	0,035	0,035	0,027	0,04
Cr, %	-	-	0,1	-
Cu, %	0,5	0,55	-	0,55
N, %	0,012	0,012	-	0,012

4 Mechanical tests

In order to determine the mechanical properties of the studied steel we have conducted tensile tests on a modernized tensile testing machine ZD10 with screw loading mechanism, which through digital control allows for precise loading (Fig. 3), in accordance with BS EN ISO 6892-1:2016. The results of these tests are needed to compare the two grades of steel, as well as a starting point for determining the change in the mechanical properties when corrosion develops.

 https://doi.org/10.1051/matecconf/201823404001

Three test pieces of each of the studied steels were tested. In order to measure the test piece extension we have used self-supporting extensometer SCHENCK with a gauge length of 25 mm. The test pieces are subjected to pure tension and the tensile force and longitudinal elongation (extension) are recorded. The set testing speed is 1050 kg/min.

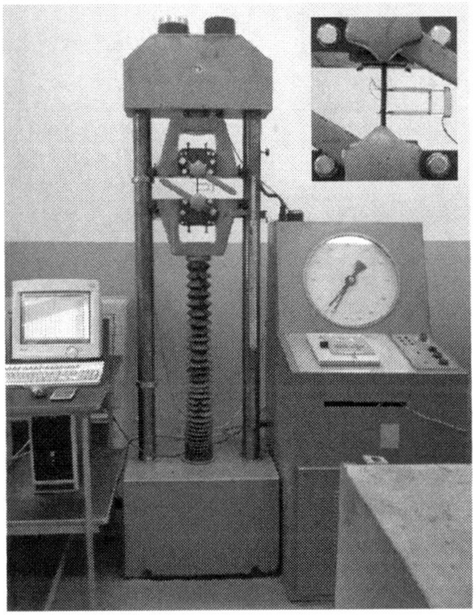

Fig. 3. Modernized mechanical tensile testing machine ZD10. The force is measured with precise 100 kN load cell, the extension with a longitudinal extensometer. The signals are recorded in a computer with special software.

Fig. 4 shows the stress-percentage extension curves of the studied steels. The damage of the test pieces has occurred in the gauge length of the extensometer. Table 2 presents the values of the following parameters determined in Fig. 4:

- Tensile strength R_m – the stress corresponding to the maximum force, which the test piece can bear during testing;
- Upper yield strength R_{eH} – the value of stress when the first decrease in the force is observed. It is determined for hot rolled steel, in which there is physical yielding;
- Proof strength, plastic extension $R_{p0,2}$ – the stress, under which the relative percentage extension is equal to 0,2%. It is determined for cold drawn steel, which does not have a physical yield limit;
- Modulus of elasticity E – the tangent of the angle between the linear part of the stress-percentage extension curve and the *x*-axis.

In Fig. 4 and Table 2 it can be seen that the two grades of steel have different stress-percentage extension curves. The differences are as follows:

- Lack of yielding in S235JRC;
- Increased by 33% yield strength of S235JRC as compared to S235JR;
- Slightly higher tensile strength (12%) of S235JRC as compared to S235JR;
- Considerably decreased percentage of total extension at fracture (61%) in S235JRC as compared to S235JR.

These differences are foreseeable and can easily be explained with the lack or presence of previous plastic deformation, as well as with the differences in the chemical composition of the tested grades of steel.

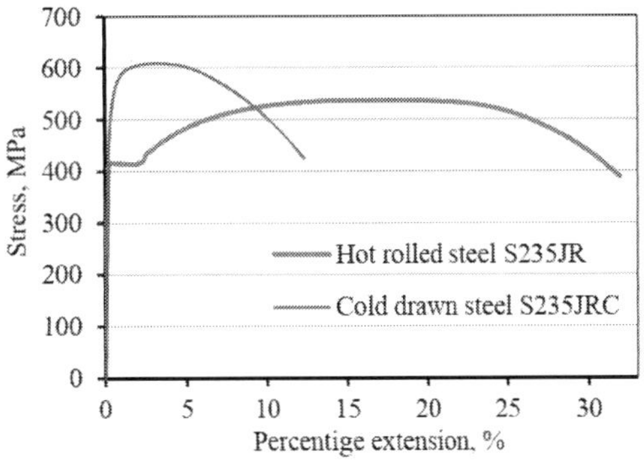

Fig. 4. Stress-percentage extension curve of S235JR and S235JRC steel under tension.

Table 2. Material constants of S235JR and S235JRC steel, determined in Fig. 4.

Value	Yield strength		Tensile strength		Elastic modulus	
Designation, unit	R_{eH}, MPa	$R_{p0,2}$, MPa	R_m, MPa		E, GPa	
Steel S235	JR	JRC	JR	JRC	JR	JRC
Test 1	416	558	537	608	206	208
Test 2	419	564	546	612	204	204
Test 3	421	554	546	606	208	212
Average	**419**	**559**	**543**	**609**	**206**	**208**

5 Corrosion behavior

Corrosion tests are carried out under the standard NACE ASTM TM0169/G31-12a procedure. The shape and size of the test pieces subjected to corrosion testing are shown in Fig.6. Three test pieces have been used for each measurement. A picture of a group of test pieces, cleaned after the experiment, is shown in Fig. 6b.

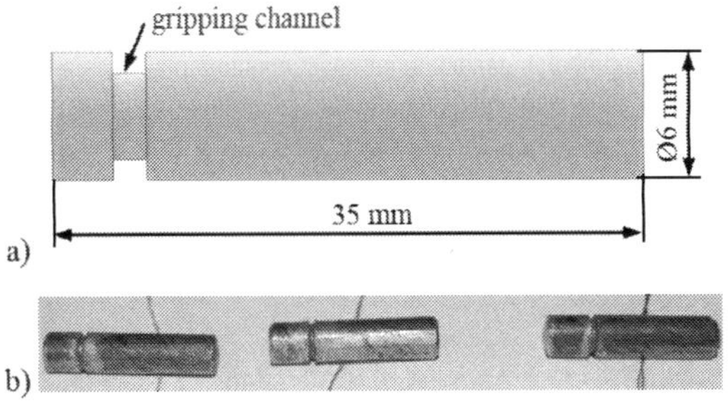

Fig. 6. Test pieces for corrosion testing. a) Overall dimensions; b) General view of the test pieces from S235JR steel, immersed in corrosive solution for 35 days.

Fig. 7 presents four groups of 3 test pieces immersed in corrosive solution. The specific colouring of the solution and the sediment from corrosion products can be seen.

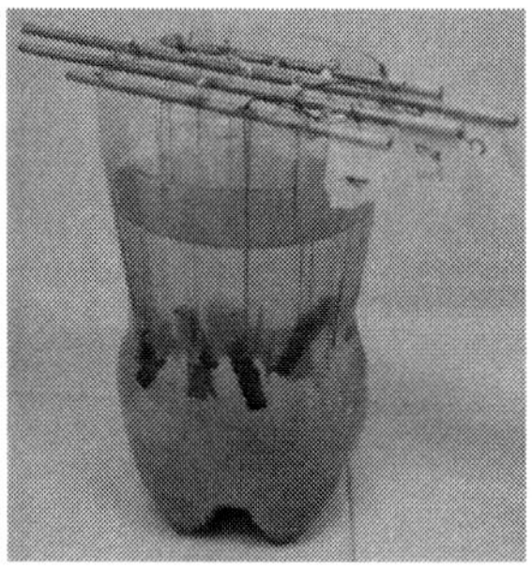

Fig. 7. Test pieces from S235JR steel, immersed in corrosive solution.

Before immersion the test pieces are subjected to the following cleaning procedure: degreasing in alkaline solution (pH 12) at 65°C for 15 min; cleaning with distilled water; submerging in acid solution (HCl + H_2SO_4); rinsing in distilled water; rinsing with ethanol to remove the residual water. After that the test pieces were weighed with an analytical balance (accuracy of 0,001g) before the immersion tests (weight $w_{b,i}$, $i = 1, 2, 3$).

The immersion test was performed in order to use the "weight loss method". A 3,5% NaCl (p.a.) water solution was used for this purpose, as it is known [6, 7] that this concentration is more aggressive to iron and carbon steel than natural seawater. All the experiments were performed in open air conditions and at room temperature (22°C).

The test pieces were immersed in the solution from 1 to 5 weeks. The products of corrosion formed on the surface of the test pieces, after the experiments, were removed mechanically with a soft material. This procedure is done very carefully so that only corrosion products are removed. After that they were rinsed in distilled water, dried with ethanol and weighed with the balance (weight $w_{a,i}$).

The percentage weight loss Δw_i between the initial and final states was obtained for each sample and the average value Δw for the three pieces was calculated as follows:

$$\Delta w = \frac{1}{3}\sum_{i=1}^{3} \frac{w_{b,i} - w_{a,i}}{w_{b,i}} 100, \%. \quad (1)$$

The percentage weight loss rate v was calculated as

$$v = \frac{\Delta w}{t}, \frac{\%}{\text{day}}, \quad (2)$$

where t, in days, is the time of exposure.

The experimentally obtained Δw and v for the two tested grades of steel are presented in Table 3 and Figures 8 and 9.

The data presented here demonstrate that the corrosion behaviour of both grades of steel in the studied interval is similar. The rate of weight loss for both grades of steel is the highest in the first week (higher in steel S235JRC) with following decrease reaching a minimum in the 3rd week, after which there is new increase. The data at the end of the experiment show a trend for a higher rate of weight loss in cold rolled steel, which can be explained with the structural changes in cold drawing and the presence of residual stress from plastic deformation in the surface layer. The trend can be easily visualized as the correlation in Figures 8 and 9 can successfully be approximated by square function. That trend will be tested for an interval of one year in a new study, which has already begun, of both grades of steel.

Table 3. Percentage weight loss Δw and percentage weight loss rate v as a function of exposure time t.

Time t, days	Steel S235JR		Steel S235JRC	
	Δw, %	v, %/day	Δw, %	v, %/day
7	0,11	0,0157	0,17	0,0243
14	0,2	0,0143	0,23	0,0164
20	0,26	0,0130	0,29	0,0145
28	0,4	0,0143	0,43	0,0154
35	0,52	0,0149	0,55	0,0166

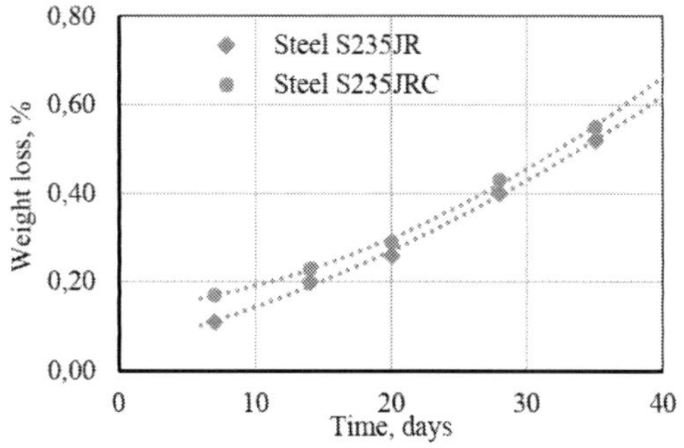

Fig. 8. Percentage weight loss Δw as a function of exposure time.

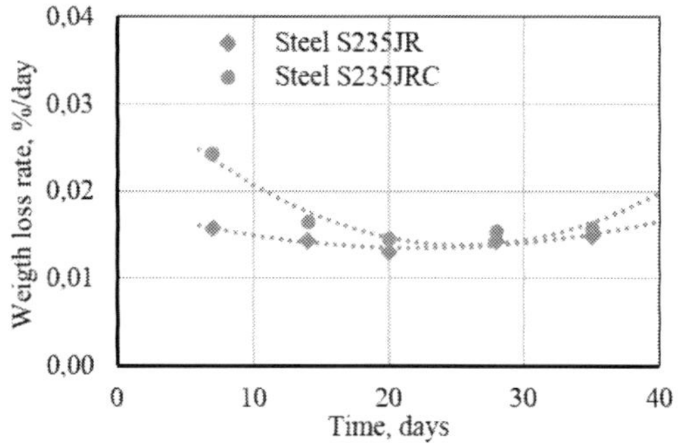

Fig. 9. Percentage weight loss rate v as a function of exposure time.

All test pieces, starting in the first week of the study, were covered with a porous layer of brown corrosion products, which is easy to clean. After it is removed the steel surface remains partially covered with black dense film composed of magnetite, which can be seen in Fig. 6b [6, 8].

The presence of a minimum in the change of the weight loss rate over time (Fig. 9) can be explained in the following way [8]: in neutral water solution and at ambient temperatures, dissolved oxygen is necessary for appreciable corrosion of iron. In air-saturated water, the initial corrosion rate may be high but diminishes over a period of days as the iron oxide film, acting as a barrier to oxygen diffusion, is formed. A study [9] explains that iron corrosion is initiated by pits and spreads out in the form of corrosion product layers, one of which is poorly adherent. According to the authors, the role of the occluded pits is negligible as compared with the porous film effect. At the corrosion potential two diffusion processes are detected: the diffusion across the layer and the mass transport in the liquid phase. The degree of coupling between the two processes is dependent on the immersion time. The oxygen transport is produced not only in the liquid phase but also through the porous layer of the corrosion products. Further increase in the weight loss rate can be related to the impact of chloride ions on the dense film formed. According to [10, 11] these ions enter the passive film and destroy the film only when a sufficiently high concentration gradient is created on the surface, which requires some time.

6 Conclusion

The paper presents the results from tension tests of two grades of steel – S235JR and S235JRC, in the shape of rods with a diameter of 6 mm, which are similar in chemical composition and designation, but differ in the way the metal has been worked and in the mechanical properties. We have determined the qualitative and quantitative differences in the stress-strain curves of both materials and have discussed the reasons for having them. Cold drawing has led to an increase in the yield limit by 33%, which allows for just as higher operation loads to be applied to the S235JRC steel rods. The S235JR hot rolled rods have considerably better plastic properties and can bear by 61% greater total deformations at fraction.

In order to determine the corrosion behaviour of the two grades of steel 15 test pieces from each grade were tested. They were immersed in 3,5% water solution of NaCl from one to five weeks. Each test piece was carefully cleaned and weighed before and after immersion in a corrosive medium. The results from the experiment have been presented, analyzed and explained. As expected the two types of material have similar corrosion behaviour but the cold drawn rods corrode faster.

The findings presented here are from the initial stage of an extensive study of the impact of corrosion on the mechanical properties of steel. The results from the tension tests will be used to be compared with the results from the mechanical tests of the test pieces immersed for a longer time in the same solution. The results of the 5-week test of corrosion behaviour indicate that the test pieces need to remain in the solution much longer for corrosion to be of an extent that will significantly affect the stress-strain curves. This led to the preparation and the start of a new experiment, where large groups of test pieces from the two types of material will be immersed in the solution from 2 to 12 months.

This work is part of research project № 182ПД0022-04 in support of PhD students funded by the Research Centre at Technical University – Sofia.

References

1. D. Dwivedi, K. Lepková, T. Becker, Carbon steel corrosion: a review of key surface properties and characterization methods, RSC Advances, **7**(8), 4580-4610 (2017)
2. A. Almusallam, Effect of degree of corrosion on the properties of reinforcing steel rods, Construction and Building Materials, **15**(8), 361-368 (2001)
3. H. Lee, Y. Cho, Evaluation of the mechanical properties of steel reinforcement embedded in concrete specimen as a function of the degree of reinforcement corrosion, International Journal of Fracture, **157**(1-2), 81-88 (2009)
4. A. Shetty, I. Gogoi, Performance enhancement of TMT rebar in accelerated corrosion, Journal of Civil Engineering Research, **2**(1), 14-17 (2012)
5. N. Tonchev, *Material science and materials technology* (in Bulgarian), VTU "T. Kbleshkov", Sofia (2010)
6. H. Möller, E. Boshoff, H. Froneman, The corrosion behaviour of a low carbon steel in natural and synthetic seawaters, Journal of the Southern African Institute of Mining and Metallurgy, **106**(8), 585-592, (2006)
7. S. Dexter, Corrosion in seawater. In: *S. Cramer, B. Covino (Edts.), ASM Handbook, 13C Corrosion: environments and industries*, ASM International, Ohio, 27-41 (2006)
8. R. Revie, *Corrosion and corrosion control: an introduction to corrosion science and engineering* (John Wiley & Sons, 2008)
9. D. You, N. Pebere, F. Dabosi, An investigation of the corrosion of pure iron by electrochemical techniques and in situ observations, Corrosion science, **34**(1), 5-15, (1993)
10. L. Li, C. Wang, S. Chen, X. Yang, Investigation of oscillatory electro dissolution processes by chloride ion perturbation, Proc. of 16th International Corrosion Congress, Beijing, China (2005)
11. L. Li, J. Luo, B. Lu, S. Chen, Effect of interface chloride ion perturbation on oscillatory electrodissolution. Electrochimica acta, **50**(16-17), 3524-3535 (2005)

https://doi.org/10.1051/matecconf/201823404002

Using statistical methods to determine the load-bearing capacity of rectangular CFST columns

Glib Vatulia[1*], *Yevhen* Orel[1], *Maryna* Rezunenko[2], and *Nataliia* Panchenko[2]

[1]Ukrainian State University of Railway Transport, Structural Mechanics and Hydraulics Department, 7 Feuerbach Sq., 61050 Kharkiv, Ukraine
[2]Ukrainian State University of Railway Transport, Applied Mathematics Department, 7 Feuerbach Sq., 61050 Kharkiv, Ukraine

Abstract. In the current practice of construction and design of transport facilities, structures with external reinforcement are commonly used which effectively resist compression. The use of steel-concrete and composite structures enables us to reduce material consumption and cost of structures significantly. There are a few established approaches used to evaluate the load-bearing capacity of steel-concrete structures under axial and eccentric compression, each being based on the initial prerequisites, which underlie the calculation formulas. In this paper, the functional relationship of the value of the maximum load-bearing capacity of rectangular concrete-filled steel tubular (CFST) columns under axial compression with the random eccentricity is plotted. A regression model is proposed based on the methods of mathematical statistics, which allows for the evaluation of the impact of geometrical and physical characteristics of rectangular CFST columns on the value of their load-bearing capacity. The correspondence of the obtained model to the experimental data, as well as the significance of the regression parameters are confirmed by Fisher and Student criteria.

1 Introduction

Due to the rapid development of scientific and technological progress, almost any structural solutions of buildings and structures can be implemented. The introduction into the practice of construction of composite structures, the elements of which are made of reinforced concrete and/or composite reinforced concrete, reduction in material, power consumption and the duration of construction works can be achieved [1-4].

Over the past four decades, researchers from the United States, Japan, China, Russia, Ukraine and other countries have conducted multiple experimental studies on the exhaustion of load-bearing capacity and the assessment of stress-strain state of steel-concrete (tubular concrete) columns.

The priority issue of structure designing is ensuring the strength and reliability of the structures, which should be solved, first of all, on the basis of calculation methods that satisfy all the requirements, according to numerous studies in this field [5-13]. The authors propose to use regression analysis as one of these methods – a statistical research method that allows for evaluation of the degree of the impact of independent factors on the dependent variable.

2 Purpose and methods of the study

The purpose of this study is to plot a functional dependence of the maximum load-bearing capacity of rectangular CFST columns under axial compression with random eccentricity on various physical and geometric characteristics.

Experimental data obtained by researchers from different countries over the past decades were used as a material for the statistical analysis of the load-bearing capacity of columns.

Such a methodology has already been used by the authors to find the maximum load-bearing capacity of various circular CFST columns [14], as well as long square columns [15].

The method of correlation regression analysis with the parameters listed below was used to solve the problem:

- tube thickness $t \geq 2{,}0$ mm;
- ratio of cross-sectional dimensions $0{,}5 \leq b/h \leq 1$ at $90 \leq b \leq 325$ mm;
- ratio $L/b \leq 4$, at $300 \leq L \leq 1220$ mm;
- yield strength of steel $200 \leq \sigma_y \leq 490$ MPa;
- modulus of elasticity of steel and concrete $1{,}68\times10^5$ MPa $\leq E_s \leq 2{,}17\times10^5$ MPa and $2{,}0\times10^4 \leq E_c \leq 4{,}4\times10^4$ MPa, respectively;
- prism strength of concrete $16 \leq f_{c,prizm} \leq 110$ MPa.

A correlation analysis of the effect of each parameter on the value of the bearing capacity has been conducted. However, it was shown that the combined influence of several factors should be taken into account for plotting an adequate mathematical model. Thus, the following regression equation (1) could be obtained:

$$\overline{N} = a_0 + \sum_{i=1}^{k} a_i x_i, \qquad (1)$$

* Corresponding author: glebvatulya@gmail.com

 https://doi.org/10.1051/matecconf/201823404002

where $\overline{N}$ is approximate maximum value of CFST columns load-bearing capacity,

$$x_1 = b\,h\,f_{c,prizm}\,,\; x_2 = L\sigma_y\,,\; x_3 = t^3\sigma_y\,E_s\,,\; x_4 = \frac{(b-t)t^2}{E_c}$$

The adequacy of the obtained model is proven by the coefficient of determination R^2 being 0,998, as well as the mean approximation error:

$$\Delta = \frac{1}{n}\sum_{i=1}^{n}\frac{\left|N_{\exp} - N_{theor}\right|}{N_{\exp}}100\% \approx 5{,}92\%, \qquad (2)$$

where N_{theor} is value of CFST columns load-bearing capacity calculated using formula (1); N_{exp} – experimental value of load-bearing capacity.

Microsoft Office Excel software was used for the calculations.

Fig. 1 presents the experimental and calculated data of the values of the load-bearing capacity of columns.

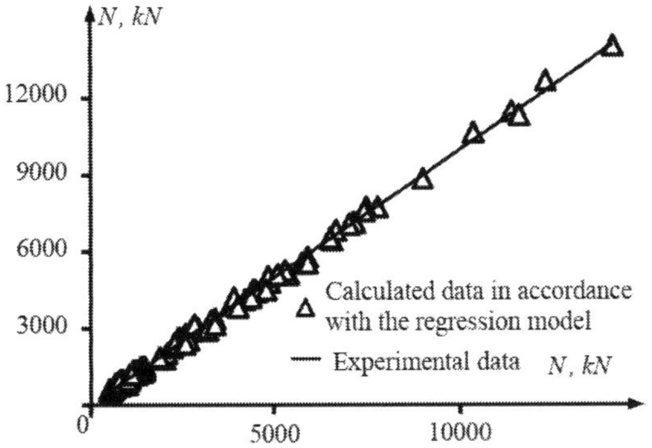

Fig. 1. Load-bearing capacity of columns.

3 Analysis of the results

The equation for multifactor regression was obtained on the basis of 134 observations of rectangular CFST columns [7, 10, 13]:

$$\overline{N} = -68{,}332 + 7{,}19\times10^{-4}x_1 + 4{,}76\times10^{-3}x_2 + \\ +1{,}34\times10^{-5}x_3 + 4{,}31x_4. \qquad (3)$$

To answer the question, how much the obtained regression coefficients may differ from the exact values, confidence intervals for each coefficient were found with a probability $P = 0{,}95$:

$$-107{,}687 \le a_0 \le -28{,}978;$$
$$7{,}01\times10^{-4} \le a_1 \le 7{,}38\times10^{-4};$$
$$4{,}473\times10^{-3} \le a_2 \le 5{,}042\times10^{-3};$$
$$1{,}13\times10^{-5} \le a_3 \le 1{,}54\times10^{-5};$$
$$3{,}875 \le a_4 \le 4{,}745.$$

The authors have analyzed the obtained equation using the Fisher and Student criteria, which confirm the correspondence of the obtained model to the experimental data and the significance of the regression parameters.

In Table 1, the correlation coefficients of the factors were presented.

Table 1. Correlation matrix.

	$\overline{N}$	x_1	x_2	x_3	x_4
$\overline{N}$	1,000	0,926	0,765	0,595	0,538
x_1	0,927	1,000	0,522	0,433	0,501
x_2	0,765	0,522	1,000	0,638	0,408
x_3	0,595	0,433	0,638	1,000	0,634
x_4	0,538	0,501	0,408	0,634	1,000

Analysis of residuals (N_{exp}–N_{theor}) using Pearson's criterion shows that they are normally distributed with the zero mathematical expectation and corrected mean square deviation S = 133,953.

Fig. 2 shows the bar graph of residuals. All values of the sample lie within [–3S; 3S], while 72% residuals belong to [–S; S], 95% are within [–2S; 2S].

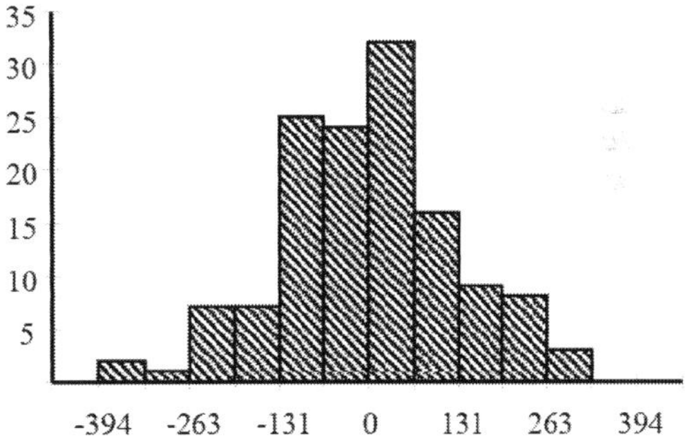

Fig. 2. Bar graph of residuals.

Dependence of the absolute error $\frac{\left|N_{\exp} - N_{theor}\right|}{N_{\exp}}$ on individual parameters is shown in Fig. 3.

The analysis of the regression model factors shows that the main contribution to the value of the load-bearing capacity of CFST columns is given by the first summand $bhf_{c,prizm}$ (Fig. 5).

Such dependence is well approximated with the linear function $y = 0{,}0012x + 818{,}65$ and gives the coefficient of determination $R^2 = 0{,}8619$.

Thus, it can be stated that 86% of the change of the load-bearing capacity is due to the combined impact of the cross-sectional area and the value of the prism strength of the concrete, but this is insufficient to obtain a more accurate profile of the behaviour of columns, which leads to the need to take into account other characteristics (length, yield strength of steel, modulus of elasticity of steel and concrete) and their combined influence.

The plotted model enables us to estimate the degree of effect of the parameters of rectangular CFST columns on the value of the bearing capacity. Fig. 4 below shows the dependence of the increase in the load-bearing capacity (%) on an increase in the tube thickness by 10% for various characteristics of columns.

Fig. 4 illustrates that an increase in the tube thickness of 10% in some cases can cause an increase in the load-bearing capacity of the column of up to 17%.

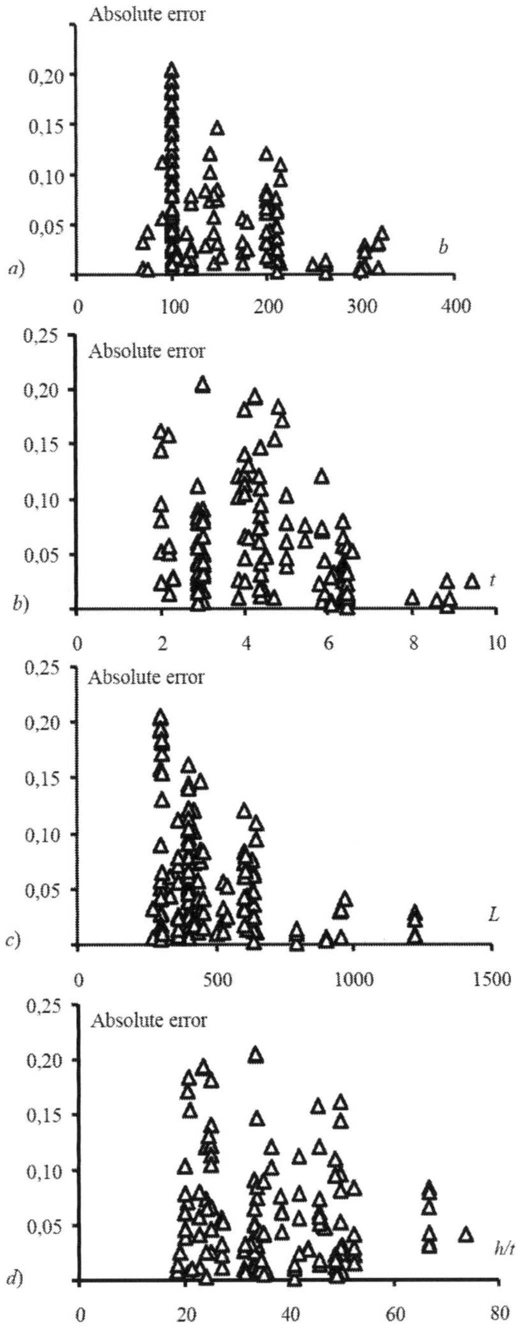

Fig. 3. Dependence of the absolute error of the regression model: *a*) on *b*; *b*) on tube thickness *t*; *c*) on column length *L*, *d*) on *h/t* value.

The adequacy of the proposed model is confirmed by the authors' comparative analysis of the error in the calculated value of the load-bearing capacity of columns calculated using the proposed method and the one obtained using Eurocode 4 calculation procedure.

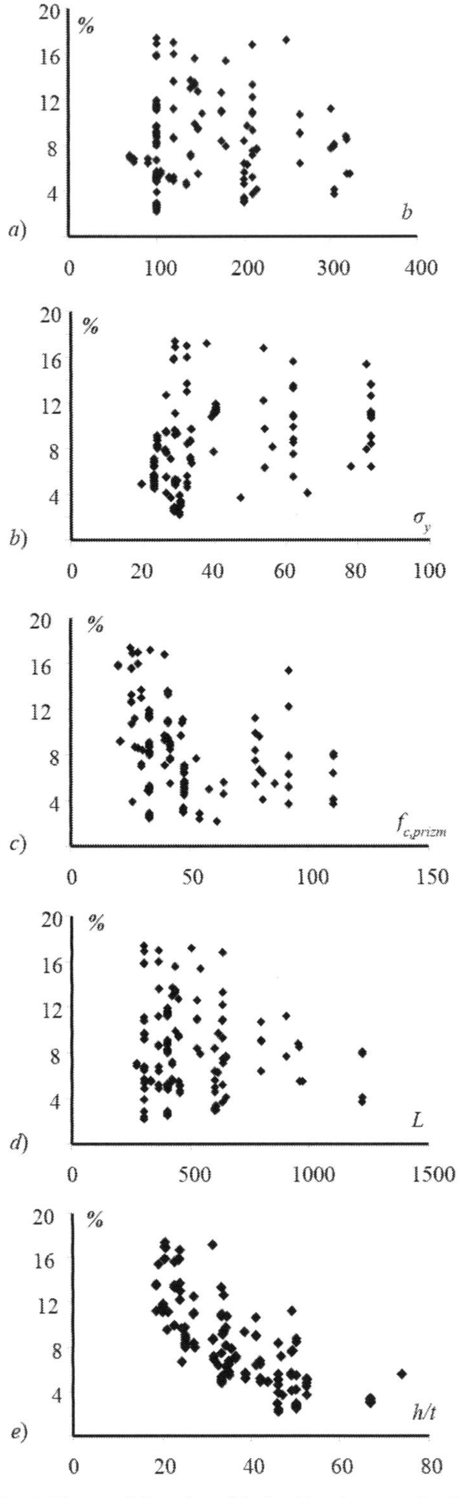

Fig. 4. Change of the value of the load-bearing capacity (%) when the tube thickness is increased by 10%.

https://doi.org/10.1051/matecconf/201823404002

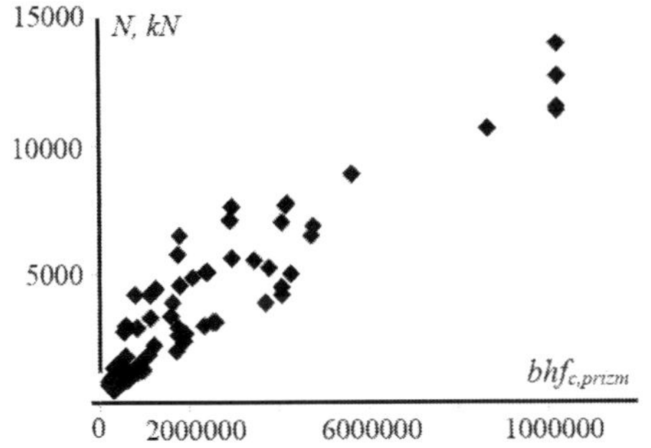

Fig. 5. Dependence between the maximum load-bearing capacity and $bhf_{c,prizm}$.

In Fig. 6, the triangles correspond to the error obtained using the statistical analysis methods, while the squares – to those of Eurocode 4.

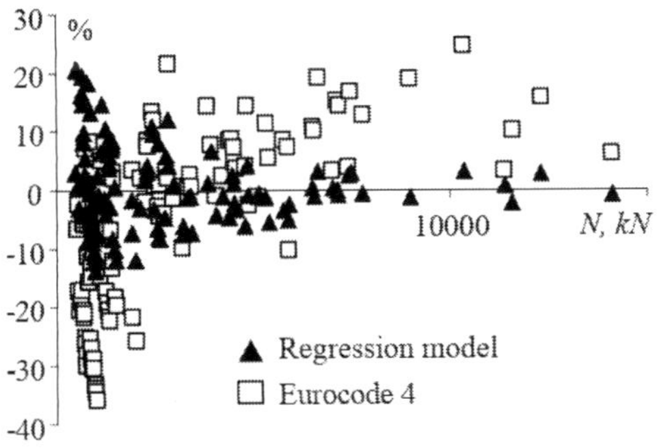

Fig. 6. Error of computation (%).

4 Conclusions

As a result of the study, a functional dependence of the maximum value of the load-bearing capacity of rectangular CFST columns under axial compression has been obtained. The obtained dependence has been analysed using the methods of mathematical statistics.

The comparative analysis of the theoretical and experimental results confirms a sufficient reliability of the obtained model within 6%.

The authors performed a comparative analysis of the calculated value of rectangular CFST columns load-bearing capacity with the value calculated using the methods of Eurocode 4.

The proposed model helps us to predict the value of the load-bearing capacity using different initial parameters together with columns optimization.

References

1. E. Chikhladze, G. Vatulia, Y. Kitov, P. Basis for calculation and design of composite and steel-concrete constructions, Transport of Ukraine, 104 (2006)
2. T. Yamamoto. Experimental study of the size effect on the behaviour on concrete filled circular steel tube columns under axial compression, J. Struct. Constr. Eng., Transactions of AIJ, **561**, 237-244 (2002)
3. M. Mursi, B. Uy, M. Bradford, Interaction buckling of concrete filled columns using high-strength steel, *ASSCCA'03 International Conference Advances in Structures (ASCCS-7)*, Sydney, Australia, 863-869 (2003)
4. L. Han, T. Zhong, L. Wei, Effects of sustained load on concrete-filled hollow structural steel columns, Journal of Structural Engineering ASCE 9, 1392-1404 (2004)
5. A. Krishan, A. Melnitchuk, Strength and deformability of short tube confined concrete columns of square cross section, News of KSUAE, **3**(29), 6-50 (2014)
6. L. Storozhenko, D. Ermolenko, O. Lapenko, Tube confined concrete, ASMI, **LLC**, 306 (2010)
7. I. Nishiyama, S. Morino, K. Sakino, H. Nakahara, *Summary of Research on Concrete-Filled Structural Steel Tube Column System Carried Out Under The US-JAPAN.* Cooperative Research Program on Composite and Hybrid Structures, 176, (2002)
8. *Design of composite steel and concrete structures – Part 1-1: General Rules and Rules for Bridges*, EN 1994-1: 2001. Eurocode 4, Brussels, 96 (2006)
9. M. Xinbo, S. Zhang, C. Goode, Comparison of design methods for circular concrete filled steel tube columns in different codes, *8th International Conference on Steel-Concrete Composite and Hybrid Structures, Harbin*, Harbin University of Science and Technology, 30-37 (2006)
10. A. Kuranovas, D. Goode, A. Kvedaras, S. Zhong, Load-bearing capacity of concrete-filled steel columns, Journal of Civil Engineering and Management, **15**(1), 21-33 (2009)
11. Y. Hao, Y. Lan, Research on the axial bearing capacity of concrete-filled rectangular steel tube short column, Applied Mechanics and Materials, **368**, 1710-1717 (2013)
12. O. Seraya, D. Demin, Linear regression analysis of a small sample of fuzzy input data, Journal of Automation and Information Sciences, **44**(7), 34-48 (2012)
13. G. Vatulia, D. Petrenko, M. Novikova, Experimental estimation of load-carrying capacity of circular, square and rectangular CFTS columns, Scientific Bulletin of National Mining University, **6**, 97-102 (2017)
14. G. Vatulia, M. Rezunenko, Ye. Orel, D. Petrenko, Regression equations for circular CFST columns carrying capacity evaluation, *DYN-WIND'2017, MATEC Web of Conferences*, **107**, 00051 (2017)
15. G. Vatulia, M. Rezunenko, D. Petrenko, S. Rezunenko, Evaluation of the carrying capacity of rectangular steel-concrete columns. Civil and Environmental Engineering, **14**, 76-83 (2018)

 https://doi.org/10.1051/matecconf/201823404003

Experimental research of steel-concrete load bearing components of bridge column piers

Nataliia Smirnova[1], *Sergiy* Bugayevskiy[2], *Andrii* Ihnatenko[2], *Olena* Synkovska[2,*], and *Maksym* Kovalov[3]

[1]Kharkov National Automobile and Highway University, Road Building and Maintenance Department, 25 Yaroslava Mudrogo st., 61002 Kharkiv, Ukraine
[2]Kharkov National Automobile and Highway University, Department of Bridges, Constructions and Structural Mechanics, 25 Yaroslava Mudrogo st., 61002 Kharkiv, Ukraine
[3]Ukrainian State University of Railway Transport, Structural Mechanics and Hydraulics Department, 7 Feierbakh Sq., Kharkiv, Ukraine

Abstract. A new constructive design of a bridge column pier, which is a load-bearing component of bridges and overpasses as well as of complex multilevel interchanges underlying a transport infrastructure, is proposed. The design is based on a steel mesh casing that is made from a solid-web plate by means of cutting meshes and further expansion of the solid-web plate using non-waste technology. Experimental research data on the bearing capacity and deformability of the cylindrical steel-concrete load bearing components of bridge column piers are provided. They are compared with research data on load bearing components having a solid casing. To make the experimental picture complete, analogous research data on cylindrical concrete load bearing components are provided. This experiment includes static and low-cycle loadings of samples under axial compression mentioned above. On the basis of the information obtained, the effort and travel fields are formed, the critical forces of structure buckling are determined. The analysis of the rationality of the cylindrical steel-concrete load bearing component of bridge column piers is carried out, using the experimental research data. To verify the research data, which are obtained in the course of the study of the critical forces of cylindrical steel-concrete load bearing component buckling, the experimental results are compared with theoretical ones.

1 Introduction

An important constituent of transport infrastructure is the engineering constructions such as bridges, overpasses and complex multilevel interchanges. However, taking into consideration their multicomponent structure, each separate component needs close scrutiny. This paper deals with a bridge column pier.

To build bridge piers, in addition to traditional concrete and reinforced concrete units, various steel-concrete load bearing units are used. As a rule, they are tubes that are filled with concrete. An advantage of a tube-concrete structure is a decrease in concrete consumption 1,5-2 times, as a result, the weight of the structure decreases by 1,8-3 times and working hours halve because there is no rebar placement and no removal of the formwork [1,2,3]. However, steel consumption increases (the percentage of the reinforcement of tube-concrete structures is 3-20%, in this case it is 4,88%); there is a reduced bond between casing and concrete due to the smooth surface of a steel tube as well as high susceptibility of a steel solid casing to mechanical damage and corrosion as a concrete cover is absent.

A steel-concrete class, in addition to tube-concrete units, also includes various systems having confinement (lateral) reinforcement such as radial rings, hoops, spiral winding etc. [4,5]. It should be noted that this reinforcement has both advantages and disadvantages. The advantages are the increase of the bearing capacity, metal and concrete saving. The disadvantages are high labour intensity to manufacture the structures and heavy longitudinal deformation of the structures.

The analysis of the disadvantages of steel-concrete load bearing units having a solid steel casing (tube-concrete units) and the advantages and disadvantages of load bearing units having lateral reinforcement enables us to offer the solution of the cylindrical steel-concrete load bearing unit of a pier having a mesh case (Fig. 1) [6,7].

This casing is a steel plate of the St3PS grade having a mesh with equally distributed cells (Fig. 2).

This plate is made using non-waste technology by cutting and further expanding the areas of a solid steel plate-workpiece using a special press [8]. In this case the thickness is 4 mm. This treatment enables us to decrease the weight of a mesh in comparison with a solid plate of the same dimensions, saving the required strength parameters (Table 1).

In this research we use the PVL 406mesh; steel saving is 52%; the reinforcement percentage of a mesh casing is 2,33%.

It should be noted that in terms of strength properties a steel-concrete structure having a mesh case is between an analogous reinforced concrete structure and an analogous tube-concrete one, integrating their best qualities.

* Corresponding author: iglema@meta.ua

https://doi.org/10.1051/matecconf/201823404003

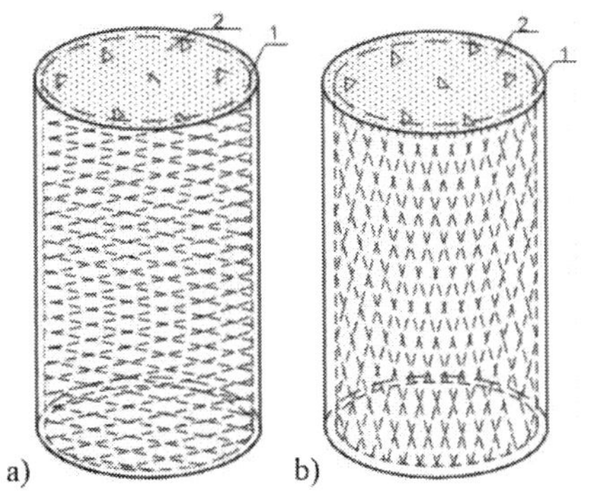

Fig. 1. Steel-concrete load bearing unit having a mesh as casing: a) cells are located across the generatrix; b) cells are located along the generatrix.

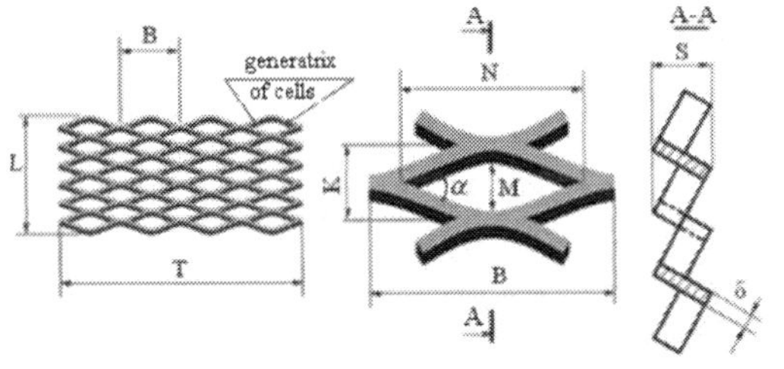

Fig. 2. Main geometric characteristics of an expanded metal mesh.

So, a mesh case carries out the function of dispersed reinforcement in addition to its main function due to volumetric rhombic cells. It differs dramatically from externally similar composite structures that are made by means of layer-by-layer winding of steel cords as a net or spiral. At the same time, an improved bond between the plate and a concrete core, even without special anchors, as well as increased corrosion resistance and fireproof of a structure due to a concrete cover are the features of a structure (Fig. 1). In turn, a variation on both the plate size and the angle of generatrix inclination enables us to set the necessary stiff characteristics to a structure [7, 9]. And it has been proven theoretically that a rational approach to optimally compress the concrete core of a mesh casing is a direction of cells across the generatrix of a steel-concrete load bearing unit [9]; it is experimentally proven in this paper.

2 Methodology of research

Thus, in order to verify the rationality of the structure of the steel-concrete load bearing unit, experimental research is conducted in Kharkov National University of Construction and Architecture [10]. Taking into account the features of the operation of piers in transport infrastructure, the programme of the testing includes central static (STcS) and low-cycle (MTcS) compression of the samples of four types: steel-concrete samples having a solid casing, steel-concrete samples having a mesh casing when cells are located along and across generatrix, and concrete samples without reinforcement (Fig. 3).

All the casings are cylindrical iron cover that is made from solid or mesh steel sheets. Solid casing is made from steel sheet *St3PS2* having a thickness of 4 mm by analogy with the thickness of the plate-workpiece of a mesh plate [8]. The joints of the casing sheets of solid and mesh casings with cells along generatrix are rolled and butt-welded; the joints of the casing sheets of a mesh casing with cells across generatrix are welded using an overlay narrow plate (Fig. 4a).

Concreting of the samples under research is made in two stages: first, the steel-concrete samples having solid and mesh casings with cells along the generatrix and the concrete samples without reinforcement and second, the steel-concrete samples having a mesh casing with cells across generatrix and the concrete samples without reinforcement. The ends of the steel-concrete samples of all types as well as the concrete samples without reinforcement are aligned with the casing; the horizontal level is under strict control. All the samples are stored in laboratory conditions for 28 days at an air temperature of (20±5)°C. The ends of samples are covered by a layer of wet sand to prevent them from contraction cracks.

The concreting of steel-concrete samples having a mesh casing with cells along and across generatrix as well as concrete samples without reinforcement is made using a disposable formwork for monolithic construction such as AMICOTUBE (Fig. 4b). The concreting of the steel-concrete samples having a solid casing needs no additional formwork.

The main power unit to test the building structures is the P-500hydraulic press. The upper base plate of the press is fixed on a movable traverse beam and it has a self-setting spherical bearing to guarantee the accuracy of application of a given compression effort. The loading

Table 1. Steel consumption per 1 m^2 of the casing of a steel-concrete load bearing unit.

Solid casing		Mesh casing (expanded metal mesh)				Steel saving coefficient, %	Weight reduction of casing, %
Thickness of plate-workpiece, mm	Weight, kg/m^2	Plate designation	Thickness of plate-workpiece, mm	Overall thickness *S*, mm	Weight, kg/m^2		
4	31,40	PVL 406	4	12,7	15,7	52,00	34,00
5	39,25	PVL 506	5	13,0	16,4	58,22	39,58
		PVL 508	5	16,8	20,9	46,75	31,79
		PVL 510	5	20,5	24,7	37,07	25,20
6	47,10	PVL 606	6	13,4	17,3	63,27	43,02
		PVL 608	6	17,1	21,9	53,50	36,38
		PVL 610	6	20,8	26,0	44,80	30,46

 https://doi.org/10.1051/matecconf/201823404003

of all the samples takes place using two additional elements – steel disks having a thickness of 30 mm with a diameter equal to the diameter of a concrete core.

During testing measurement of the longitudinal deformations and relative lateral movements (relative to the radius of a certain sample) is made with the help of four clock-type indicators that are fixed on a sample and four digital indicators that are fixed in the middle of the sample height (Fig. 5). The data of a measuring system are read 15 minutes later after having loaded the tested sample with a programmed loading, i. e. after complete stabilisation of the numerical readings of devices. [10].

Loading on all the samples is transferred due to the strict alignment of a structure, a press axis and corresponding additional devices. The tests, which include static axial loadings and low-cycle axial ones, are made until the samples break.

Loading is carried out by means of the phase-by-phase application of load (F, kN) from zero to a set level (including the moment of destruction), and a phase varies within a range of 100-400 kN depending on the type of sample. During low-cycle loading we have full load relief after each phase. The programme of tests predetermines the record of loading cycles.

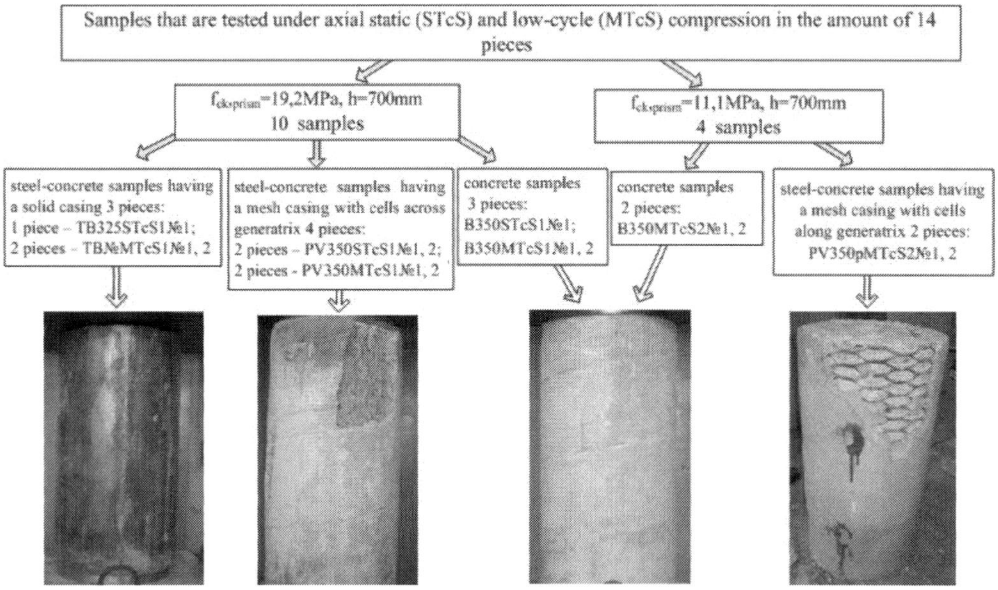

Fig. 3. Structure chart of experimental research of steel-concrete samples of various types and concrete samples without reinforcement under axial static and low-cycle compression.

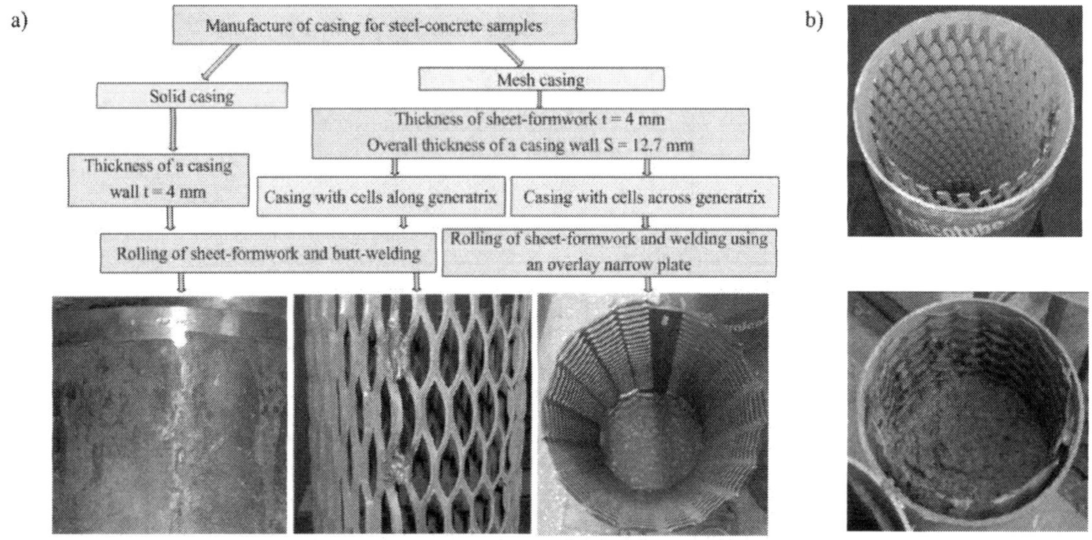

Fig. 4. Features of sample manufacture: a) joining of various steel sheets by welding; b) concreting of samples using a disposable formwork for monolithic construction AMICOTUBE.

 https://doi.org/10.1051/matecconf/201823404003

Fig. 5. Installation scheme of a system to measure longitudinal deformations and lateral movements.

3 Results and discussion

In the course of experimental research the breaking loads of the samples of all types are recorded, their values are recorded in Table 2. Table 2 also contains the results of the theoretical research of the bearing capacity of the cylindrical steel-concrete units having a solid casing and a mesh casing where cells are located along and across the generatrix of structure, using the methodology that is presented in papers [9, 11]. Hence, the theoretical calculation of the bearing capacity of a cylindrical bearing structure having a solid casing as well as a mesh casing is performed as follows:

$$N_{estimated} = A_C\, f_{ck} + \varphi_{sv}\, A_{as,red}\, f_{tk} + \left(\frac{\delta_{as,red}}{R_m} + \mu_s\right) f_{yk}\, A_c, \quad (1)$$

where $N_{estimated}$ is the bearing capacity of a cylindrical steel and concrete component; A_c, $A_{as,red}$ are the square of the lateral section of a concrete core and the reduced square of a steel casing respectively; f_{ck} is characteristic concrete strength under compression; f_{tk}, f_{yk} are the characteristic steel strength value under tension and the value of steel strength on a fluidity limit correspondingly; φ_{sv} is the coefficient that takes into account the decrease of steel-concrete strength due to a welding seam; $\delta_{as,red}$ is the reduced thickness of a steel casing wall; μ_s is the reinforcement coefficient of the lateral section of a cylindrical steel-concrete bearing component $\mu_s = A_a / (A_a + A_{as,red})$; R_m is the mean radius of the steel casing of a cylindrical steel-concrete component.

The bearing capacity increase coefficient is calculated using the ratio between a steel-concrete structure and a corresponding concrete sample without reinforcement.

Table 2. Breaking loads as the results of tests and theoretical calculations.

№	Sample type	Sample designation	Experimental bearing capacity, kN	External diameter of casing, mm	Characteristic strength of concrete in compression $f_{ck,prism}$, MPa	Characteristic strength of steel in tension f_{tk}, MPa	Characteristic strength of steel on limit of yielding flow f_{yk}, MPa	Theoretical carrier capacity, kN	Discrepancy between theoretical and experimental data, ε %	Coefficient of the increase of bearing capacity
1	Solid casing	TB325CTcS1№1	4110	325	19,2	435,6	235,15	4015,15	-2,36	1,67
2		TB325MTcS1№1	3960	325	19,2	435,6	235,15	4015,15	+1,39	1.61
3		TB325MTcS1№2	4020	325	19,2	435,6	235,15	4015,15	-0,12	1,63
4	Mesh casing with cells along generatrix	PV350STcS1№1	1800	325	19,2	365,1	222,77	1748,46	-2,95	1,16
5		PV350STcS1№2	1640	325	19,2	365,1	222,77	1748,46	+6,61	1,06
6		PV350MTcS1№1	1850	325	19,2	365,1	222,77	1748,46	-5,81	1,20
7		PV350MTcS1№2	1750	325	19,2	365,1	222,77	1748,46	-0,09	1,13
8	Concrete	B350STcS1№1	2340	350	19,2	-	-	-	-	-
9		B350MTcS1№1	1800	350	19,2	-	-	-	-	-
10		B350MTcS1№2	2350	350	19,2	-	-	-	-	-
11	Concrete	B350MTcS2№1	1490	350	11,1	-	-	-	-	-
12		B350MTcS2№2	1440	350	11,1	-	-	-	-	-
13	Mesh casing with cells across generatrix	PV350pMTcS2№1	2250	325	11,1	365,1	222,77	2298,36	+2,10	1,61
14		PV350pMTcS2№2	2360	325	11,1	365,1	222,77	2298,36	-2,68	1,67

The results of the experimental research of the bearing capacity of the tested samples confirm the proposed solution of the steel-concrete load bearing unit having a mesh steel casing where cells are located across a structure generatrix for structure piers in transport infrastructure. Thus, in the samples where the cells are located along generatrix the coefficient of the bearing capacity increase is approximately 1,05; it is half of the increase of the bearing capacity in the samples having a mesh casing with cells across the generatrix. In addition, the concentration of the values of the critical efforts of buckling confirms the adequacy of the experimental research. The comparison of the experimental and theoretical bearing capacity values indicates a discrepancy of about 5%. It is worth noting that both the steel-concrete sample in a casing that is made from a solid steel sheet and the sample in a mesh casing with cells across generatrix have approximately equal increase in bearing capacity (by 1,61-1,67 times). At the same time, a mesh casing needs half the iron of a solid casing.

The results of the research are displayed in the diagrams (Fig. 6) where the correlation between the sample longitudinal deformations and the level of low-cycle loading under axial compression are shown.

So, the diagrams of the correlation between the longitudinal deformations and the level of external loading indicate that under low-cycle axial compression both a solid steel casing and a mesh casing with cells along the generatrix decrease the longitudinal deformations of the samples approximately two times as compared to an analogous concrete unit, whereas a mesh casing with cells across generatrix does not influence the value of longitudinal deformations.

In addition, the diagrams of the casing between relative lateral movements and the level of low-cycle loading under axial compression are made for the same samples (Fig. 7).

The diagrams of the casing between the relative lateral movements and the level of low-cycle axial compression indicate that the steel-concrete samples having a mesh casing with cells along the generatrix have movements that are decreased by 1,5 times, and the samples having cells across the generatrix demonstrate movements that are decreased by 4 times as compared with analogous movements of the concrete sample.

4 Conclusions

The results, which are obtained in the experimental and theoretical research, enable us to put into practice the proposed structures of cylindrical steel-concrete column piers for engineering constructions in transport infrastructure. The use of the structure, which has cells across the generatrix, enables us to save 52% of steel as compared with a solid steel casing of analogous thickness. At the same time, bearing capacity increases by 1,6 times as compared with concrete samples without reinforcement.

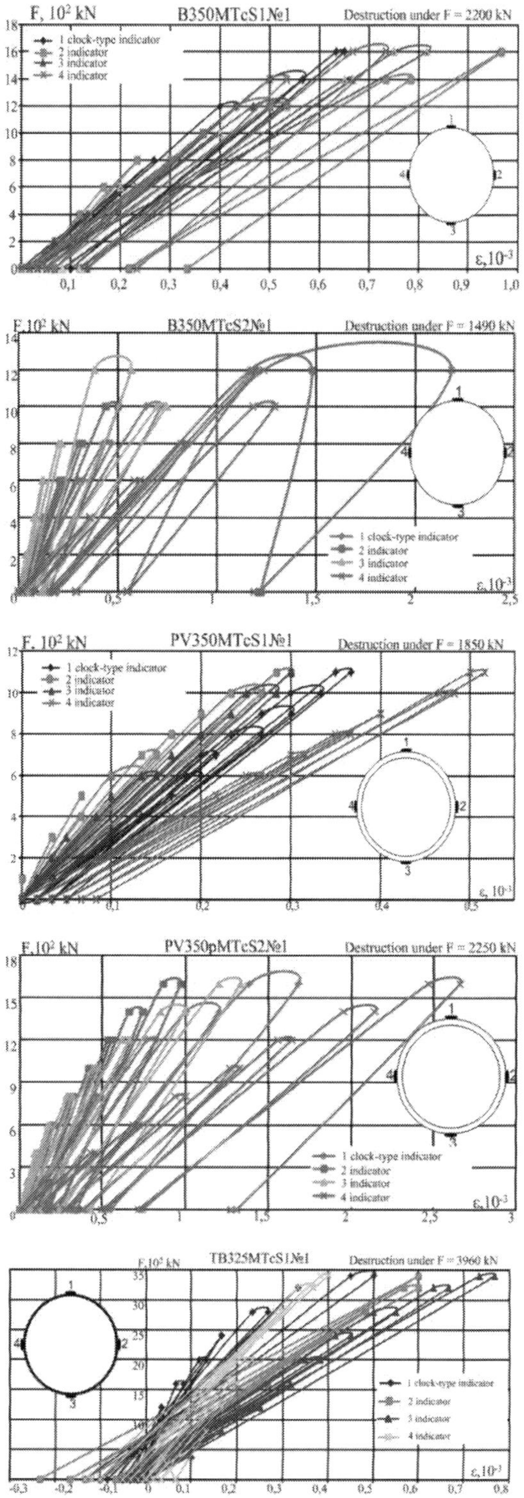

Fig. 6. Correlations between the longitudinal deformations of tested samples and the level of low-cycle axial loading.

https://doi.org/10.1051/matecconf/201823404003

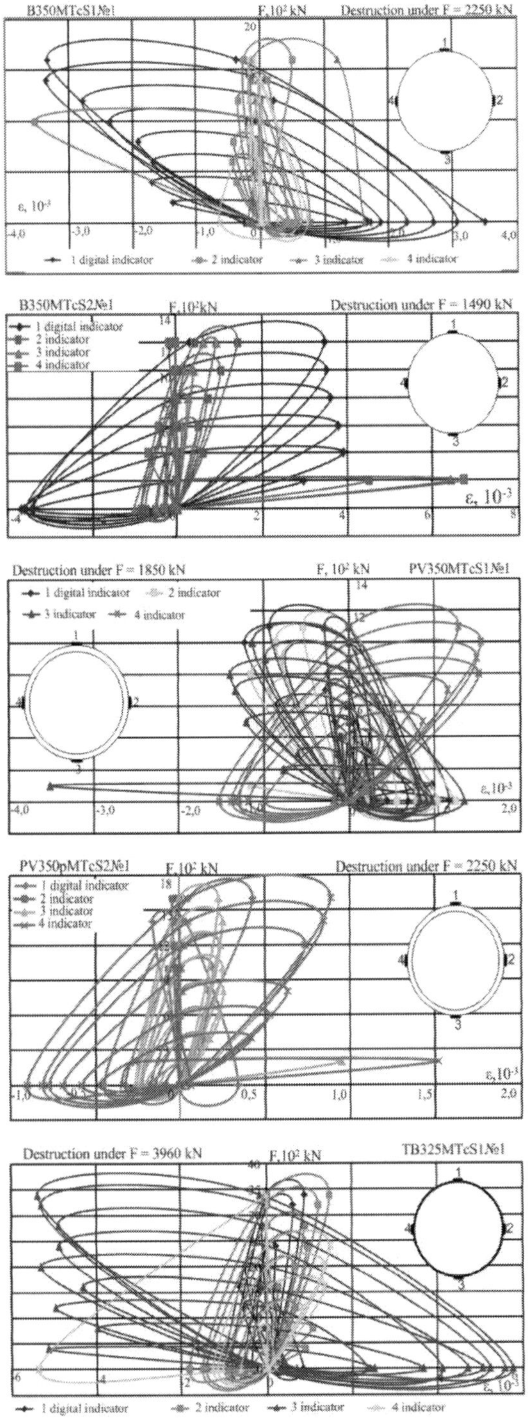

Fig. 7. Correlation between the relative lateral deformations of tested samples and the level of low-cycle axial loading

5 References

1. L. Storozhenko, D. Ermolenko, O. Lapenko, *Tube-concrete* (in Russian), Poltava, 306 (2010)
2. M. Lachemi, K. Hossain, V. Lambros, Axial load behavior of self-consolidating concrete-filled steel tube columns in construction and service stages, ACI Structural Journal – January-February 2006, 38-47 (2006)
3. E. Chihladze, M. Verevicheva, Operation of cylindrical steel-concrete columns under static short-term loading (in Russian), Construction, Material Science, Mechanical Engineering, Collection of works of PGASA, **35**(4), 47-54 (2005)
4. V. Gnedovskiy, *Confinement reinforcement of reinforced concrete structures* (in Russian), 126, Leningrad (1981)
5. W. Yi, P. Li, S. Kunnath, Experimental studies on confinement effect of steel hoops in concrete columns, ACI Structural Journal – January-February 2012, 3-10 (2012)
6. V. Shmukler, O. Synkovska, O. Petrova, M. Shutkin, B. Shutkin, Steel-concrete column (in Russian), Useful model patent UA93491E04C3/30, Ukraine (2014)
7. V. Shmukler, Yu. Klimov, N. Burak, *Lightened frame systems* (in Russian), Kharkov, 336 (2008)
8. Expanded steel mesh. Specification (in Russian), GOST 8706-78, Moscow, 11 s (1978)
9. O. Sinkovskaya, A. Ignatenko, Peculiarities of carrying capacity evaluations of cylindrical CFST columns with new type casing, Transbud-2017, MATEC Web of Conferences, 116, 02031 (2017)
10. O. Synkovska, Rational structures of steel-concrete load-bearing units of bridge footing of new type, Concrete and reinforced concrete – view on the future: III All-Russian conference in concrete and reinforced concrete (in Russian), Moscow, 328-329 (2014)
11. O. Synkovska, E. Surzhan, Cylindrical load-bearing steel-concrete units of building supports and structure footing of new type, their bearing capacity (in Russian), Collection of works of UkrDUZT, **149**, 169-178 (2014)

Modelling of motorway bridge spans under modernization with consideration of rheological properties of the materials

Alexey Lobiak[1*], *Andrii* Plugin[2], *Larisa* Kravtsiv[1], and *Oksana* Kovalova[3]

[1]Ukrainian State University of Railway Transport, Structural Mechanics and Hydraulics Department, Feuerbach sq. 7, 61050 Kharkiv, Ukraine
[2]Ukrainian State University of Railway Transport, Building Materials and Structures Department, Feuerbach sq. 7, 61050 Kharkiv, Ukraine
[3]Ukrainian State University of Railway Transport, Manage freight and commercial work Department, Feuerbach sq. 7, 61050 Kharkiv, Ukraine

Abstract. The paper presents a computer modelling technique for modernization of bridgework operations by building-up a mounted reinforced concrete slab. It implements the technique of the evolutionary transformation of a model in one calculation cycle with redistribution of forces between the elements of the built-up section, and consideration of the impact of elastoplastic and rheological properties of the materials. Consideration of the concrete creep implies the application of the generalized kinetic curve of prolonged deformation and phenomenological deformation development equations based on the colloid-chemical concept of the prolonged concrete deformation mechanism. The creep control was implemented through new structural coefficients which determined the structure of matrix interlayers between the sand grains, and mortar between the crushed stone grains. The technique proposed was realized in the program complex "LIRA-SAPR" based on the building information modelling (BIM) and the finite element method (FEM). The multistage modelling technique was shown by an example of calculation of a motorway bridge slab span within the transport structures under modernization along the Lev Landau Avenue in Kharkiv (Ukraine).

1 Introduction

According to the State Agency of Automobile Roads of Ukraine, the total length of the road network in Ukraine is over 172000 kilometers. And in the long term we are going to witness both dynamic construction of new roads and modernization of existing ones in order to handle much heavier traffic flows which may lead to greater loads on engineering structures including bridges and overpasses. At the same time, we constantly notice the overrunning of the building norms and basic loads with which new and existing bridgeworks should conform. In turn, change in the building norms and increase in the loads from transportation means may require modernization of spans and bring their loading capacity into conformity with the requirements. Besides, bridgeworks constantly suffer from various defects and damages due to the climatic conditions and unavoidable physical depreciation.

At present, in Ukraine a great many existing road bridgeworks do not satisfy the modern requirements for carrying capacity, traffic area dimensions and engineering state. One of the most common modernization techniques for bridge spans is the building-up of a mounted solid concrete reinforced slab. The advantage of this reinforcement technique is the monolithic action of the prefabricated span elements in transverse direction, which increases the carrying capacity and eliminates "the key effect". For slab spans of road bridges this problem is most acute as in operation their key connections often damage.

The efficiency of span modernization with a mounted slab is calculated in the design stage, and, first of all, depends on the model adequacy. However, the existing building norms of Ukraine [1] do not provide this design procedure which complicates the application of the modernization method. The paper deals with practical experience in finite-element modelling of the span operation through an example of road overpass considering the operation stages from the current state to latest service life periods after modernization. The technique proposed implements the evolutionary transformation of a model in one calculation cycle with consideration of the impact of elastoplastic and rheological properties of the materials.

2 A model of the span operation

2.1 Problem statement

The effect of the increased carrying capacity of a span by building-up a mounted slab depends on the technological maintenance diagrams accepted and existing technical state of the structure. Taking into account that the mounted slab in modernization is put into operation for the second part of the constant and the temporary loads, the calculation was done non-linearly considering

* Corresponding author: Lobiak@ukr.net

changes in the structure. And at each stage of the evolutionary changes the calculation should be done for a corresponding construction diagram considering the mounted elements. For this case the design procedure considered three mounting stages (Fig. 1) and redistribution of forces as a result of the structural changes and concrete creep. The effect of the concrete creep was considered on the basis of a method developed by the authors and proven to be effective in earlier research into composite [2] and concrete structures [3].

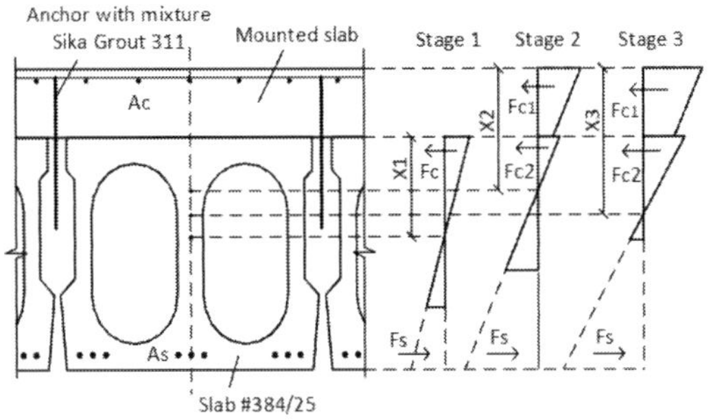

Fig. 1. Stages of the span operation.

The first stage involves the loading of the existing span elements with the first part of the constant loads consisting of the weight of the initial and built-up elements (a mounted slab), as well as the forces of the reinforcement pre-stress. The second stage puts the mounted slab into operation, and the composite section bears the additional loads from the weight of the road pavement, sidewalk blocks and road guards. The third stage implies the combination of constant and temporary loads. Here, the carrying capacity is assessed separately for the existing elements of the span and the reinforcement elements.

Generally, the design of transport facilities at present assumes the application of industrial program complexes as the basic computer modelling tool. The method proposed was fully realized with the program complex LIRA-SAPR [4]. The extensive library of finite elements (FE), support of the non-linear material behavior and powerful analyzing tools provided the needed functionality in the numerical analysis of reinforced structures.

2.2 The concrete creep

Numerous experiments demonstrate that concrete-based structures under long-term loads suffer from non-elastic deformations, which can exceed the initial conditionally elastic ones several times. But for bridge works of reinforced concrete and composite structures the problem of concrete creep is most acute. Thus, the problems of long-term deformation forecasting for bridge works, as well as synchronous redistribution of loads in the composite sections of spans, are rather urgent.

The existing creep theories widely use two approaches in forecasting long-term deformations. According to the first one, material behavior is based on mechanical systems made of combination of elements intended for mechanical property modelling (module of elasticity and viscosity coefficient). The second approach, the phenomenological one, is based on experimental data establishing dependencies between deformations and time (creep curves) or stress and time (relaxation curves). The existing phenomenological theories, despite the success achieved, do not accurately consider concrete creep deformations. Particularly, the most general hereditary theory of concrete ageing connects residual deformations with ageing only, excluding the part of residual deformation which is the result of the stresses and change in the concrete structure.

The Building Norms of Ukraine, harmonized with Eurocode 2 [5], are based on the simplest approach establishing dependencies between deformations and time with the creep coefficient $\varphi(t, t_0)$. According to this method, ultimate deformations of creep depend on concrete strength, age and relative moisture, and creep development in time t depends on the relative moisture and section size:

$$\varepsilon_{c,t} = \sigma_c / E_c \left[1 + \varphi_0 \beta(t, t_0)\right], \tag{1}$$

where $\beta(t, t_0)$ is coefficient describing the development of creep over time; t_0 – age of the concrete at the time of the first load.

The method realizes 90% of concrete creep deformations in a life span of 20-25 years. Despite its advantages, this method cannot reveal the actual mechanism of long-term deformations of concrete, and thus, does not allow us to control the mechanism and influence the ultimate values of creep deformations.

Since alternative methods to consider the creep can be applied with special software, we propose a method based on the colloidal chemical representation of the mechanism of long-term concrete deformations [3].

The theory is based on a more accurate representation of the kinetic concrete deformation curve (Fig. 2). On the basis of the findings of the study, the equations of deformation development were obtained depending on four stages of concrete performance: compression in the conditionally elastic setting, the stage of high-rate creep (1), ordinary creep, consisting of nonlinear (2) and linear (3) parts, as well as long-term creep (4).

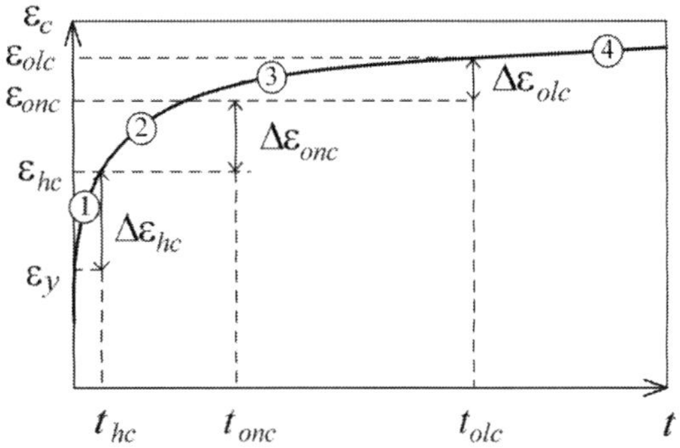

Fig. 2. The generalized kinetic concrete creep nomogram.

 https://doi.org/10.1051/mateconf/201823404004

In accordance with Fig. 2, the pattern of the long-term deformation of the concrete under compression and the general concrete deformation equation take the form:

$$\varepsilon_c = \varepsilon_y + \Delta\varepsilon_{hc} + \Delta\varepsilon_{onc} + \Delta\varepsilon_{olc} + \Delta\varepsilon_{1c}. \quad (2)$$

The specific equations simulating deformation kinetics at each of the above stages:

$$\varepsilon_{c,hc}(t) = \varepsilon_y + \Delta\varepsilon_{hc}(1 - e^{-t/T_r^{hc}});$$

$$\varepsilon_{c,onc}(t) = \varepsilon_{hc} + \Delta\varepsilon_{onc}(1 - e^{-t/T_r^{onc}}); \quad (3)$$

$$\varepsilon_{c,olc}(t) = \varepsilon_{onc} + k_{olc}(t - t_{onc});$$

$$\varepsilon_{c,lc}(t) = \varepsilon_{olc} + k_1(t - t_{olc}),$$

where T_r^{hc}, T_r^{onc} are relaxation time according to the Tertsagi's dispersion phase model on the exponential deformation stages; k_{olc}, k_1 – kinetic coefficients reflecting a rate of general (on the linear section) and long-term creep deformation.

In view of the latter, the expression for determining the creep coefficient at any concrete deformation stage takes the form:

$$\varphi(t) = \lfloor \varepsilon_c(t) - \varepsilon_{c,y} \rfloor / \varepsilon_{c,y}. \quad (4)$$

The total relative concrete deformations, corresponding to the completion of the high-rate (ε_{hc}) and ordinary (ε_{onc}, ε_{olc}) creep stages, are determined on the basis of the hypothesis about the behaviour mechanism of submicroscopic structure of the concrete consisting of electrically charged particles (globules) of hydrosilicate gel when it is exposed to compressing load. It is assumed that ultimate creep deformations are determined by gel deformations due to compression of a double electric layer, water filtration from the compressed areas of the construction into those stretched according to the pressureless water penetration law, while the long-term creep kinetics depends on the scale factor during filtration of water squeezing from the gel.

Creep deformations can be controlled through new structural coefficients determining the structure of cement stone layers between the sand and mortar grains between chip grains [3]:

$$\alpha = V_{csm} / V_{emp}^{BS} = \left(\frac{S}{\rho^S} + \frac{C}{\rho^C} + \frac{W}{\rho^W} \right) \Big/ \left(\frac{BS}{\rho_{pour}^{BS}} \cdot S_{emp}^{BS} \right);$$

$$\mu = V_{cs} / V_{emp}^S = \left(\frac{C}{\rho^C} + \frac{W}{\rho^W} \right) \Big/ \left(\frac{S}{\rho_{pour}^S} \cdot S_{emp}^S \right). \quad (5)$$

The best coefficients α_{opt} and μ_{opt} were obtained under the condition of the fullest contact, which decreases disjoining pressure and, respectively, compressive deformations and hair cracks.

3 Realization of the method

3.1 Constructive solution

The design calculation of an overpass within the transport structures under modernization along the Lev Landau Avenue in Kharkiv (Ukraine) is an example of a multistage modelling technique proposed. The span consists of 31 pre-stressed slabs of a length of 12m and a thickness of 600 mm with oval hallows. The carrying structures are made for loads H-30 (the load from a convoy of 50-ton trucks) and HK-80 (the wheel load from a convoy of 80-ton trucks) in accordance with a standard project of series 3.503-12 inventory # 384/25 "Soiuzdorproekt". The ultimate bending moment of the section in accordance with the standard design is M_u = 541.3 kNm with the consideration of the partial loss of concrete strength being M_{uf} = 520.66 kNm. According to the routine inspection of the technical state and calculations, the experts recommended refurbishment of the overpass due to its disabled state (State 5 by Classification [6]).

The span modernization involved (Fig. 1) the installation of a solid reinforced concrete mounted slab with a minimal thickness of 150 mm put into operation with the existing span slabs with anchors. The anchors were fixed in restored monolithic sections of the seams with the self-leveling cement epoxy mixture SikaGrout 311. The slab was installed without preliminary unloading of the spans.

3.2 The design diagram

The following hypotheses were taken for designing a finite element model of the span.

1. The existing span slabs were modelled with universal spatial axial FEs considering physical non-linearity. The mounted slab was modelled with FEs of thin shallow shells as a biomaterial system of isotope physically-nonlinear materials.
2. The flexibility of links between the sectional elements was provided with resilient filling pieces of constant rigidity in shearing, which physically modelled the anchor devices. The pre-stress forces of the reinforcement were set with a group of FEs "hung" to the rods with rigid filling pieces.
3. The structural non-linearity of the system was revealed with the design method "Mounting" which considered the evolutionary transformations. Besides, each building stage had the finite elements specified for it. And the load corresponding to this stage was calculated. For finite elements included for several stages all the efforts were summed up.
4. The concrete strength and stress-strain behaviour were characterized by parabolical and linear stress-strain diagrams $\sigma_c - \varepsilon_c$ according to the grade, and also by long-term deformation diagrams $\varphi(t) - t$. In order to consider the reinforcement steel, the diagram $\sigma_s - \varepsilon_s$ without the yield line was applied.

The span model was made according to statistically determinate diagram and hinged edges.

Generally, the design diagram was drawn up of materially nonlinear FEs of slabs simulating operation of the mounted slab, FE grid frameworks, a group of rigid inserts to transfer forces of pre-tension and a group of flexible inserts for "hanging" precast elements of the existing span (Fig. 3).

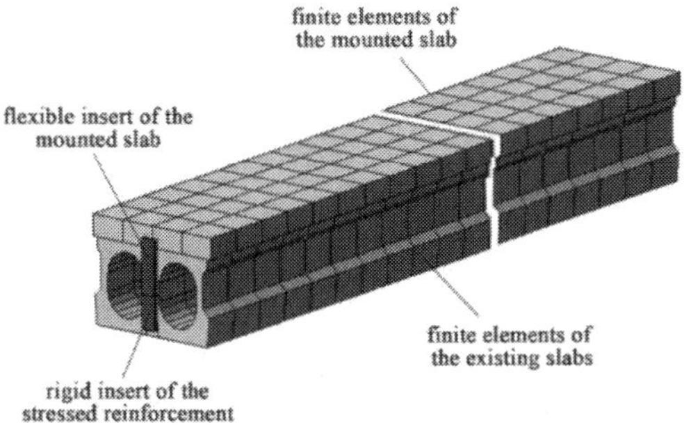

Fig. 3. A fragment of the span model.

The triangulation diagram was obtained as the solution of a separate task to seek a compromise between time and calculation accuracy.

The calculation was done using the iterative and incremental method by three mounting stages considering the creep at two first stages (Fig. 1). The number of loading steps at each mounting stage was 50, and the increase at each step was irregular. The shear rigidity along the contact of the composite section was defined by the number and types of anchors.

The creep calculation considered the relative humidity RH = 74% and the age of concrete of the mounted slab to be 28 days. The creep nomograms obtained according to [5] and the method proposed are presented in Fig. 4. The software specified the dependency proposed in the description of the stress-strain properties of the concrete as the axis set of a piecewise linear diagram.

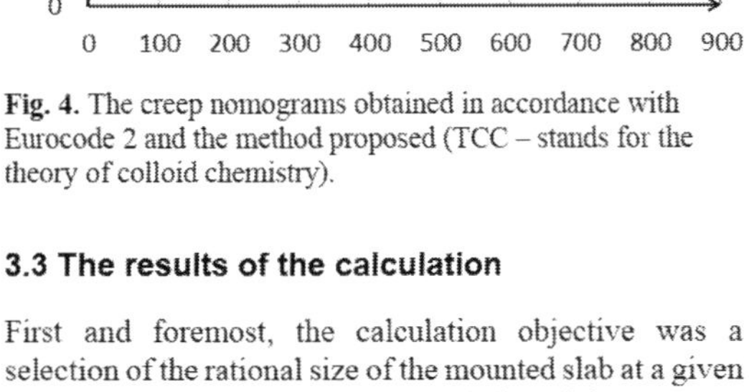

Fig. 4. The creep nomograms obtained in accordance with Eurocode 2 and the method proposed (TCC – stands for the theory of colloid chemistry).

3.3 The results of the calculation

First and foremost, the calculation objective was a selection of the rational size of the mounted slab at a given concrete grade providing due operability of the overpass (State 3 by Classification [6]) at minimal added dead weight of the span.

The operational states of bridges presented in the Ukrainian Building Norms are classified by a safety indicator according to the reliability evaluation model put forward by A.R. Rzhanitsin [7]

$$\beta = \frac{\gamma - 1}{\sqrt{V_q^2 + \gamma^2 V_r^2}}. \qquad (6)$$

Formula γ presents the mathematical expectation of the generalized safety coefficient, which, in this case, can be represented as the relation of the ultimate moment M_u to the design moment of the existing span elements M_{sl}

$$\gamma = \frac{M_u B_r}{M_{sl} B_q}, \qquad (7)$$

where V_q and V_r are generalized coefficients of influences and resistances

$$B_r = \frac{1}{1 - 1.64V_r}; \; B_q = \frac{1}{1 + 1.64V_q}. \qquad (8)$$

The generalized coefficients of the variations are defined by the general formulae of quadratic deviation of the sum of mutually independent variables taken by statistical models of loading and resistance of the system. Thus, the adopted statistical models are presented in Tables 1 and 2.

Table 1. The statistical model of effects.

Random factor	V_q
Dead weight of the span	0.033
Bogie load A-11	0.17
Uniform load A-11	0.24
Roadway and pavement load	0.17
Pedestrian load	0.24
Cross-section moment	0.0229
Cross-section area	0.0237
Eccentricity of pre-stress forces	0.0167
Shrinkage and creep effect	0.03

Table 2. The resistance statistical model.

Random factor	V_r
Cross-section moment	0.03
Cross-section area	0.0237
Concrete strength of the mounted slab	0.082
Mounted slab reinforcement	0.035
Concrete strength of span slabs	0.117
Span reinforcement	0.05
Influence line length	0.0465
Uncertainty factor of the model	0.02

In order to define the mathematical expectation of the generalized safety factor the span was loaded with various combinations of temporary loads by diagrams H-30 (according to Building Specifications СН 200-62), A-11 (the load from a convoy of two-axle 22-ton trucks) and HK-80 (according to State Building Norms in Production ДБН В.1.2-15:2009). The reliability was evaluated by the maximum design moment in the existing span slabs M_{sl} and was equal to 514.8 kNm (Table 3).

Table 3. Results of the calculation.

Force	Constant loads	A-11	HK-80	H-30
	kNm			
M_{init}	323.9	602.2	483.3	524.4
M_{tot}	366.3	643.4	532.3	591.7
M_{sl}	351.7	514.8	453.5	493.2
M_{msl}	14.6	128.6	78.8	98.5

The results of the calculation are presented in the paper as force lines, obtained by the method of finite elements built for the vulnerable section of a span (Fig. 5). Generally, the vulnerable section was situated in the middle of the slab and searched by changing the temporary loads along and across the traffic way. The summary data of the calculation with a consideration of a 900-day long-term creep are given in Table 3. The maximum bending moment in slabs for the initial structural state M_{init} was defined as a solution to a separate task. The bending moment to the mounted slab M_{msl} was defined as the difference between the total moment M_{tot} and the moment to the existing slabs M_{sl}.

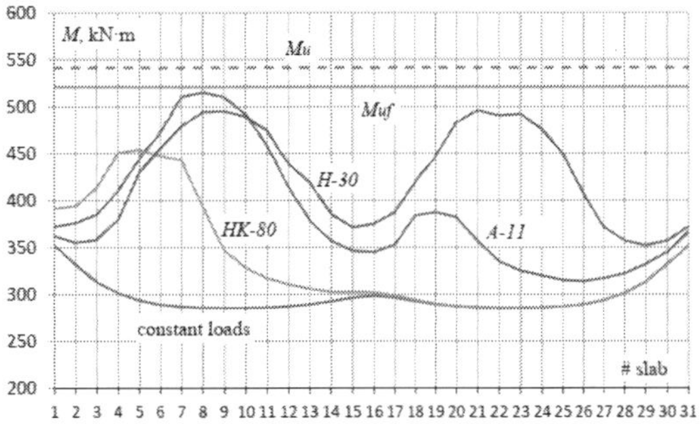

Fig. 5. The force lines (moments) for the existing span slabs after the modernization.

On the basis of the load calculation made and statistic models taken, the authors calculated the safety factor. In this case the values were: V_q = 0.4243; V_r = 0.1679; B_q = 0.5896; B_r = 1.3799; γ = 2.4607; β = 2.4665. Later, by means of the tables of Laplace transforms, the reliability was evaluated as equal to P = 0.9931. Thus, the probability of a trouble-free operation of the overpass corresponded to Operational State 3.

The result obtained corresponded to a thickness of the mounted slab of 150 mm, which was obtained by stepwise approximation. The diagram and parameters of the mounted slab reinforcement were set for each design cycle according to the acting loads.

As the calculation demonstrated, consideration of evolutionary changes in the span model led to a lower impact of the mounted slab on the carrying capacity of the modernized structure. A fraction of the total bending moment from all constant loads and sustained by a mounted slab was 4%, the same from the load A-11 according to the diagram (with consideration of the constant load and passersby) was 20%, from the load HK-80 it was 15%, and from the load H-30 it was 17%.

4 Conclusion

The modelling method proposed made it possible to establish the real carrying capacity of the overpass after modernization and assign the most effective reinforcement structure satisfying the present loads of HK-80 and A-11.

The taken approach to consider the creep included the control algorithm through the structural properties of concrete. It was established that long-term processes in the composite section influenced the stress-strain behavior. The efficiency of the modernization method does not decrease with time, and depends on a rational concrete mix and the optimum correlation of the structural parameters.

References

1. Bridges and pipes. Rules of designing: DBN V.2.3-14:2006. K.: Ministerstvo budivnitstva, arhitekturi ta zhitlovo-komunalnogo gospodarstva, 359 (2006)
2. A. Lobiak, G. Vatulia, Ye. Orel, Simulation of performance of circular CFST columns under short-time and long-time load. *6th International Scientific Conference "Reliability and Durability of Railway Transport Engineering Structures and Buildings" – Transbud 2017*. MATEC Web of Conferences, **116**, 02036 (2017)
3. A.N. Plugin, A.A. Plugin, O. Kalinin et al., *Fundamentals of theory of solidification, strength, fracture and durability of Portland cement, concrete, and structures of them* (in Russian), Naukova dumka, Kyiv, 3, 287 (2012)
4. A. Gorodetsky, Computer models of structures, Fact, Kyiv, 394 (2007)
5. Design of Concrete Structures – Part 1-1: General Rules and Rules for buildings EN 1992-1-1. Eurocode 2, Brussels, 227 (2004)
6. Instructions on assessment and forecasting the operational condition of road bridges (in Ukrainian), DSTU-N B.V.2.3-23:2013. K.: Minregion Ukrayiny, 54, (2013).
7. A. Rzhanitsyin, *Reliability calculation theory for structures* (in Russian), 239 (Stroyizdat, Moskva, 1978)

An experimental and numerical study of forces and residual stresses in AISI 316L stainless steel joints due to conventional and pulse gas tungsten arc welding

Ivan Muhtarov[1*], *Teofil* Iamboliev[2], *Veselin* Tsonev[3]

[1]Visteon Electronics Bulgaria, 90 Tsarigradsko Shosse Blvd., Capital Fort Building, 1000 Sofia, Bulgaria
[2]Technical University – Sofia, Plovdiv Branch, "Mechanical Technologies" Department, 25 Tsanko Diustabanov St., 4000 Plovdiv, Bulgaria
[3]Technical University – Sofia, "Strength of Materials" Department, 8 Kliment Ohridski Blvd., 1000 Sofia, Bulgaria

Abstract. This paper aims to present the results of a numerical and experimental study of the temperature field, internal forces and the residual stresses in 2 mm thick autogenous welds of AISI 316L stainless steel produced by continuous and pulse current gas tungsten arc welding. A special experimental device was used to measure the temperature and the internal forces due to the welding. The welds were qualified for internal and external weld imperfections according to ISO 15614-1. FEM software ANSYS® Multyphysics™ was applied in order to solve the thermal and mechanical problems. Normal residual stresses were measured by the hole-drilling strain gauge method in the continuous current weld. The peak value of the longitudinal stress was 80 % of the base metal yield stress. The magnitude of the numerically obtained residual stress values was found to be 16 % to 19 % above the measured one in the longitudinal and transverse direction, respectively. The experimental device used in this study allowed for a real time measurement of forces far from the weld seam. On the basis of the correspondence between the calculated and measured forces the numerical results were verified. Therefore, this device might open up new possibilities for determining thermo-mechanical material data.

1 Introduction

Stainless steel is a desired material for a variety of structural engineering applications. Austenitic stainless steels, for example 316L, combine excellent mechanical properties, corrosion and heat resistance ranging from cryogenic up to high temperatures. They are used as equipment in chemical plants, power plants, food industry, etc. [1].

Continuous current gas tungsten arc welding (CC-GTAW) is a key technology used for joining thin plates or thin-wall pipes in equipment manufacturing. It is known for its flexibility, providing a stable concentrated arc and narrow heat-affected zone with minimal distortion of the weld joint. If the thickness of the components to be joined is of up to 3 mm, a filler metal is not required and the process can be applied autogenously, hence, providing an equal chemical composition of the weld metal and the base metal [2, 3].

The pulse current gas tungsten arc welding (PC-GTAW) is a variant of the constant current welding process. It is the changing of welding current from peak amperage to a background amperage rapidly and repeatedly. This process is useful whenever welding heat input must be minimized or penetration accurately and repetitively controlled [2, 4-6]. This is particularly advantageous for welding thin sheet or when welding the root pass of a multilayer weld in thicker sections. The superior mechanical properties (higher yield strength and hardness), preferred microstructures in the weld metal region (very fine equiaxed grains with higher amount of precipitates) and favorable residual stress field in the weld metal region (large magnitude of compressive stress) as well as less warpage are the reasons for better performance of the PC-GTAW joints and the advantages of using PC-GTAW over the conventional GTAW process for welding of stainless steels [4-6]. It has been estimated that the average current needed for PC-GTAW is by 10 to 15 % less than the average current used for GTAW [5].

During welding, a non-uniform heat introduction in the components results in a non-uniform temperature field. After the completion of welding and cooling to room temperature, residual stresses are built up, causing distortion of the weld joint. This may sometimes lead to a premature failure. There are several factors affecting the extent of the distortion and residual stresses. The most important of them are the welding parameters, the thickness of the components to be joined, the degree of restraint, and the welding sequence, etc. [2, 3, 6-9].

Experimentally residual stresses may be measured by means of the hole drilling method [10, 11], ultrasonic methods [12] as well as X-ray and neutron diffraction methods [13, 14]. Numerical simulation by means of a

* Corresponding author: i_muhtarov@tu-sofia.bg

finite element code is an efficient tool for determining the residual distribution stress within the weld [15-17]. Its practical application may be limited by missing or not precisely determined material data dependent on temperature [16, 18].

There is a great number of studies measuring and modelling residual stress and distortion as a result of pulse gas metal arc welding [19-27]. However, the number of works dealing with experimental and numerical determination of the residual stress and distortion in PC-GTA welds is low. In an experimental work on the effect of PC-GTAW on the residual stress of 304 and 310 stainless steel welds the hole drilling strain-gauge method has been applied. It was found out that the magnitude of the residual stress can be reduced when welding with high pulse frequency, large pulse spacing, high base current/pulse current amplitude and base time/pulse time duration ratios. Higher residual stresses were determined in the 310 welds than in the 304 welds due to the difference in the thermophysical properties of these alloys [6].

Most of previous work has been focused on the technological welding aspects using austenitic [5, 6, 9, 21, 22] and ferritic [17, 19, 24, 25] stainless steels as well as aluminum alloys [20]. Less attention has been paid to numerical studies of the residual stresses and distortions of these steel alloys subjected to GTAW.

In view of the above, this paper presents a numerical and experimental study of the temperature field, internal forces and the residual stresses in thin plates of 316L subjected to autogenous CC-GTAW and PC-GTAW.

2 Experimental procedure

2.1 Material

Square-edge plates sized 200x115x2 mm for butt joints of stainless steel alloy AISI 316L were cut. The chemical composition of this alloy is listed in Table 1. After degreasing, the plates were tack-welded so that the welding gap was approximately zero.

Table 1. Chemical composition of steel alloy 316L.

Element	%
C	0,017
Si	0,54
Mn	1,76
S	0,002
P	0,034
Cr	16,9
Ni	10,1
Mo	2,06
N	0,045

2.2 Experimental device

An experimental device was developed to include the welded samples in a statically indeterminate structure, allowing measurement of internal forces and temperature in selected points. The setup is shown in Fig. 1. A wide flange standard beam 1 was used as a holding structure. The plates 4 to be joined were mounted on it by means of clamps 2 and 3. The cylindrical part 5 of the clamp 3 was a hollow section of 1,5 mm wall thickness. After machining the clamp was hardened and tempered up to 45 HRC. Resistance strain gauges of type LK11O 3/350 with gauge factor k=2,00 made by HBM GmbH were glued onto the surface of the hollow section and covered with a mechanically protecting layer. The maximum permissible temperature of the glue and the strain gauges was 160 °C. In order to prevent sensor overheating when welding under high heat input per unit length, there was a water cooling system of the clamp represented by the inlet and outlet pipes 13. All measuring circuits were executed as full Wheatstone bridges. Cable 6 connected the strain gauges to an amplifier and further, to a PC. Hence, clamp 3 and cylindrical part 5 acted as a combined dynamometer.

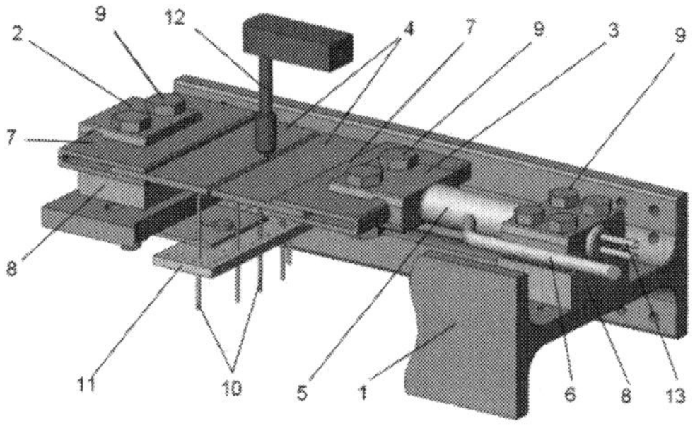

Fig. 1. Experimental device: 1 – steel profile; 2, 3 – clamps; 4 – plates to be joined; 5 – cylindrical part; 6 – cable; 7 – fitting plates; 8 – distance plates; 9 – bolt; 10 – thermocouples; 11 – supporting plate; 12 – welding torch; 13 – inlet and outlet cooling pipes.

Distance plates 8 helped to adjust the vertical position of the sample. They were clamped using fitting plates 7 of suitable thickness. The clamp constraint was provided by pressure achieved after screwing bolts 9.

The principal axis orientation of the dynamometer and the positions of the strain gauges are visible in Fig. 2. The internal forces due to welding were measured by the dynamometric clamp 3 as follows: the torque moment M_x around axis x, the bending moments, M_y and M_z around axes y and z, respectively, as well as the normal force N_x along axis x.

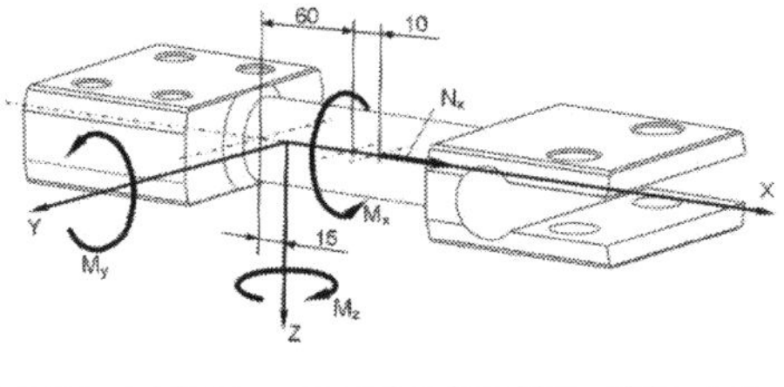

Fig. 2. Internal forces measured and positions of the strain gauges.

Prior to welding, the dynamometer was calibrated in order to convert the measured strain into relevant forces. The calibration was accomplished using appropriate

 https://doi.org/10.1051/matecconf/201823404005

appliances in accordance with ISO 376. The total error of the dynamometer was within 0,5 %. The interference between the measuring channels was less than the common limit of 5 % in engineering practice.

The insulated NiCr-Ni thermocouples 10 of diameter 3 mm and precision class 2 providing ±0,75 % error within the temperature range 333-1100 °C were used to measure the temperature in selected points at the weld seam. They were supported by a plate 11 (Fig.1), and fixed in the blind drilled holes of the tacked plates prior to welding. The temperature was measured along the welding direction in 4 points (T1, T2, T3, T4) and away from the edge in 2 points (T5, T6) according to Fig. 3.

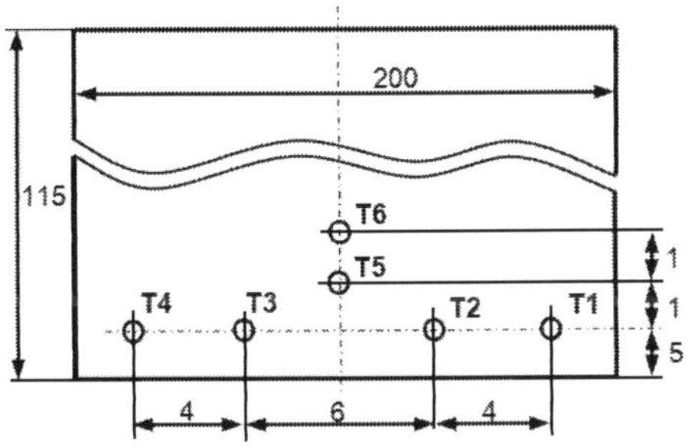

Fig. 3. Location of points for temperature measurement.

2.3 Experimental layout

The experimental layout is shown in Fig. 4. The strain gauge signals from dynamometer 1 were processed by a digital multichannel strain gauge amplifier 2. Further, PC 3 with appropriate software was used to record the signals and control the amplifier 2. Thermocouples 4 were connected to processing unit 5 via shielded cables. An interface adapter transformed the signals in an appropriate format for the PC. The PC was used to save and visualize the thermal cycles.

2.4 Welding conditions

Both the CC-GTAW and the PC-GTAW were applied autogenously without backing. The following welding conditions were applied:

- CC-GTAW: current 100 A, arc voltage 9,7 V, welding velocity 3,5 mm/s, heat input 1524,3 J/cm;
- PC-GTAW: pulse current 158 A, base current 45 A, pulse on time 0,47, frequency 3 Hz, average current 98,1 A, arc voltage 8,7 V, welding velocity 3,5 mm/s, heat input 1341,3 J/cm.

A thoriated W-electrode with a diameter of 2,4 mm was used, as well as Ar shielding with a flow rate of 6,5 l/min and 5 l/min for the face side and the root side of the weld seam, respectively. The welding machine was Kempi Mastertig 2300 allowing for a precise set up of the parameter values. The torch was moved by a carriage ensuring constant welding velocity.

The welds were qualified for internal and external weld imperfections according to ISO 15614-1 [28]. After non-destructive testing the macrostructure was studied in a weld cross section.

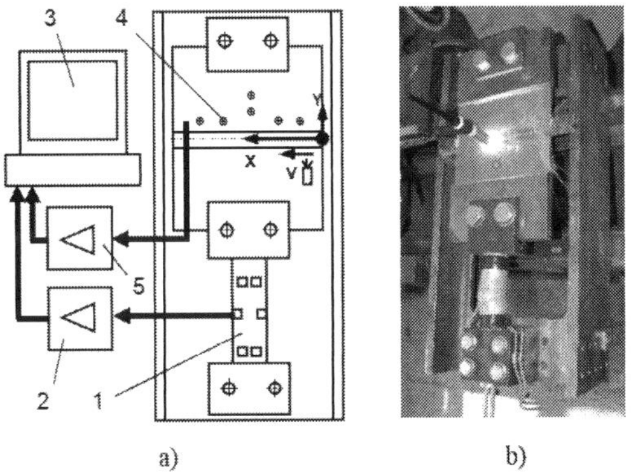

Fig. 4. Experimental layout: a) set up: 1 – dynamometer; 2 – amplifier; 3 – PC; 4 – thermocouples; 5 – processing unit; b) view during welding.

2.5 Residual stress measurement

The residual stress measurement was accomplished by means of the hole-drilling method following a standard ASTM procedure [10, 11]. It was carried out after weld cooling to room temperature and dismantling from the experimental device. Two sets of strain gauges RY61-120 were used – one of them was active and the other one was put on a neighboring plate to compensate for the temperature effect. The location of the strain gauges at measuring point SG on the CC-GTA weld sample is illustrated in Fig. 5. The hole-drilling was carried out with a special guiding device equipped with a drill of 1,3 mm diameter at 1000 rpm. The stress measurement was accomplished using appropriate HMB equipment.

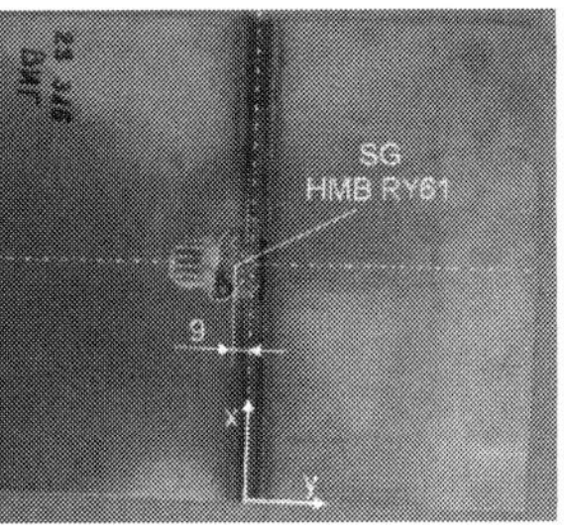

Fig. 5. Location of point SG and strain gauge rosette on a CC-GTA weld sample.

3 Computational models

FEM software ANSYS® Multiphysics™ was used to develop the computational models. They were produced as sequential decoupled analyses. This approach has already been used to study thermal fields, forces and stresses in single and multi-pass welds of various-thickness plates [24-26]. The thermo-mechanical problem was divided into two parts solved sequentially [15]. The thermal field was calculated. Thereafter, the mechanical solution was conducted determining the stresses and strains.

https://doi.org/10.1051/matecconf/201823404005

The geometrical model shown in Fig. 6 was the same for both the CC-GTAW and PC-GTAW processes. The body of the experimental device was neglected, since it was assumed to be rigid. The clamp and the dynamometer were fixed on the surfaces connected with the experimental device body. The finite elent mesh in heat-affected zone (HAZ) can be seen in Fig. 7. Prismatic solid elements of small average size were used to build a fine mapped mesh in this zone. Volumes far from the weld seam were meshed with elements of bigger size. Hence, the total number of the model elements was decreased.

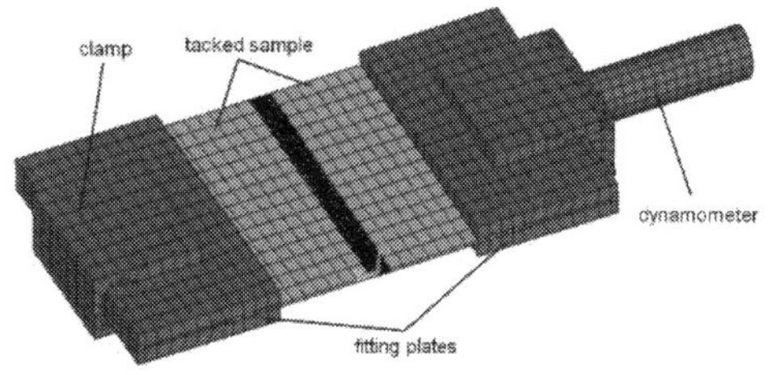

Fig. 6. Computer model of the investigated weld joints.

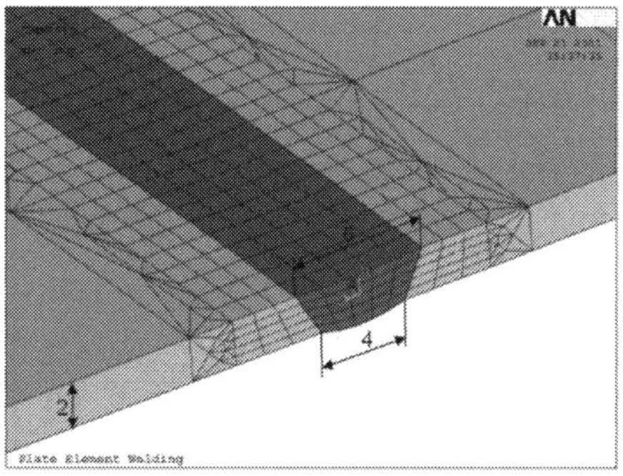

Fig. 7. Mesh in the weld heat-affected zone.

A double ellipsoidal heat source with Gaussian distribution of the thermal flux [8, 15] was applied in the model of the conventional CC-GTAW process. For the heat input in the PC-GTAW model two types of heat sources were used – a constant double ellipsoidal and a pulsing spherical source. The constant double ellipsoidal source was used to supply the heat due to the base current. The pulsing spherical source provided the heat resulting from the increased pulse current. The parameters of the sources were tuned with test solutions, until the fusion zone was properly represented in the models. The movement of the sources was realized by defining them in a local Cartesian coordinate system moving in regard to a global stationary coordinate system [15].

In the presented numerical models the weld seam formation was accomplished by an activation and deactivation technique [28, 29]. This technique was also used in order to determine the residual stresses in the welded samples after disassembling from the experimental device. This was accomplished by deactivating the fitting plates at t = 1800 s after the start of the welding process. Thereby, the samples in the model were released from the constraints. Force determination in the cylindrical section of the dynamometric clamp was done through classical integral relation between stresses and internal forces.

The data for the material thermodynamical and thermomechanical properties used for the numerical solution were adopted from [30] as shown in Table 2, where: T – temperature; c – thermal capacity; k – thermal conductivity; E – Young's modulus; ν – Poisson's ratio; σ_y – yielding stress; E_T – tangent modulus; α – thermal expansion.

Table 2. Material properties data [30].

T °C	c J/kg.K	k W/m.K	E GPa	ν -	σ_y MPa	E_T/E -	α °C^{-1}
20	442	15,0	200	0,278	318	0,014	$11,8.10^{-6}$
200	515	17,5	185	0,288	205	0,014	$12,6.10^{-6}$
400	563	20,0	170	0,298	132	0,014	$13,6.10^{-6}$
600	581	22,5	153	0,313	105	0,014	$14,8.10^{-6}$
800	609	25,5	135	0,327	77	0,014	$14,7.10^{-6}$
1000	631	28,3	96	0,342	50	1.10^{-4}	$14,0.10^{-6}$
1200	654	31,1	50	0,350	10	1.10^{-4}	$13,8.10^{-6}$
1340	669	33,1	10	0,351	10	1.10^{-4}	$13,6.10^{-6}$
1400	675	66,2	10	0,353	10	1.10^{-4}	$12,7.10^{-6}$

4 Results and discussion

4.1 Weld quality

Fig. 8 presents a view of the macrostructure of full penetration welds. The weld geometry represented a smooth weld bead profile excluding stress concentration. No weld imperfections were found in the welds, hence, the weld covered the highest quality requirements of ISO 5817, level B [28]. The weld bead area of the PC-GTAW and CC-GTAW were found to be 8,14 mm^2 and 11,96 mm^2, respectively. The difference in the area could be attributed to the 12 % lower heat input of the PC-GTAW as compared with that of the CC-GTAW.

a)

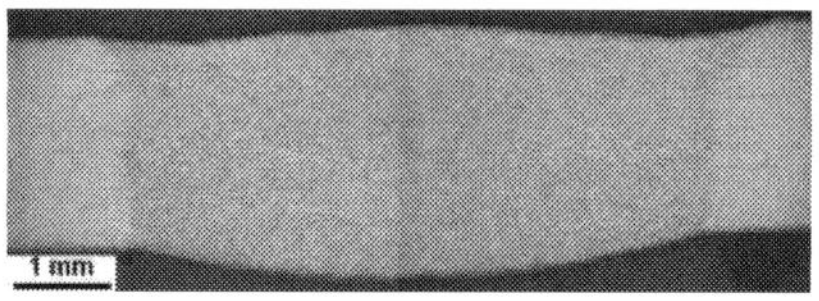

b)

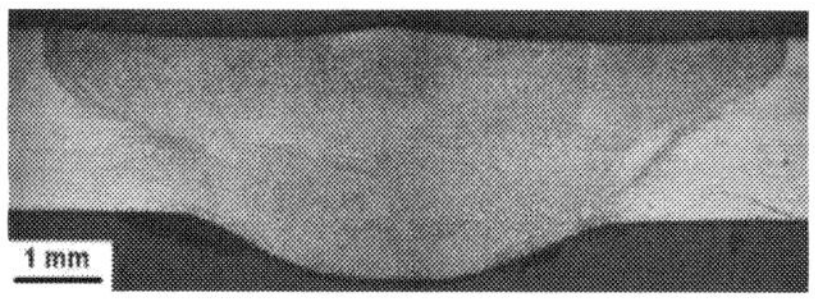

Fig. 8. Macrostructure of welds with figures for the heat input and the weld bead area: a) PC-GTAW, q = 1341,3 J/cm, A = 8,14 mm^2; b) CC-GTAW, q = 1524,3 J/cm, A= 11,96 mm^2.

4.2 Weld thermal cycles

The weld thermal cycles taken at points T2 and T6, Fig. 3, are illustrated in Fig. 9. The duration of the welding process was about 57 s. The peak temperatures at T2 to

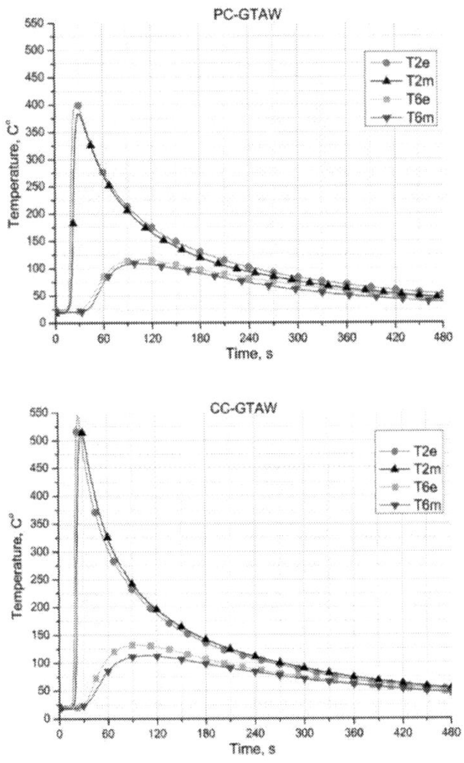

Fig. 9. Numerically (subscript *m*) and experimentally (subscript *e*) obtained thermal cycles at points T2 and T6.

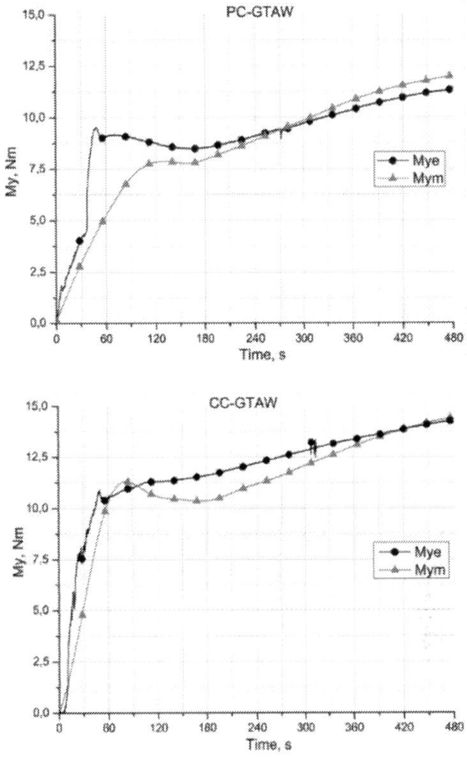

Fig. 10. Kinetics of the measured and calculated bending moment M_y.

T4 were found to be much higher than those at point T6 because of the smaller distance to the heat source during welding. After about 480 s, on cooling, the curves are brought close together corresponding to the stage of heat equalization due to heat conduction. Also numerically obtained thermal cycles are superimposed showing good correlation to the experimental curves.

The peak temperatures occurring with the PC-GTAW were lower than those with the CC-GTAW. For example, peak temperatures of 400°C and 520°C were measured at point T2 for the PC-GTAW and the CC-GTAW, respectively. The peak values at point T6 were 120°C and 133°C for each process. The lower peak temperature of the PC-GTA welds resulted in a narrower HAZ and better mechanical properties as found earlier [31].

4.3 Internal forces

The kinetics of the internal forces, Fig. 2, arising in the sample during welding and after its completion are shown in Fig. 10 to Fig. 12. Bending moment M_y originating from an angular distortion of the weld was measured to be positive as demonstrated in Fig. 10. On weld cooling, M_y increased abruptly due to an increase of the modulus of elasticity. At different stages of cooling, M_y of the PC-GTA welds was lower than that of the CC-GTA welds. For example, just after welding completion at t = 57 s M_y = 9 Nm for the PC-GTA welds and M_y = 10,5 Nm for the CC-GTA welds. Later at t = 480 s the corresponding measured values were M_y = 11,25 Nm and M_y =13,40 Nm, respectively. The modeled M_y curves are superimposed in Fig. 10. After welding completion there is a good correlation between the experiment and the model of both the PC-GTAW and CC-GTAW processes.

Similar observations were established considering the kinetics of the bending moment M_z in Fig. 11. The initial negative value of M_z could be attributed to the linear expansion of the plate edges after the welding start, which led to an angular distortion in the plane z = 0. When half of the weld length was accomplished the stiffness of the sample arose and the rate of deformation slowed down reaching a negative peak value. Even before welding completion M_z increased and became positive due to a powerful longitudinal shrinkage on cooling of the already deposited weld metal. After welding completion on further cooling the rate of the M_z rise was lower as corresponding to the lessening cooling rate of the weld. The M_z values measured in the PC-GTA welds were lower than those obtained in the CC-GTA welds. For example, at t = 480 s M_z approximated 5 Nm in the PC-GTA welds and 7 Nm in the CC-GTA welds, the former amounting to about 70 % of the latter. A good correlation between the experimentally and numerically obtained M_z values for both the PC-GTAW and CC-GTAW processes was found.

The torque moment was also time dependent and varied within the range M_x = –0,5 Nm to M_x = 1 Nm. These values were found to be small and insignificant, so an M_x diagram is not provided.

 https://doi.org/10.1051/matecconf/201823404005

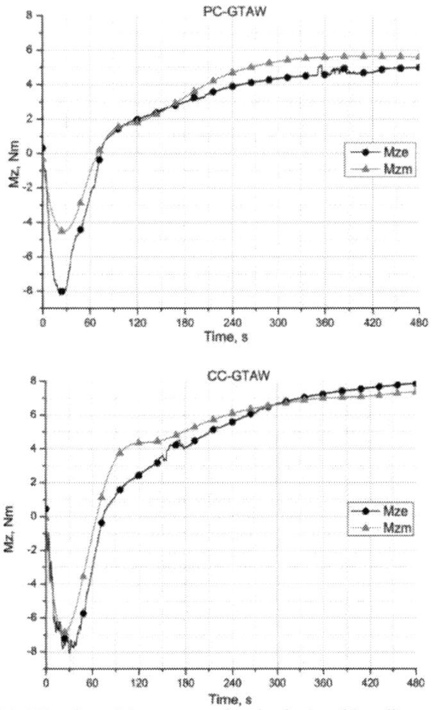

Fig. 11. Kinetics of the measured and calculated bending moment M_z.

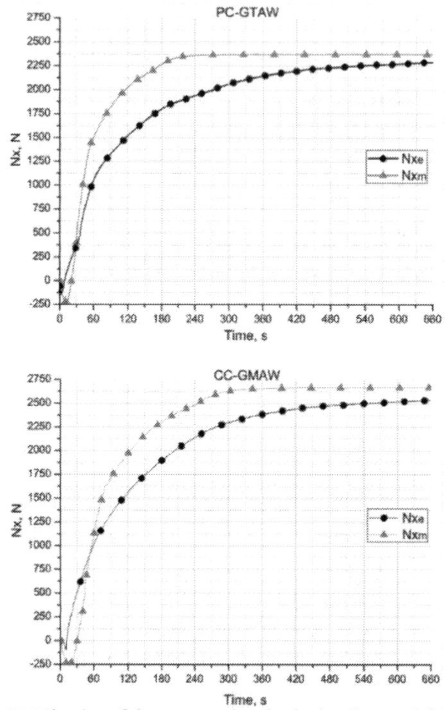

Fig. 12. Kinetics of the measured and calculated normal force N_x.

Short-time negative values of normal force N_x were established just after the beginning of the welding process, as demonstrated by the experimental curve in Fig. 12. These were to be related to a transverse expansion of the weld sample, causing pressure on the dynamometer. On cooling, the expansion turned to shrinkage of the deposited weld metal, so N_x became positive and, after a steady increase, reached its final constant value. Also N_x showed lower values for the PC-GTA welds compared with those of the CC-GTA welds as follows: at $t = 180$ s $N_x = 1750$ N for the PC-GTA welds and $N_x = 2000$ N for the CC-GTA welds; also, at $t = 600$ s, the corresponding residual values were 2250 N and 2500 N. A good correlation between the experiment and the model was reached about 600 s after the arc had extinguished.

4.4 Residual stress

As mentioned previously, the weld stresses were measured in the CC-GTA weld using the hole drilling method. The experimental results obtained are shown in Fig. 13. It can be seen that the longitudinal residual stress σ_x is positive and comes up to 222 MPa, which represents about 70 % of the base metal yield strength. In addition, the transverse positive residual stress σ_y amounted to 75 MPa, hence, a ratio $\sigma_x/\sigma_y = 2{,}96$ applied.

The normal stresses in the CC-GTAW sample were modelled from the beginning of the welding process. They were superimposed in Fig. 13. During welding there was a transient process of stress development. It was followed by a steady stress increase up to the residual stress values reached at $t = 660$ s after the welding start. The longitudinal stress $\sigma_x = 271$ MPa was higher than the transverse stress $\sigma_y = 104$ MPa and a relation applied $\sigma_x/\sigma_y = 2{,}6$. These residual stress values applied for the weld sample constrained in the experimental device according to Fig 1 and Fig. 4. After weld dismantling at $t = 1800$ s the actual residual stresses decreased to $\sigma_x = 258$ MPa and $\sigma_y = 89$ MPa causing an abrupt change of the curves as pointed by the arrows in Fig. 13. In this case, the residual stress ratio was $\sigma_x/\sigma_y = 2{,}90$, thus, approximately confirming the one of 2,96 obtained from the measurement.

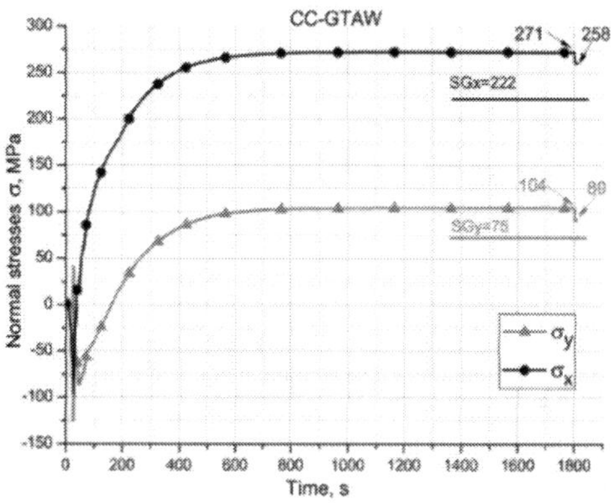

Fig. 13. Calculated and measured normal stresses in point SG of the CC-GTAW sample.

The modelled distribution of the normal residual stresses in longitudinal and transverse direction at $t = 1805$ s of the CC-GTA weld is shown in Fig. 14. The numerically obtained residual stresses were found to be 16 % to 19 % higher than the measured values. First of all, these discrepancies could be attributed to the unreliable thermomechanical material data above 900°C, which were used for the calculation presented in Table 2.

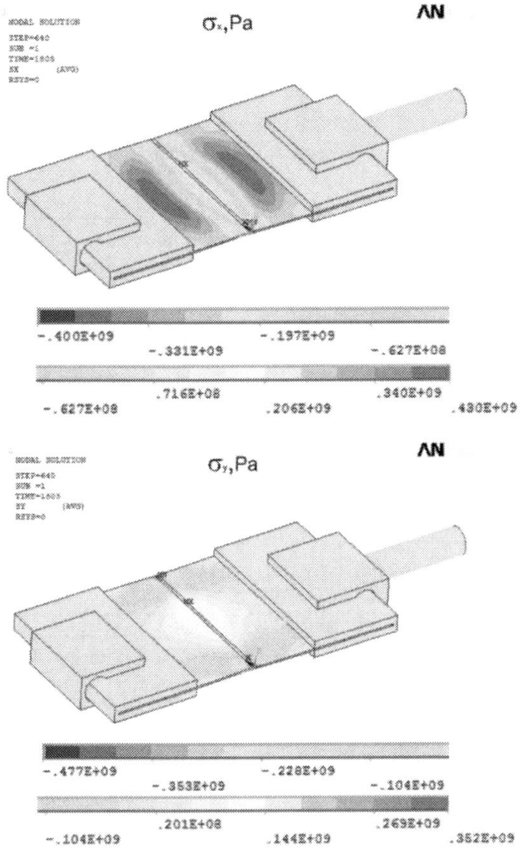

Fig. 14. Distribution of the normal residual stresses at time 1805 s for CC-GTAW (σ_x – longitudinal stress; σ_y – transverse stress).

Since the internal forces in the PC-GTA weld were lower than those of the CC-GTA weld, it might be expected that the same tendency applies for the values of the corresponding residual stresses and their distribution in the PC-GTA weld.

In the present study the measurement of the temperature and internal forces arising during welding far from the weld seam was possible because of the use of the special experimental device. The accuracy of the computational models was assessed by the relation between the calculated and measured forces. On the basis of this the results gained from the numerical models could be verified. And so, the difficulties experienced in the real time measurement of stresses and strains were removed.

It is well known that thermomechanical material data are temperature dependent and play a key role for the accuracy of the computational model. Also it is recommended that the material data in the temperature range close to the melting point of the base metal are to be adopted with alternative methods [8, 15], characterized by a fair approach to accuracy. Further on, the data should be applied over several test solutions until a good correspondence is achieved between computational and available experimental values. At the end of this procedure the computational values obtained would be very close to the real material properties data and may be used in computational models with similar solution algorithms and process parameters. Consequently, the use of the experimental device might open new possibilities for computerized determination of the real thermomechanical data of the materials subjected to welding. This procedure is to be tested in a following study.

5 Conclusions

As a result from the experimental and numerical study of the forces and residual stresses in 316L stainless steel welds produced under PC-GTAW and CC-GTAW the following conclusions can be drawn.

1. The heat input of the PC-GTAW used for the autogenous butt welds was found to be by 12 % lower when compared with that of the CC-GTAW. As a consequence the peak temperature and the internal forces of the PC-GTA welds were lower than those of the CC-GTA welds. The weld geometry and the absence of weld imperfections of both covered the highest quality requirements of ISO 5817, level B.

2. Weld thermal cycles at different points of the welds were obtained. For $t > 400$ s after the welding start there was a good agreement between the experimental and calculated temperature values. The discrepancies established at $t < 400$ s were attributed to the thermomechanical material data, which were considered unreliable within the temperature range from 900°C to the melting point of the base metal.

3. Internal forces represented by the bending moments, torque moment and the normal force were calculated far from the weld seam and measured by the special experimental device used in this study. Verification of the numerical model based on the correspondence between the calculated and measured forces was done. After an initial transient process, there was a good agreement between the experimental and numerical values. Therefore, the experimental device might open new possibilities for determining and testing thermomechanical material data.

4. Normal residual stresses were measured and modeled for the CC-GTAW in both the longitudinal and transverse direction. It was found out that the stresses were positive and the peak value of the longitudinal stress amounted to 80 % of the base metal yield stress. The magnitude of the numerically obtained residual stress values was found to be 16 % to 19 % above the measured one.

https://doi.org/10.1051/matecconf/201823404005

References

1. A. Guljaev, *Physical Metallurgy* (Mashinostroenie, Moskow, 1977)
2. H. Cary, *Modern welding technology*, 4 th ed., (Prentice Hall, Englewood Cliffs, NJ, 1998)
3. A. Zhelev, *Materials Engineering. Manufacturing processes and processability*, Sofia (2002)
4. D. Kotecki, F. Armao, *Stainless steels properties – how to weld them where to use them* (The Lincoln Electric Company, 2003)
5. A. Lakshminarayanan, K. Shanmugam, V. Balasubramanian, *Effect of autogenous arc welding processes on tensile and impact properties of ferritic stainless steel joints*, Journal of Iron and Steel Research, **16**(1), 62-68 (2009)
6. K. Tseng, C. Chou, *The effect of pulsed GTA welding on the residual stress of a stainless steel weldment*, Journal of Materials Processing Technology, 123, 346-353 (2002)
7. V. Vinokurov, *Weld stresses and deformations* (Machinostroenie, Moskow, 1968)
8. D. Radaj, *Welding residual stresses and distortion. Calculation and measurement*, (DVS Verlag, 2003)
9. T. Toyooka, T. Tsunerari, R. Ide, T. Tange, *Fatugue test of residual stress induced specimens in carbon steel*, Welding Journal, **64**(1), 29-36 (1985)
10. ASTM Standard E837-13a, *Determining residual stresses by the hole-drilling strain-gage method.*
11. Y. Arai, M. Kikuchi, T. Watanabe, M. Nagaki, *Residual stresses due to welding and its effect on the assessment of cracks near the weld interface*, International Journal of Pressure Vessels and Piping, **63**, 237-248 (1995)
12. P. Withers, H. Bhadeshia, *Residual stress. Part 1 Measurement techniques*, Materials Science and Technology, **17**, 355-365 (2001)
13. T. Holden, J. Root, V. Fidleria, R. Holt, G. Roy, *Application of neutron diffraction to engineering problems*, Material Science Forum, 27-28, 359-370 (1988)
14. M. Park, H. Yang, D. Jang, J. Kim, T. Jin, *Residual stress measurement on welded specimen by neutron diffraction*, Journal of Materials Processing Technology, 155-156, 1171-1177 (2004)
15. J. Goldak, M. Akhalaghi, *Computational welding mechanics*, (Springer Verlag, 2005)
16. C. Schwenk, M. Rethmeier, Material properties for welding simulation – measurement, analysis, and exemplary data, Welding Journal, **90**(6), 220-227 (2011)
17. M. Onsøien, M. M'hamdi, O. Akselsen, Residual stresses in weld thermal cycle simulated specimens of X70 pipeline steel, Welding Journal, 89(6), 127-132 (2010)
18. V. Tsonev, Determination of mechanical characteristics of a new alloy steel at high temperatures and static load, *BulTrans-2010 Proceedings*, 209-211 (2010)
19. P. Ghosh, K. Devakumaran, A. Pramanick, Effect of pulse current on shrinkage stress and distortion in multipass GMA welds of different groove size, Welding Journal, **89**(3), 43-53 (2010)
20. V. Goyal, P. Ghosh, J. Sainic, Analytical studies on thermal behaviour and geometry of weld pool in pulsed current gas metal arc welding, Journal of Materials Processing Technology, **209**, 1318-1336 (2009)
21. P. Ghosh, A. Ghosh, Control of residual stresses affecting fatigue life of pulsed current gas metal arc weld of high strength aluminum alloy, Metallurgical and Materials Transactions A, **35**(8), 2439-2444 (2004)
22. S. Kulkarni, P. Ghosh, S. Ray, Improvement of weld characteristics by variation in welding processes and parameters in joining of thick wall 304LN stainless steel pipe, ISIJ International, **48**(11), 1560-1569 (2008)
23. P. Ghosh, S. Gupta, H. Randhawa, Analytical studies on characteristics of vertical up bead on plate weld deposition using pulsed current GMAW, International Journal for the Joining of Materials, **12**(3), 76-85 (2000)
24. L. Lazov, I. Muhtarov, V. Tsonev, *Method for examination of forces in butt single pass arc welded joins*, Journal of the Technical University Sofia, branch Plovdiv, **14**, 179-184 (2009)
25. L. Lazov, I. Muhtarov, P. Petrov, Modeling of Temporary and Residual Stresses and Strains in Arc Welded Joints, Journal of the Technical University Sofia, branch Plovdiv, **14**, 173-178 (2009)
26. T. Iamboliev, I. Muhtarov, T. Petrov, Investigation of the temperature field and forces under GMAW of double pass butt welds, *Proceedings of the 2nd SE IIW Congress*, Sofia, 272-276 (2010)
27. L. Yajiang, W. Juan, C. Maoai, S. Xiaoqin, Finite element analysis of residual stress in the welded zone of high strength steel, Bulletin of Materials Science, **27**, 127-132 (2004)
28. ISO 15614-1:2004, Specification and qualification of welding procedures for metallic materials. Welding procedure test. Part 1: Arc and gas welding of steels and arc welding of nickel and nickel alloys
29. *ANSYS User's manual*, (Swanson Analysis Systems, Inc., Houston, 1998)
30. *Lucefin Group catalog data*, www.lucefin.com
31. T. Iamboliev, Optimizing pulse GTA welding of austenitic stainless steel 1.4404, *Proceedings of science conference, Ruse University*, **51**(2), 196-200 (2012)

Some aspects of the influence of the vertical wind speed gradient on the blade aerodynamics loads in high class wind generators

Julian Genov[1,*]

[1]Technical University of Sofia, Department of Mechanics, 8 Kl. Ohridski Blvd., 1000 Sofia, Bulgaria

Abstract. The powerful wind generators, producing electric power on the order of megawatts are with impressive sizes. The studies on the interaction between the air flow and the wind turbine usually assume the airflow as homogeneous in vertical direction. While for smaller wind turbines this assumption does not lead to significant inaccuracies, for powerful generators the vertical gradient of the wind velocity causes a dependence of the aerodynamic interaction from the angle of rotation of the propeller and leads to a significant cyclic variability of the forces and moments acting on the blades, even for a flow considered as stationary in time. The study of this problem is important for strength loads and fatigue calculations as well as for an implementation of an individual pitch-control of the blades, which could allow a synthesis of a regulation depending on the angular position of the turbine. This could contribute to higher performance and higher reliability.

1 Vertical distribution of wind speed

The change of the wind speed in the vertical direction depends on a number of factors such as: the degree of stability which depends of the vertical convection processes, the turbulence of the flow, the extent of its thrust, atmospheric pressure, the nature of the terrain and the presence of bulkheads with their characteristics, etc. From the basically used two representations for this distribution, which are based on power and logarithmic functional dependences [1], in the present study is preferred the logarithmic model, derived from Prandtl's mixing length theory and which is based on physical and meteorological studies and more detailed takes into an account the factors noted above. The dependence formula is:

$$v_\infty(z) = \frac{v^*}{k}\left[\ln\left(\frac{z-d}{z_0}\right) - \psi(z, z_0, L)\right], \text{m/s} \qquad (1)$$

where: $v^* = \sqrt{\tau_s/\rho}$, m/s – shear (friction) velocity of the air flow, that describes shear-related motion of vertical transference of unit mass air due to the turbulence [2];

τ_s, Pa – shear stress of flow;

$\rho = p(z)Mg/RT(z)$, kg/m^3 – mass density, depending according the Clapeyron's law from temperature and altitude;

$p(z) = p_0\left(\frac{T(z)}{T_0}\right)^{\frac{gM_g}{Rh_t}}$, Pa – atmospheric pressure at level z;

p_0 – atmospheric pressure at the zero level (for sea level p_0 = 101325 Pa);

T_0, K – temperature at the zero level,

h_t – dry air adiabatic lapse rate (temperature gradient with average value 0,0065 K/m),

g = 9,81 m/s^2 – gravitational acceleration,

R = 8,31447 J/(mol K) – universal gas constant,

M_g – molar mass (0,02896 kg/mol for dry air),

$T(z) = T_0 - h_t z$, K – temperature at height z;

$k \in [0{,}35 \div 0{,}42]$ von Kármán constant for the logarithmic distribution (the best estimate for the von Kármán constant for turbulent flow is found to be 0,4 [3]);

d – zero-plan displacement, define the height at which the mean velocity is zero due to large obstacles;

z_0 – aerodynamic roughness length is a corrective measure to account for the effect of the roughness of a surface on wind flow.

Table 1. Values of the parameters considering the influence of unevenness on the ground.

Ground surface condition	d, m	z_0, m
Water surface	0,009	0,002
Marshy or snowy without plants and obstacles	0,023	0,005
Flat grassy surface with small obstacles	0,138	0,03
Low crops occasional obstacles with a width to high ratio x/H>20	0,466	0,1
Low crops occasional obstacles with a width to high ratio 15<x/H<20	1,175	0,25
Parkland, bushes, numerous obstacles	2,235	0,5
Large forests, barriers suburbs of settlements	4,467	1,0
City centres with high- and low- rise buildings	>10	>2

$$\psi = \begin{cases} \ln\left((1+x^2)(1+x)^2\right) - 2\arctan(x) - 0{,}5086, & \text{for}\zeta \in (-2 \div 0) \text{ unstable flow} \\ -6\zeta, & \text{for}\zeta \geq 0 \text{ stable flow} \end{cases}$$

$$\zeta = \frac{z}{L}, x = (1 - 16\zeta)^{0{,}25}$$

$L = v^{*3}/\left(k\frac{g}{T_v}\Phi_H\right)$, m – Monin-Obukhov length, height where turbulence is generated more by buoyancy than by

* Corresponding author: j_genov@mail.org

https://doi.org/10.1051/matecconf/201823404006

wind shear – a criteria for fluid flow stability and turbulent kinetic energy;

$T_v = (1 + 0{,}61w)T$ – virtual temperature. It is the temperature that a dry air would have if its pressure and density were equal to those of actual moist air;

w – water mixing ratio, it is the ratio of moisture to the amount of dry air in the flow (often $w = 3{,}5\times10^{-3}$),

$\Phi_H = -h/(\rho c_p)$, Km/s – kinematic surface heat flux,

$h = 10{,}45 - v_\infty + 10\sqrt{v_\infty}$ W/(m^{2}K) – heat transfer coefficient characterizes the convective heat transfer across a unit area;

c_p – specific heat capacity, for the operating temperature range $c_p \approx 1005$ J/kg/K.

In the current study it is used the logarithmic wind distribution with the speed of $v_\infty(z_t) = 11{,}4$ m/s at the height of the turbine axis $z_t = 90$ m with temperature of 298 K (25 °C) and grassy terrain. The values of the parameters and the calculated values according to the above dependencies are given in Table 2.

Table 2. Values of the parameters for the calculation of logarithmic wind speed distribution

L	Φ_H	ρ	T_0	T_v	c_p
m	Km/s	kg/m^3	K	K	J/(kgK)
59,9	–0,03	1,23	298	290	1005
k	h	v^*	τ_s	d	z_0
-	W/(m^{2}K)	m/s	Pa	m	m
0,4	33,09	0,28	0,097	0,138	0,03

On Fig. 1 is shown the vertical wind speed distribution. As a result of the vertical gradient, the difference in the turbine disk speeds between the lower level $z = 27$ m and the upper $z = 153$ m (for a radius of the turbine disk $R = 63$ m) is about 10 m/s: 6,367÷15,98 m/s.

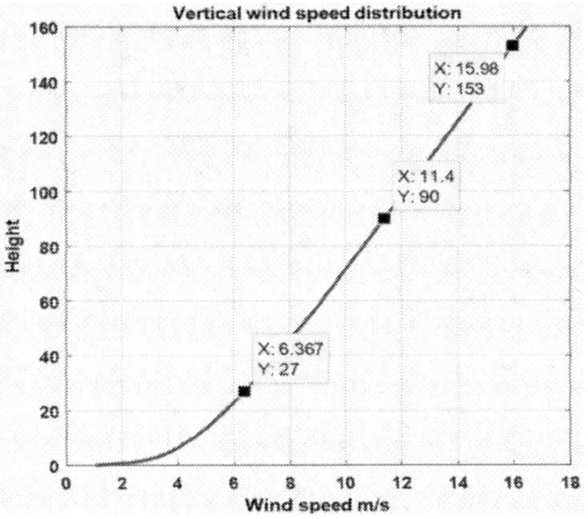

Fig. 1. Vertical distribution of wind speed.

2 Mathematical model of the aerodynamic interaction between the wind and the turbine in considering the vertical wind velocity gradient

The classical approach of the Blade element momentum (BEM) theory considers the fluid field as homogeneously [4-9]. In them for all blades, the dynamic interaction with the wind flow is the same and also does not depend on the azimuth angle which determines the position of each of them in the cross section – Fig. 2. But for the high class wind turbines, due to the inflow vertical speed gradient, it is observes significant different of the speed in the turbine disk and respectively causes different angles of attack for each of blades. For this reason the generated power and aerodynamic forces are function of the azimuth angle even for time-stationary wind flow. The common used methodology for calculation of induction coefficients [6] is not valid and is needs to be some refined.

In [10-15] some semi-analytical or numerical investigations in this or similar sense are made.

Many software realizations [16-26] in ANSYS Fluent, SolidWorks, Adams, Abacus, as well specialized products as FAST, AeroDyn, HAWC2 and others give also possibilities for modelling the aerodynamic loads in non homogenously wind field by using CFD computation, numerical-analytical calculations and approximations.

The aim of this investigation is to give one analytical model which among other approaches may be useful for the purposes of wind turbine control, loads and fatigue analysis, optimization, and others.

The elementary turbine disk area represented by the polar coordinates r and γ (azimuth angle) – Fig. 2, is:

$$\Delta A_d(r,\gamma) = r\Delta r\Delta\gamma.$$

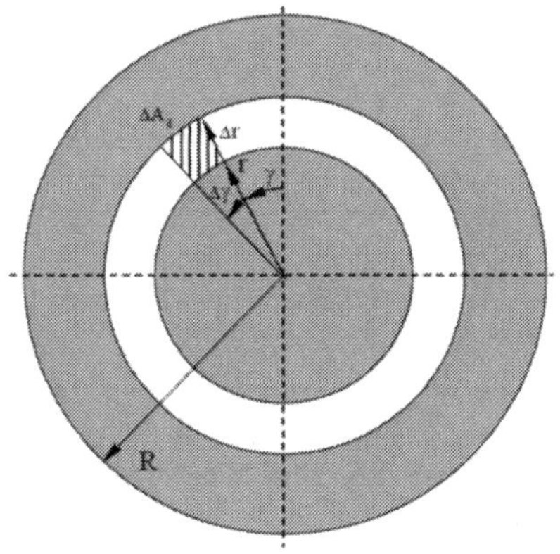

Fig. 2. Representation of the blade element area by polar frame of reference.

The dependencies for the axial thrust forces F_{Thrust} and the torque M_{wt} acting on that elementary area, on the corresponding ring of the turbine disk are:

$$\Delta F_{Thrust}(\varphi_{wt}, r, \gamma) =$$

$$= \frac{d}{dt}[\Delta Q_m(v_\infty(r,\gamma) - v_w(\varphi_{wt}, r, \gamma))]F_c(\varphi_{wt}, r) =$$

$$= \rho\Delta A_d(r,\gamma)v_d(\varphi_{wt}, r, \gamma)\big(v_\infty(r,\gamma) - v_w(r,\gamma)\big)F_c(\varphi_{wt}, r) =$$

$$= 2\rho a(\varphi_{wt}, r)\big(1 - a(\varphi_{wt}, r)\big)F_c(\varphi_{wt}, r) r v_\infty^2(r,\gamma)\Delta\gamma\Delta r,$$

$$\Delta F_{Thrust}(\varphi_{wt}, r) = \int_{\gamma=0}^{2\pi} \Delta F_{Thrust}(\varphi_{wt}, r, \gamma)d\gamma =$$

$$= 2\rho a(\varphi_{wt}, r)\left(1 - a(\varphi_{wt}, r)\right)F_c(\varphi_{wt}, r)r\Delta r.$$

$$\cdot \int_{\gamma=0}^{2\pi} v_\infty^2(r,\gamma)\, d\gamma$$

 https://doi.org/10.1051/matecconf/201823404006

$$F_{\text{Thrust}}(\varphi_{wt}) = \int_{r=0}^{R} \Delta F_{\text{Thrust}}(\varphi_{wt}, r)\, dr =$$
$$= 2\rho \int_{r=0}^{R} a(\varphi_{wt}, r)\big(1 - a(\varphi_{wt}, r)\big) F_c(\varphi_{wt}, r) r. \quad (2)$$
$$.\int_{\gamma=0}^{2\pi} v_\infty{}^2(r, \gamma)\, d\gamma dr\,.$$

$$\Delta M_{wt}(\varphi_{wt}, r, \gamma) = F_c(r, \varphi_{wt}) \frac{d}{dt} \Delta I(\varphi_{wt}, r, \gamma)\omega_d(\varphi_{wt}, r) =$$
$$= \rho r v_d(\varphi_{wt}, r, \gamma) r^2 \omega_d(\varphi_{wt}, r) F_c(\varphi_{wt}, r) \Delta r \Delta\gamma =$$
$$= 2\rho\omega_{wt} r^3 a'(\varphi_{wt}, r)(1\text{-}a(\varphi_{wt}, r)) v_\infty(r, \gamma).$$
$$.F_c(\varphi_{wt}, r)\Delta\gamma\Delta r,$$

$$\Delta M_{wt}(\varphi_{wt}, r) = \int_{\gamma=0}^{2\pi} \Delta M_{wt}(\varphi_{wt}, r, \gamma) d\gamma =$$
$$= 2\rho\omega_{wt} a'(\varphi_{wt}, r)\big(1 - a(\varphi_{wt}, r)\big) F_c(\varphi_{wt}, r) r^3.$$
$$.\int_{\gamma=0}^{2\pi} v_\infty(r, \gamma) d\gamma \Delta r,$$

$$M_{wt}(\varphi_{wt}) = \int_{r=0}^{R} \Delta M_{wt}(\varphi_{wt}, r) dr =$$
$$= 2\rho\omega_{wt} \int_{r=0}^{R} a'(\varphi_{wt}, r)\big(1 - a(\varphi_{wt}, r)\big) F_c(\varphi_{wt}, r) r^3 \,.(3)$$
$$.\int_{\gamma=0}^{2\pi} v_\infty(r, \gamma) d\gamma\, dr.$$

where: φ_{wt} – blade azimuth angle measurement from vertical position in counter clockwise direction;
Q_m – mass flow rate;
ρ – mass density;
I – moment of inertia of the fluid in the turbine disk;
v_∞ – inflow speed;
v_w – flow speed in wake;
v_d – flow speed in turbine disk;
$a = (v_\infty - v_d)/v_\infty$ – axial induction factor;
$a' = 0{,}5\ \omega_d/\omega_{wt}$ – tangential induction factor;
ω_d – angular velocity of the flow in the plane of the disk;
$F_C = F_{TL}F_{Root}$ – flow correction factor [5],
$F_{TL} = \frac{2}{\pi}\arccos\left(e^{\frac{N}{2\sin(\phi)}\left(\frac{R}{r}-1\right)}\right)$ – Tip-Loss Prandtl-Glauert correction;
$F_{Root} = \frac{2}{\pi}\arccos\left(e^{-\frac{N}{2\sin(\phi)}\left(1-\frac{R_{Hub}}{r}\right)}\right)$ is a correction that considers the vortices at the blade's base around the turbine hub;
N is the number of the blades; R_{Hub} is the hub radius.

On Fig. 3 are shown angular, velocities and forces that describe the aerodynamic interaction for given blade cross-section.
$\beta(\varphi_{wt}, r) = \beta_c(\varphi_{wt}) + \beta_0(r)$ – pitch angle for the section;
$\beta_0(r)$ – constructive pitch angle for the section;
$\beta_c(\varphi_{wt})$ – blade pitch angle set by the system for active pitch control;
$\phi(\varphi_{wt}, r) = \text{arctg}(v_d(\varphi_{wt}, r)/v_b(\varphi_{wt}, r))$ – inflow angle for the section;
$\alpha(\varphi_{wt}, r) = \phi(\varphi_{wt}, r) - \beta(\varphi_{wt}, r)$ – angle of attack.
$v_b(\varphi_{wt}, r)$ – a tangential component of the relative flow velocity for the section;
$v_{rel}(\varphi_{wt}, r)$ – relative (effective) flow velocity for the section;
$F_{Torque}(\varphi_{wt}, r)$ – torque force;
C_{Torque}, C_{Thrust}, C_L, C_D – torque, thrust, lift and drag coefficients of the section.

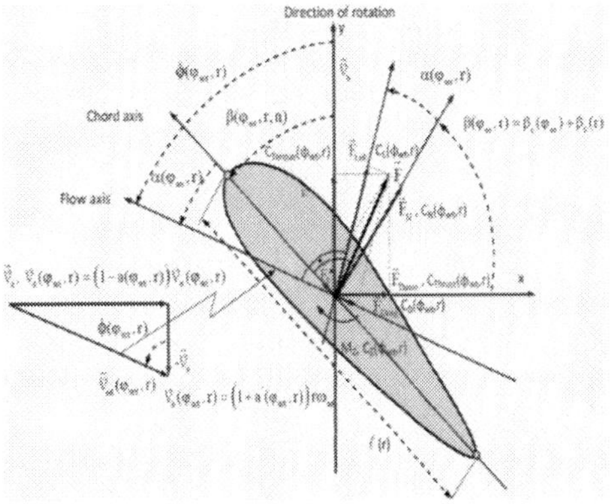

Fig. 3. Wind stream and induced forces acting on blade elements.

$$\begin{cases} v_b(\varphi_{wt}, r) = r\omega_{wt}\left(1 + a'(\varphi_{wt}, r)\right); \\ v_{rel}(\varphi_{wt}, r) = \frac{(1-a(\varphi_{wt}, r))}{\sin(\phi(\varphi_{wt}, r))} v_\infty(\varphi_{wt}, r); \\ \phi(\varphi_{wt}, r) = arctg\left[\frac{(1-a(\varphi_{wt}, r))v_\infty(\varphi_{wt}, r)}{r\omega_{wt} r(1+a'(\varphi_{wt}, r))}\right] \end{cases}$$

$$\begin{matrix} F_{\text{Thrust}}(\varphi_{wt}, r) = C_{\text{Thrust}}(\varphi_{wt}, r) F(\varphi_{wt}, r) \\ F_{\text{Torque}}(\varphi_{wt}, r) = C_{\text{Torque}}(\varphi_{wt}, r) F(\varphi_{wt}, r) \end{matrix} \quad (4)$$

where:

$$F(\varphi_{wt}, r) = 0{,}5\rho\ell_T(r) v_{rel}{}^2(\varphi_{wt}, r) F_c(\varphi_{wt}, r);$$

$$C_{Thrust}(\varphi_{wt}, r) = C_L\left(\alpha(\varphi_{wt}, r)\right)\cos\left(\phi(\varphi_{wt}, r)\right) +$$
$$+ C_D\left(\alpha(\varphi_{wt}, r)\right)\sin\left(\phi(\varphi_{wt}, r)\right);$$

$$C_{Torque}(\varphi_{wt}, r) = C_L\left(\alpha(\varphi_{wt}, r)\right)\sin\left(\phi(\varphi_{wt}, r)\right) -$$
$$- C_D\left(\alpha(\varphi_{wt}, r)\right)\cos\left(\phi(\varphi_{wt}, r)\right);$$

$$\ell_T(\varphi_{wt}, r) = \ell(r)\cos\left(\beta(\varphi_{wt}, r)\right).$$

For the blade it is obtain:

$$\begin{cases} a(\varphi_{wt}, r) = \frac{K(\varphi_{wt}, r)}{1+K(\varphi_{wt}, r)} \\ a'(\varphi_{wt}, r) = \frac{\omega_d(\varphi_{wt}, r)}{2\omega_{wt}} = \frac{K_{Torque}(\varphi_{wt}, r)}{\tilde{\lambda}(r) K_{Thrust}(\varphi_{wt}, r)} a(\varphi_{wt}, r) \end{cases} \quad (6)$$

where: $K(\varphi_{wt}, r) = \frac{\ell_T(\varphi_{wt}, r)}{4r\int_{\gamma=0}^{2\pi} V_\infty{}^2(r, \gamma) d\gamma} K_{Thrust}(\varphi_{wt}, r)$,

$$\tilde{\lambda}(r) = r\omega_{wt}\int_{\gamma=0}^{2\pi} V_\infty(r, \gamma) d\gamma \,/\, \int_{\gamma=0}^{2\pi} V_\infty{}^2(r, \gamma)\, d\gamma.$$

The iterative cycle for determining the aerodynamic characteristics for each blade element is:

- Set Initial values for the induction coefficients (most commonly);
- Calculate $\phi(\varphi_i, r)$ и $\alpha(\varphi_i, r)$, i=1,..N for the current values of $v_\infty(\varphi_i, r)$ and ω_{wt} (note for individual pitch control $\beta_c(\varphi_{wt}) \Rightarrow \beta_c(\varphi_i)$;
- Determine the values of $C_L(\alpha(\varphi_i, r))$ and $C_D(\alpha(\varphi_i, r))$ via the calculated angles, i=1,..N, using experimentally obtained data for the corresponding airfoil;
- Calculate the corrections $F_c(\varphi_i, r)$, i=1,..N;
- Obtain new values for the induction coefficients;
- Iterations continue until reaching a convergence.

 https://doi.org/10.1051/matecconf/201823404006

3 Numerical results

The obtained dependences are tested for prototype wind turbine NREL5MW [27] with reference operating wind speed 11,4 m/s and angular speed of the turbine 12 rpm. The hub radius is R_{hub} = 1,5 m, the blade length is L = 61,5 m, and the tower is with height of H = 90 m.

The types of the airfoils along the blades, with their main characteristics and lift and drag coefficients as function of the angle of attack are given in [27].

Results of numerical simulation are shown on Figures 4 to 12.

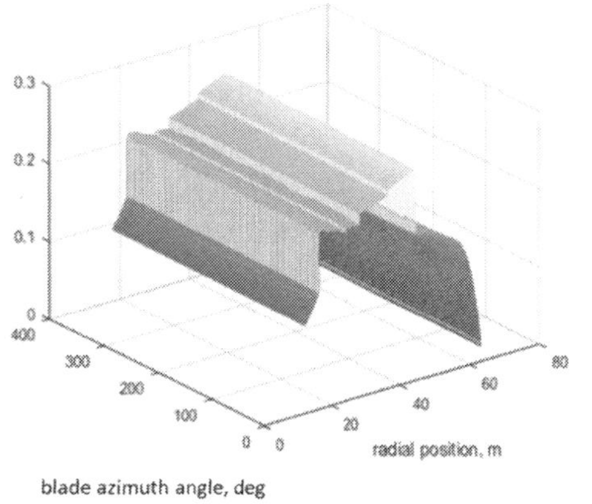

Fig. 4. Axial indication factor.

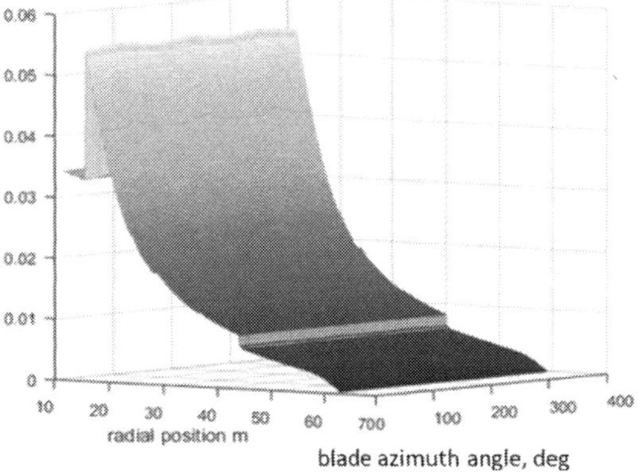

Fig. 5. Tangential induction factor.

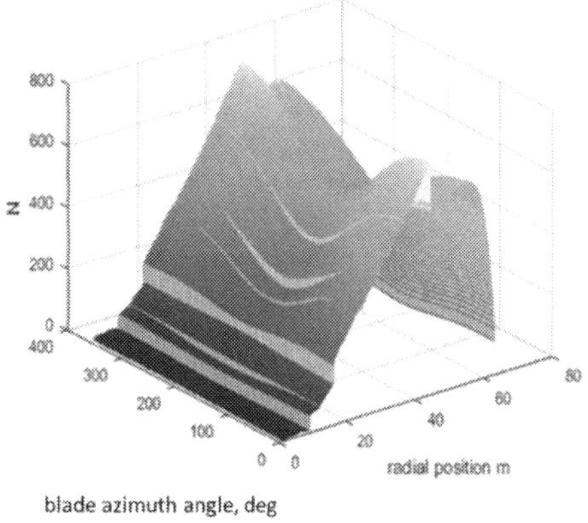

Fig. 6. Blade 1 Thrust forces distribution.

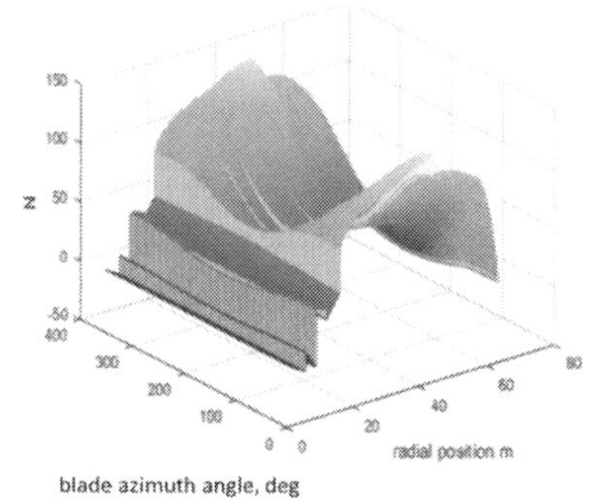

Fig. 7. Blade 1 Torque force distribution.

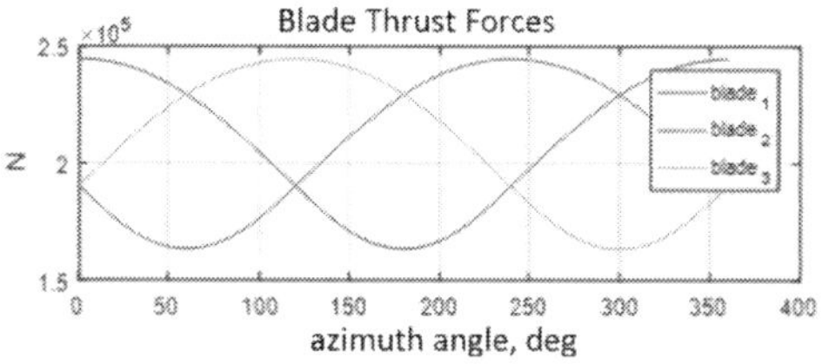

Fig. 8. Blade Thrust Forces.

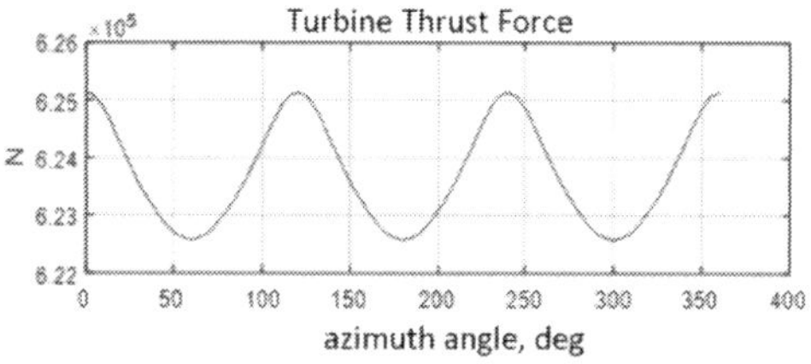

Fig. 9. Turbine Thrust Forces.

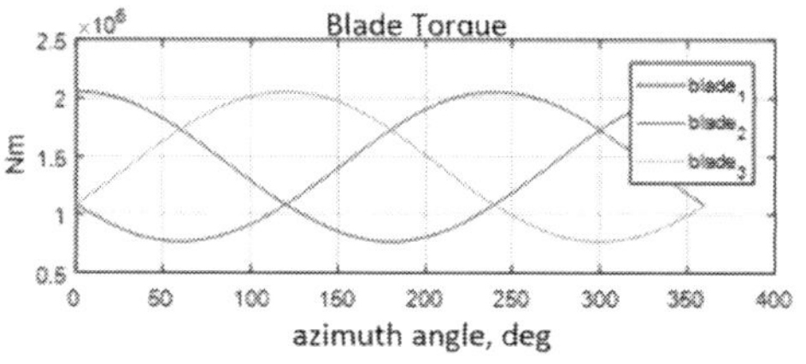

Fig. 10. Blade Torque.

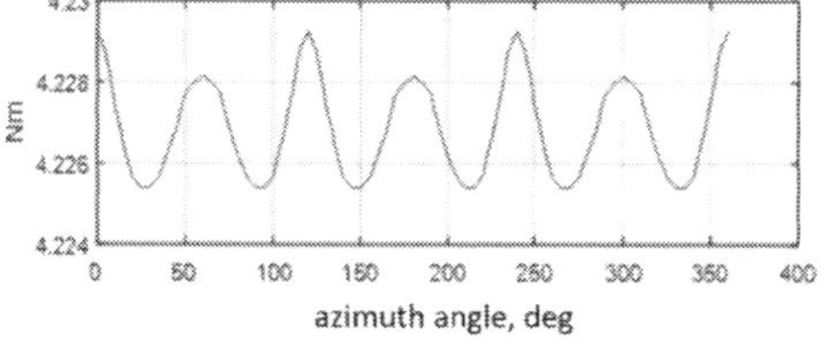

Fig. 11. Turbine Torque.

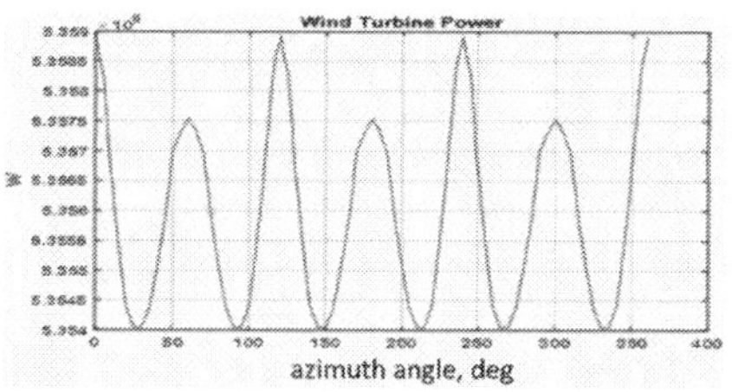

Fig. 12. Wind Turbine Power.

 https://doi.org/10.1051/matecconf/201823404006

The pulsations in the total axial force, torque and power generation have small amplitude with respect to their mean values: $F_{Thrust} = 6{,}233 \pm 1{,}28$ kN, $M_{wt} = 4227 \pm 1{,}92$ kNm, $P_{wt} = 5{,}356$ MW.

The calculations show that the variation in the load of the individual blades as a function of the angle of rotation due to the vertical gradient of the wind speed is very significant. For example for F_{Thrust} amplitude is 40,6 kN – nearly 20% of its mean value $\tilde{F}_{Thrust}$=208 kN.

These values are aggregated for the whole blade. For the some elements in the middle part of the blade they are significant bigger.

4 Conclusion

The publication presents the BEM Theory adaptation in cases when the vertical speed gradient of the wind flow must be taken into an account. In those cases the turbine blades have different aerodynamic interactions with the flow and the classical BEM approach is not accurate. Here is proposing the solution of this problem what is applied and verified for a prototype of high class.

References

1. L. Landberg, *Meteorology for Wind Energy – An Introduction*, (John Wiley & Sons Ltd., 2016)
2. S. Smith, et al., Sea surface wind stress and drag coefficients: The hexos results, Boundary-Layer Meteorology, **60**, 109-142 (1992)
3. Y. Zhang, J. Ma, Z. Cao, The Von Karman constant retrieved from CASES-97 dataset using a variational method, Atm. Chem. and Phys. **8**, 7045-7053 (2008)
4. W. Rankine, W. Froude, On the mechanical principles of the action of propellers, Transactions of the Institution of Naval Architects, **6**, 13-39 (1865)
5. L. Prandtl, Applications of modern hydrodynamics to aeronautics, Translated by the NACA, Technical Report №116, NACA Annual Report 7, 157-215 (2003)
6. R. Wilson, P. Lissaman, *Applied aerodynamics of wind power machines*, National Science Foundation, Oregon State University, USA, 109 (1974)
7. J. Anderson, *Fundamentals of aerodynamics*, (McGraw-Hill Inc., 1991)
8. M. Hansen, *Aerodynamics of wind turbines – Third edition*, (Taylor &Francis Group, 2015)
9. P. Jamieson, *Innovation in wind turbine design*, 297 (John Wiley & Sons Ltd., 2011)
10. K. Cox, A. Echtermeyer, Structural design and analysis of a 10MW wind turbine blade, Energy Procedia, **24**, 194-201 (2012)
11. J. Conway, Analytical solutions for the actuator disc with variable radial distribution of load, J. Fluid Mech., **297**, 327-355 (1995)
12. J. Conway, Exact actuator disk solutions for nonuniform heavy loading and slipstream contraction, J. Fluid Mech., **356**, 235-267 (1998)
13. T. Wang, *Unsteady aerodynamic modelling of horizontal axis wind turbine performance*, PhD Dissertation, University of Glasgow, 299 (1999)
14. R. Bontempo, *The nonlinear actuator disk method as applied to open and ducted rotors*, PhD Dissertation, University of Naples Federico II, 153 (2014)
15. T. Sant, *Improving BEM-based aerodynamic models in wind turbine design codes*, PhD Dissertation, Delft University – Wind Energy Institute, 447 (2007)
16. A. Miller et al., Review of Computer-Aided Numerical Simulation in Wind Energy, Renewable & Sustainable Energy Reviews, **25**, 122-134 (2013)
17. M. Buhl, A. Manjock, A comparison of wind turbine aeroelastic codes used for certification, NREL/CP-500-39113, 15 (2006)
18. M. Sprague, J. Jonkman, B. Jonkman, FAST modular wind turbine CAE tool: Nonmatching spatial and temporal meshes, *SciTech 2014 National Harbor,* Maryland Jan. 13-17, 2014, 25, (2014)
19. M. Sprague, J. Jonkman, B. Jonkman, FAST modular framework for wind turbine simulation: New algorithms and numerical examples, *AIAA SciTech 2015: 33rd Wind Energy Symposium*, Kissimmee, Florida Jan. 5-9, 2015, 26 (2015)
20. J. Jonkman, FAST open-sources platform for wind turbine multy – physics engineering modeling, *NAWEA Symposium 2017*, 68 (2017)
21. B. Jonkman, L. Kilcher, *TurbSim User's Guide: Version 1.06.00*, NREL, U.S. Department of Energy, 97 (2012)
22. G. Bir, User's Guide to MBC3 (Multi-blade coordinate transformation utility for 3-bladed wind turbines), National Renewable Energy Laboratory, 19 (2008)
23. T. Larsen, A. Hansen, *How 2 HAWC2, the user's manual*, Risø-R-1597(ver. 4-6) (EN), TU of Denmark Roskilde, Denmark, 113 (2015)
24. T. McCoy, Advanced methods for development of wind turbine models for control design, XCEL RDF Project CWO2, Global Energy Concepts, LLC, 50 (2003)
25. Cv. Velkova, J. Minchev, Numerical flow simulation of a SCREW propeller, Machines Mechanics, ISSN 0861-9727, **118**, 1-7 (2017)
26. S. Ning, et al., Development and validation of a new blade element momentum Skewed-Wake model within AeroDyn, *AIAA Science and Technology Forum and Exposition*, 2015 Kissimmee, Florida, 19 (2015)
27. J. Jonkman, S. Butterfield, W. Musial, G. Scott, Definition of a 5-MW reference wind turbine for offshore system development, Technical Report NREL/TP-500-38060, 75 (2009)

A synthesis of a blade pitch control for large wind turbines, taking into account the functional dependence of the torque generated in the blades from the azimuth angle

Julian Genov[1,*]

[1]Technical University – Sofia, Department of Mechanics, 8 Kl. Ohridski Blvd., 1000 Sofia, Bulgaria

Abstract. In multi-MW (1MW and up) wind generation systems a variable speed - variable pitch (VS-VP) control technology is usually adopted to improve speed and torque characteristics and to obtain the optimal energy, which means to obtain the maximum power at low wind speed and the rated power at high wind speed. The power generated by an individual blade of the wind turbine changes considerably during rotation due for the change of wind speed in vertical direction, which decreases the angles of attack in the low blade positions and which leads to a reducing the generated energy. In the paper basing on BEM theory modification for none-homogenously fluid field and an optimization procedure is obtained an individual variable pitch control law as function of wind speed values and blades positions.

1 Introduction

The vertical speed gradient leads to significant changes of the torque generated from one blade during rotation. This is due to the fact that in the lower blade positions the wind speed decreases considerable, this leads to reducing the angles of attack and as a result of this, the torque diminishes. Therefore, if an individual blade pitch control is applied, it is possible to increase power to a certain extent, especially for wind speeds below the nominal. The determination of this control is complicated because the blades are building from sections with different profile and respectively different lift and drag coefficients. Thus if the angle of attack exceeded a certain value the aerodynamic interaction may get worse. Moreover due to the continuity of the fluid flow, the induction coefficients are in dependence with aerodynamic state of all blades, which is a prerequisite for additional complexity in determination of the optimum angle of attack in the given position of the blade. For a solution of this problem, basing on the adaptation of the classical blade element momentum (BEM) theory to the accounting of the vertical gradient of the wind speed, in this research an optimization approach for the determination of the individual pitch control law is proposed. The approach is realised for a set of discrete positions of the turbine and then the control dependence is obtained of by interpolation between them. The obtained results for a known prototype of wind turbine show simultaneously the validity of the BEM theory modification and of the propose approach for synthesis of an individual blade pitch control.

2 Pitch control synthesis

In [1-9] is investigated a pitch control synthesis that provided efficiency and high quality of produced electricity as well as maintaining the lowest possible dynamic loads in some elements of the wind generator's system.

In present paper this synthesis is treated in an aspect of obtaining both the maximum power at lower than the nominal wind speeds and the rated power at higher then nominal wind speeds, taking in account the fluid speed gradient in vertical direction. For the first purpose is investigated an individual form of pitch angle control and for the second is using collective pitch control form.

2.1 Wind speed distribution

The proposed algorithm is presented for the prototype wind turbine of high class - NREL5MW [10]. For the vertical wind speed distribution is preferred the logarithmic model [11]. For the parameters values defined in [12], the wind distribution as a function of the speed in the hub axis at a height of 90 m, and height z has the form:

$$v_\infty(z)=0{,}059v_\infty(z=90)\left[\ln\left(\frac{z-0{,}138}{0{,}03}\right)+0{,}1z\right],\frac{\mathrm{m}}{\mathrm{s}} \quad (1)$$

On Fig. 1 are shown vertical distributions for in the hub axis wind speeds: 3÷10, 10.5, 11, 11.4 m/s.

* Corresponding author: j_genov@mail.org

https://doi.org/10.1051/matecconf/201823404007

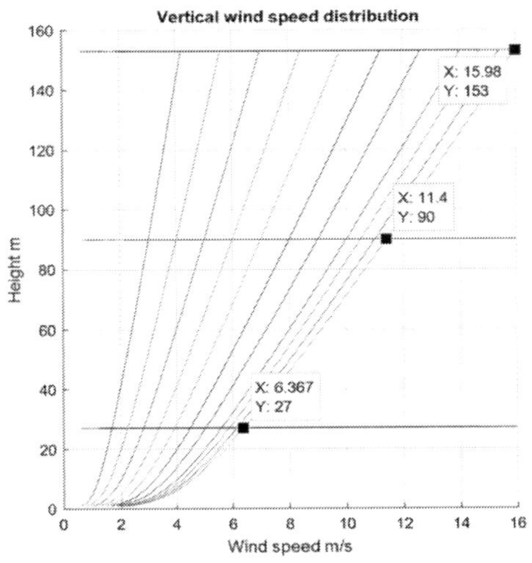

Fig. 1. Vertical distribution of wind speed.

2.2 BEM theory adaptation with taking in account the vertical wind speed gradient

The BEM theory is presented in many sources [13-23]. In the most of them the fluid field is considered as a homogeneous i.e. the differences of the wind speed in a vertical direction are neglected. Thus for all blades, the dynamic interaction with the wind flow is an identical and does not depend on the azimuth angle. For the turbines with standard size this is acceptable but for the high class it is observes significant different of the vertical speed distribution in the turbine disk. This causes that, for the given position each of the blades have a considerably different aerodynamic interaction with the wind. Respectively the aerodynamic forces that act on the blade become a function of the blade azimuth angle and varying even if the longitudinal turbulence of the wind velocity is not considered. The common used methodology for calculation of induction coefficients is needs to be revision. In [14, 16-19, 21, 23] some semi-analytical or numerical investigations in this or similar sense are made.

In the next text are presented the results obtained by the author in adaptation of BEM theory for the case of nonhomogeneous vertical wind speed distribution.

On Fig. 2 are shown the elements of the aerodynamic excitations in the plane of the blade cross-section.

The designations shown in the figure are as follows:

r – distance from the hub axes;

φ_{wt} – blade azimuth angle (zero in vertical upward direction);

$\omega_{wt} = \dot{\varphi}_{wt}$ – angular velocity of the turbine rotation;

$v_\infty(\varphi_{wt}, r)$ – inflow speed; $v_d(\varphi_{wt}, r)$ – flow speed in turbine disk;

$v_b(\varphi_{wt}, r) = r\omega_{wt}\big(1 + a'(\varphi_{wt}, r)\big)$ – tangential component of the relative flow velocity for the section;

$v_{rel}(\varphi_{wt}, r) = \frac{(1-a(\varphi_{wt},r))}{\sin(\phi(\varphi_{wt},r))} v_\infty(\varphi_{wt}, r)$ – relative (effective) flow velocity;

$a = (v_\infty - v_d)/v_\infty$ – axial induction factor;

$a' = 0{,}5\omega_d/\omega_{wt}$ – tangential induction factor;

ω_d – angular velocity of the flow in the disk plane;

$\beta(\varphi_{wt}, r) = \beta_c(\varphi_{wt}) + \beta_0(r)$ – section pitch angle;

$\beta_0(r)$- construction pitch angle for the section;

$\beta_c(\varphi_{wt})$ – blade pitch angle set by the system of the active control;

$\phi(\varphi_{wt}, r) = \mathrm{arctg}(v_d(\varphi_{wt}, r)/v_b(\varphi_{wt}, r))$ – inflow angle for the section;

$\alpha(\varphi_{wt}, r) = \phi(\varphi_{wt}, r) - \beta(\varphi_{wt}, r)$ – angle of an attack;

$F_{Torque}(\varphi_{wt}, r)$ – torque force, $F_{Thrust}(\varphi_{wt}, r)$ – thrust forces, $M_{wt}(\varphi_{wt}, r)$ - torque

$C_{Torque}, C_{Thrust}, C_L, C_D$ – torque, thrust, lift and drag coefficients of the section.

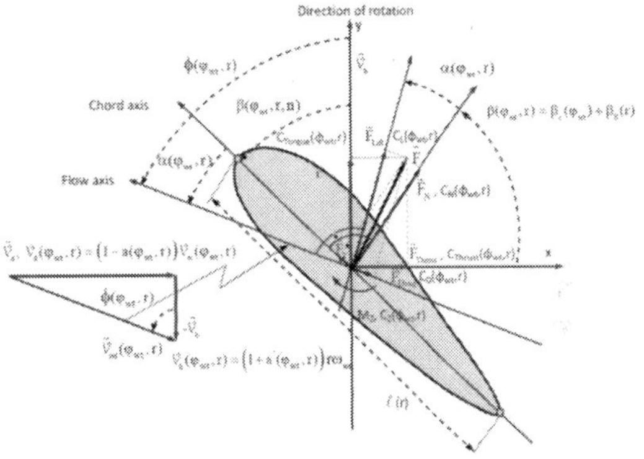

Fig. 2. Wind stream and induced forces acting on blade elements.

The following dependencies are in effect:

$$F_{\text{Thrust}}(\varphi_{wt}, r) = C_{\text{Thrust}}(\varphi_{wt}, r)F(\varphi_{wt}, r)$$
$$F_{\text{Torque}}(\varphi_{wt}, r) = C_{\text{Torque}}(\varphi_{wt}, r)F(\varphi_{wt}, r) \quad (2)$$

where:

$$F(\varphi_{wt}, r) = 0{,}5\rho\ell_T(r)v_{rel}{}^2(\varphi_{wt}, r)F_c(\varphi_{wt}, r);$$

$$C_{Thrust}(\varphi_{wt}, r) = C_L\big(\alpha(\varphi_{wt}, r)\big)\cos\big(\phi(\varphi_{wt}, r)\big) + \\ +C_D\big(\alpha(\varphi_{wt}, r)\big)\sin\big(\phi(\varphi_{wt}, r)\big);$$

$$C_{Torque}(\varphi_{wt}, r) = C_L\big(\alpha(\varphi_{wt}, r)\big)\sin\big(\phi(\varphi_{wt}, r)\big) - \\ -C_D\big(\alpha(\varphi_{wt}, r)\big)\cos\big(\phi(\varphi_{wt}, r)\big),$$

$\ell_T(\varphi_{wt}, r) = \ell(r)\cos\big(\beta(\varphi_{wt}, r)\big)$ – projection of the chord of the blade cross-section $\ell(r)$,

$F_c = F_{TL}F_{Root}$ – flow correction factor [20],

$F_{TL} = \frac{2}{\pi}\arccos\left(e^{-\frac{N}{2\sin(\phi)}\left(\frac{R}{r}-1\right)}\right)$ – Tip-Loss Prandtl-Glauert correction;

$F_{Root} = \frac{2}{\pi}\arccos\left(e^{-\frac{N}{2\sin(\phi)}\left(1-\frac{R_{Hub}}{r}\right)}\right)$ – is a correction that considers the vortices at the blade's base around the turbine hub;

N is the number of the blades;

R_{Hub} is the hub radius.

For the blade, it is obtained:

$$F_{\text{Thrust}}(\varphi_{wt}) = \\ =0{,}5\rho\int_{r=R_{Hub}}^{R} \ell_T(\varphi_{wt}, r)\big(1-a(\varphi_{wt}, r)\big)^2 K_{Thrust}(\varphi_{wt}, r)dr$$

$$M_{wt}(\varphi_{wt}) = 0{,}5\rho \int_{r=R_{Hub}}^{R} \ell_T(\varphi_{wt}, r)\left(1 - a(\varphi_{wt}, r)\right)^2 r K_{Torque}(\varphi_{wt}, r)dr \qquad (3)$$

where:

$$K_{Thrust}(\varphi_{wt}, r) = \sum_{i=1}^{N}\left[C_L\left(\alpha(\varphi_i, r)\right)cotg\left(\phi(\varphi_i, r)\right) + \right.$$
$$\left. + C_D\left(\alpha(\varphi_i, r)\right)\right]\Upsilon(\varphi_i, r);$$
$$K_{Torque}(\varphi_{wt}, r) = \sum_{i=1}^{N}\left[C_L\left(\alpha(\varphi_i, r)\right) - \right.$$
$$\left. - cotg\left(\phi(\varphi_i, r)\right)C_D\left(\alpha(\varphi_i, r)\right)\right]\Upsilon(\varphi_i, r);$$
$$\Upsilon(\varphi_i, r) = \frac{v_\infty{}^2(\varphi_i, r)F_C(\varphi_i, r)}{\sin\left(\phi(\varphi_i, r)\right)}.$$

The induction factors are:

$$\begin{cases} a(\varphi_{wt}, r) = \dfrac{K(\varphi_{wt}, r)}{1 + K(\varphi_{wt}, r)} \\ a'(\varphi_{wt}, r) = \dfrac{\omega_d(\varphi_{wt}, r)}{2\omega_{wt}} = \dfrac{K_{Torque}(\varphi_{wt}, r)}{\tilde{\lambda}(r)K_{Thrust}(\varphi_{wt}, r)}a(\varphi_{wt}, r) \end{cases} \qquad (4)$$

where:

$$K(\varphi_{wt}, r) = \frac{\ell_T(\varphi_{wt}, r)}{4r\int_{\gamma=0}^{2\pi} V_\infty{}^2(r, \gamma)\, d\gamma} K_{Thrust}(\varphi_{wt}, r),$$

$$\tilde{\lambda}(r) = r\omega_{wt}\int_{\gamma=0}^{2\pi} V_\infty(r, \gamma)d\gamma \,/\, \int_{\gamma=0}^{2\pi} V_\infty{}^2(r, \gamma)\, d\gamma.$$

2.3 Obtaining of pitch control law

The relation between torque and angular velocity of the turbine is shown of Fig. 3.

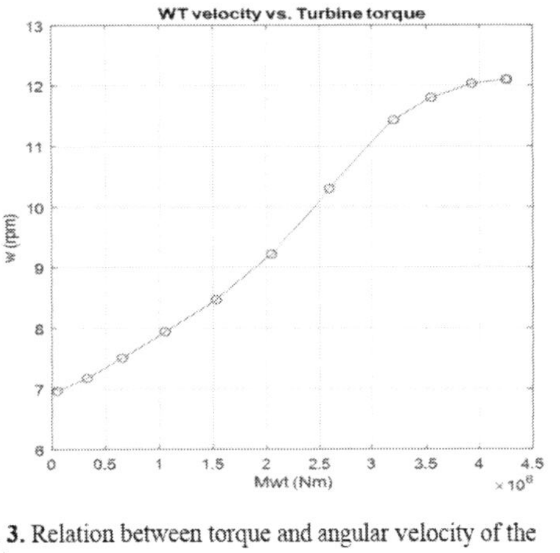

Fig. 3. Relation between torque and angular velocity of the turbine

The results, from the numerical modelling, for the angular turbine velocities ω_{wtr}, average thrust force – F_{Thrust}^{av}, torque – M_{wt}^{av} and power – P_{wt}^{av} values, without using pitch control, obtained for values between the minimal and nominal wind velocity, related to the hub axis level – v_∞^{Hub}, are shown in Table 1.

The proposed control law is individual pitch for wind velocity under the nominal value and collective for over the nominal.

The pitch angle has two components:

$$\beta(\varphi_{wt}, r) = \beta_c(\varphi_{wt}) + \beta_0(r) \qquad (2)$$

where: $\beta_0(r)$ is the construction pitch angle for the blade section;

$\beta_c(\varphi_{wt})$ blade pitch angle set by the system for active pitch control.

Table 1. The results from the modelling without pitch control

v_∞^{Hub}	ω_{wtr}	F_{Thrust}^{av}	M_{wt}^{av}	P_{wt}^{av}
m/s	tr/min	kN	MNm	MW
3	6.97	37.0	0.053241	0.038860
4	7.18	106.7	0.325716	0.244902
5	7.51	154.4	0.658100	0.517560
6	7.94	206.7	1.054802	0.877041
7	8.47	262.2	1.530413	1.357440
8	9.22	329.7	2.044856	1.974342
9	10.30	415.0	2.594580	2.798549
10	11.43	511.8	3.204408	3.835507
10.5	11.80	558.0	3.544856	4.380354
11	11.99	598.5	3.925783	4.945617
11.4	12.10	625.6	4.254310	5.390675

The torque of the blade elements as function of the azimuth angle and the element position by the length of the blade for the different wind speed from 3 to 25 m/s is shown of Fig. 4.

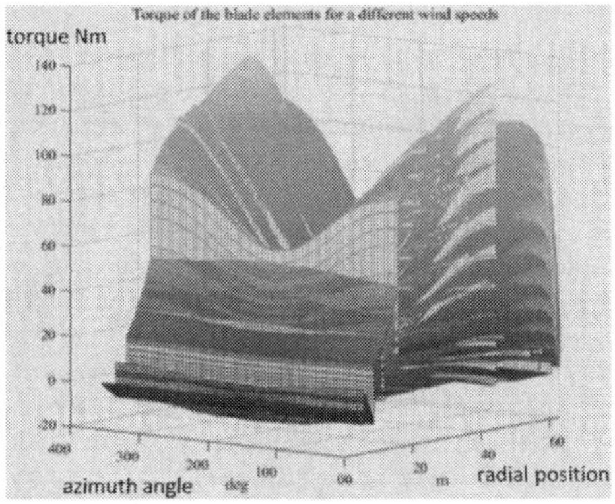

Fig. 4. Torque of the blade elements as function of the blade azimuth angle and the element position for wind speed from 3 to 25 m/s, without using pitch control.

By using the described models of the wind and of the aerodynamic interaction, for fixed azimuth angles $\varphi_{wti} = 0°$, $30°$ and $60°$ measured from the upper vertical position, an optimization procedure is used applied for all discrete values of the wind velocities v_{∞_j} in the working range from 3 to 25 m/s.

The object functions for the velocities under the nominal are defined by mean of the wind turbine torque for the corresponding discrete angle of rotation φ_{wti}. The procedures of optimization are realized consecutively for three of the positions of the turbine:

$$\varphi_{wt_1} = 0^0, \varphi_{wt_2} = 30^0, \varphi_{wt_3} = 60^0,$$

and for these positions, as an optimization parameters are obtain the pitch control angles of the three blades:

$$\beta_{i,1}, \beta_{i,2}, \beta_{i,3}, \text{ for } \varphi_{wt_i}, \text{ i=1,2,3,}$$

 https://doi.org/10.1051/matecconf/201823404007

witch maximise the turbine torque at the corresponding position.

This is repeated for all discrete values of the wind speed referred on level of turbine axis – v_{∞}^{Hub}:

$$3 \div 10;\ 10{,}4;\ 11;\ 11{,}4\ \text{m/s}.$$

The used constraints for the optimization parameters are of the type Lower-Upper boundaries.

In this way the optimization can be described mathematically as:

$$\left\{ J_i(\mathbf{X}_i) = \max_{X_i \in D} M_{wt}(\varphi_{wt_i}, v_{\infty_j}, \mathbf{X}_i), \mathrm{i} = 1,2,3 \right\} \Bigg|_{j=1,..11} \quad (3)$$

$$\mathbf{X}_i = \left[\beta_{i,1}, \beta_{i,2}, \beta_{i,3}\right],$$

$$\mathrm{D} = \left\{ \mathbf{X} \in \left| \mathbf{X}_i^{Lower} \le \mathbf{X} \le \mathbf{X}_i^{Upper} \right\} \in \mathrm{P}^{3}_{\le 0}$$

$$\varphi_{wt_i}\Big|_{i=1,2,2} \in \left\{ \varphi_{wt_1} = 0^0, \varphi_{wt_2} = 30^0, \varphi_{wt_3} = 60^0 \right\}$$

$$v_{\infty_j} \in \{3 \div 10;\ 10{,}4;\ 11;\ 11{,}4\}\ \mathrm{m/s}.$$

The pitch control angles values in the other positions in the turbine disk, located through 30 degrees, are obtained taking in an account theirs symmetry relative to the aerodynamic interaction. Then through a polynomial interpolation of the obtaining data for these discrete positions, are calculated the pitch angles values of the blades, for any other intermediate turbine position.

The object functions, for the all velocities of the wind over the nominal, is referred to the condition that the average values of the torques for a one full rotation of the turbine M_{wt}^{av} should be equal to the nominal value of 4.25431 MNm (see Table 1). In this case the optimized parameter is the collective pitch angle, which is not treated as function of the azimuth angle and it is only a function of the wind speed. The mathematical presentation of this optimization task is:

$$J_j(\beta_j) = \min_{\beta_i \in D} \left| M_{wt}\left(v_{\infty_j}, \beta_j\right) - M_{wt}(v_{\infty}^{\mathrm{n}}, 0) \right| \Bigg|_{j=12,..25} \quad (4)$$

$$\mathrm{D} = \left\{ \beta \in \left| 0 \le \beta \le \beta_j^{Upper} \right\} \in \mathrm{P}_{\ge 0}$$

$$v_{\infty_j} \in \{12, 13, ..25\}\mathrm{m/s}.$$

The optimization is realised by the Quasi-Monte Carlo method based on the using of the Sobol sequences [24]. It is generated cover of 150 points in the definition area, for each position. This number of points proves to be sufficient for the localization of the optimum.

The obtained results are given in Table 2.

Table 2. Results of modeling with pitch control - individual for wind velocity under the nominal value and collective for over the nominal.

v_{∞}	Pitch angle β	ω_{wt}	F_{Thrust}^{av}	M_{wt}^{av}	P_{wt}^{av}
m/s	deg	tr/min	kN	MNm	MW
3	polynomial approximation	6.97	37.0	0.057231	0.0389
4	--//--	7.18	106.7	0.395435	0.2449
5	--//--	7.51	154.4	0.7242781	0.5176
6	--//--	8.00	238.6	1.105343	0.9254
7	--//--	8.54	306.9	1.584803	1.4169
8	--//--	9.37	385.1	2.128162	2.0880
9	--//--	10.53	483.0	2.703402	2.9800
10	--//--	11.6	594.0	3.338278	4.0550
10,5	--//--	11.92	564.1	3.708268	4.6285
11	--//--	12.08	691.5	4.105221	5.1942
11,4	0,00	12.10	625.6	4.254310	5.3907
12	2,75	12.10	542.5	4.254371	5.3908
13	5,51	12.10	471.3	4.255132	5.3917
14	7,66	12.10	419.4	4.253839	5.3901
15	9,48	12.10	380.4	4.255933	5.3927
16	11,13	12.10	348.5	4.254364	5.3907
17	12,64	12.10	322.7	4.256122	5.3930
18	14,07	12.10	299.8	4.250831	5.3863
19	15,42	12.10	280.5	4.251241	5.3868
20	16,71	12.10	263.6	4.252468	5.3883
21	17,95	12.10	248.7	4.252856	5.3888
22	19,14	12.10	235.5	4.257366	5.3946
23	20,30	12.10	223.3	4.253881	5.3901
24	21,42	12.10	212.4	4.254932	5.3915
25	22,51	12.10	202.6	4.254687	5.3912

The individual pitch control as function of the blade azimuth angle, for different wind speeds is shown on Fig. 5.

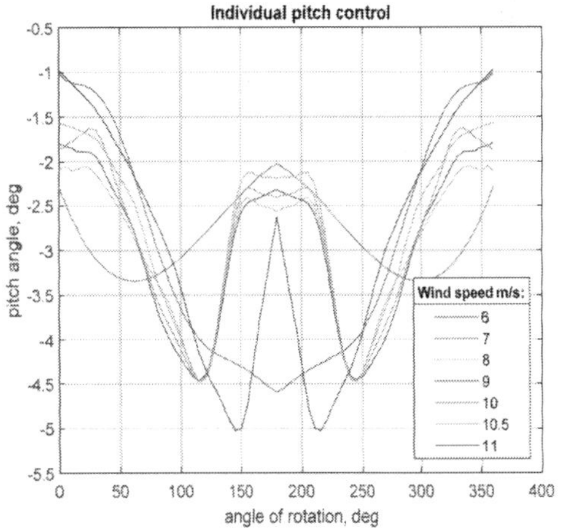

Fig. 5. Individual pitch control as function of angle of the turbine rotation.

The collective pitch control as function of the wind speeds on the level of the hub axis is shown on Fig. 6.

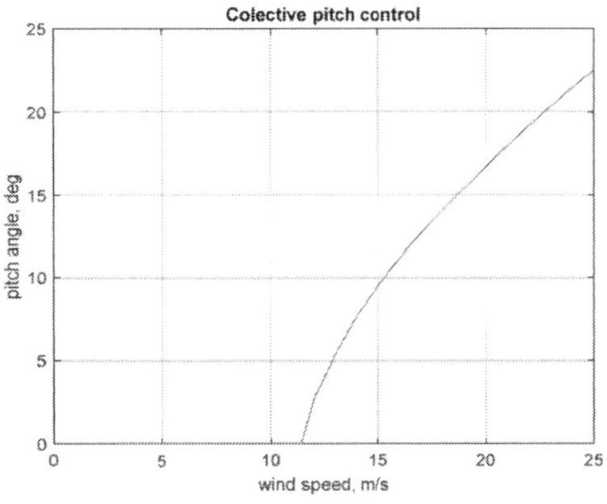

Fig. 6. Collective pitch control as function of the wind speed on the level of the turbine axis.

The pitch control as function of the wind speed at the level of hub axis and the angle of blade azimuth is shown on Fig. 7.

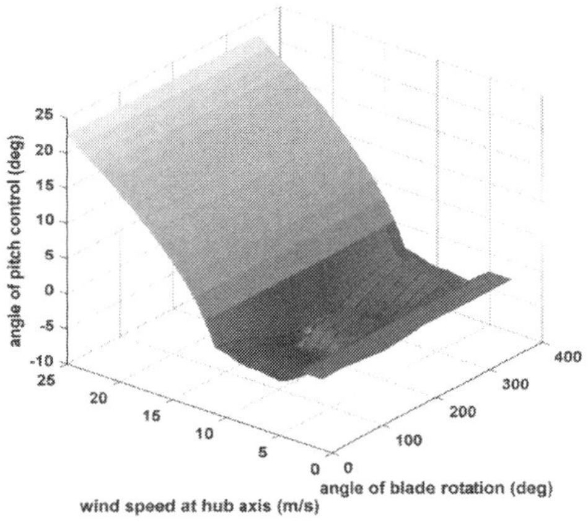

Fig. 7. Pitch control as function of the wind speed and the angle of blade rotation.

On Fig. 8 is shown the turbine power with and without pitch control and on Fig. 9 the thrust forces of the blade elements as function of the azimuth angle and the element position by the length of the blade for the different wind speed in range from 3 to 25 m/s.

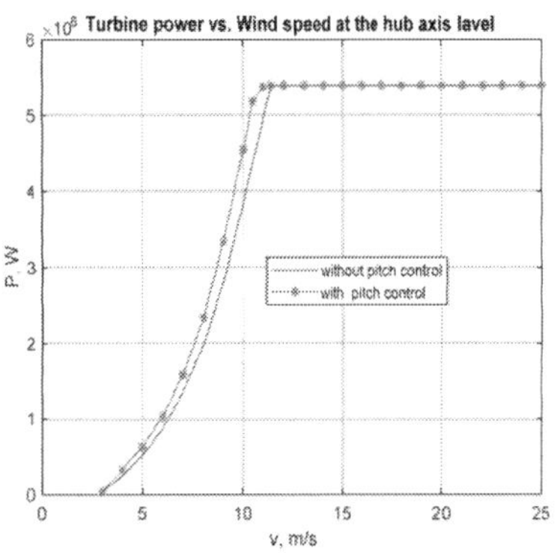

Fig. 8. Turbine power without and with proposed pitch control.

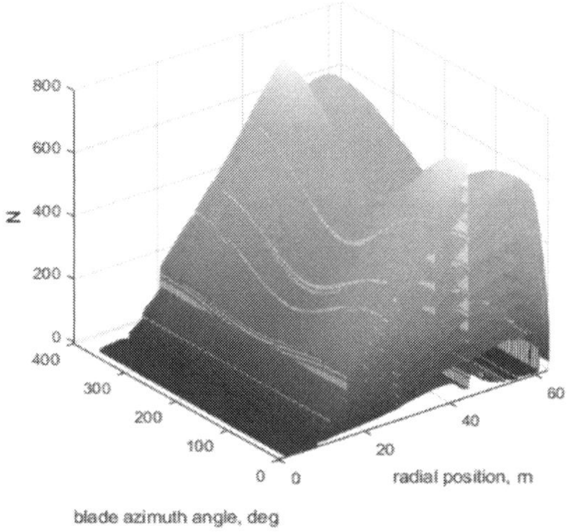

Fig. 9. Thrust forces of the blade elements as function of the azimuth angle and the element position for wind speed from 3 to 25 m/s, with using pitch control.

This is not very large, but is a tangible improvement of the power characteristics.

3 Conclusion

The paper presents an optimal synthesis of a new type pitch control of the wind turbines of high class that takes into account the effects due to the vertical gradient of the wind velocity. It is based on proposed modification of BEM theory. The obtained from the simulation results show that they are some better in comparison of these without of such type control. The theory may to be applied ease for the synthesis of the blade pitch control when the longitudinal turbulence of the wind velocity is also included in the model.

References

1. E. Hooft; P. Schaak; T. Engelen, Wind turbine control algorithms, DOWEC-F1W1-EH-03-094/0, Petten, Nederland (2003)
2. M. Hansen, A. Hansen, T. Larsen, S. Øye, P. Sørensen, P. Fuglsang, Control design for a pitch-regulated, variable speed wind turbine, Risø National Laboratory Information Service Department, Denmark, 85 (2005)
3. K. Stol, W. Zhao, Al. Wright, Individual blade pitch control for the controls advanced research turbine, Transactions of the ASME Journal of Solar Energy Engineering, **128**, 498-505 (2006)
4. F. Bianchi, H. Battista, R. Mantz, *Wind turbine control systems principles - Modelling and gain scheduling design*, 205 (Springer-Verlag London Ltd., 2007)
5. I. Munteanu, A. Bratcu, N. Cutululis, E. Ceanga, *Optimal control of wind energy systems - Towards a global approach*, 283 (Springer-Verlag London Ltd., 2008)

6. A. Wright, L. Fingersh, *Advanced control design for wind turbines Part I: Control design, implementation, and initial tests*, *Technical Report* NREL/TP-500-42437, U.S. Department of Energy, 148 (2008)
7. C. Bottasso, A. Croce, *Advanced control laws for variable-speed wind turbines and supporting enabling technologies*, Scientific Report DIA-SR 09-01, Dept. of Aerospace Engineering, Polytechnics of Milano, 41 (2009)
8. J. Genov et al., Modeling and control of wind turbine tower vibrations, AIP Conference Proceedings, **1293**(1), 30-38 (2010)
9. J. Genov, G. Venkov, B. Gilev, Modeling and control of a wind turbine, Information Technologies and Control, **2**, 15-22 (2012)
10. J. Jonkman, S. Butterfield, W. Musial, G. Scott, Definition of a 5-MW reference wind turbine for offshore system development, Technical Report NREL/TP-500-38060, 75 (2009)
11. L. Landberg, *Meteorology for wind energy – an introduction*, (John Wiley & Sons Ltd., 2016)
12. J. Genov, I. Angelov, Influence of the vertical wind speed distribution on the dynamic load of wind turbines of high class (in bulgarian), *BulTrans-2016 Proceedings*, 136-140 (2016)
13. W. Rankine, W. Froude, On the mechanical principles of the action of propellers, Transactions of the Institution of Naval Architects, **6**, 13-39 (1865)
14. R. Wilson, P. Lissaman, *Applied aerodynamics of wind power machines,* National Science Foundation, Oregon State University, USA, 109 (1974)
15. J. Anderson, *Fundamentals of aerodynamics*, (McGraw-Hill Inc., 1991)
16. J. Conway, Analytical solutions for the actuator disc with variable radial distribution of load, J. Fluid Mech., **297**, 327-355 (1995)
17. J. Conway, Exact actuator disk solutions for nonuniform heavy loading and slipstream contraction, J. Fluid Mech., **356,** 235-267 (1998)
18. T. Wang, *Unsteady aerodynamic modelling of horizontal axis wind turbine performance,* PhD Dissertation, University of Glasgow, 299 (1999)
19. R. Bontempo, *The nonlinear actuator disk method as applied to open and ducted rotors*, PhD Dissertation, University of Naples Federico II, 153 (2014)
20. M. Hansen, *Aerodynamics of wind turbines - Third edition*, 187 (Taylor&Francis Group, 2015)
21. T. Sant, *Improving BEM-based aerodynamic models in wind turbine design codes*, PhD Dissertation, Delft University – Wind Energy Institute, 447 (2007)
22. P. Jamieson, *Innovation in wind turbine design*, 297 (John Wiley & Sons Ltd., 2011)
23. Cv. Velkova, J. Minchev, Numerical flow simulation of a SCREW propeller, Machines Mechanics, ISSN 0861-9727, **118,** 1-7 (2017)
24. I. Sobol, Uniformly distributed sequences with an addition uniform property, USSR Computation Maths. Math. Phys., **16**, 236-242 (1976)

https://doi.org/10.1051/matecconf/201823404008

Dynamic response of the tower of a NREL5MW wind turbine generator

Ivo Angelov[1,*]

[1]Technical University – Sofia, Department of Mechanics, Bulgaria, 8 Kl. Ohridski Blvd., 1000 Sofia, Bulgaria

Abstract. Considering the complexity of the aerodynamic interaction between a non-homogeneous, non-stationary wind field and a wind turbine, one can easily establish that the constructive elements of the turbine will be exposed to dynamic loads, which will eventually lead to forced motion and particularly to forced vibrations. One of the most significant elements of every wind turbine is the tower of the generator. The tower is subjected to forced vibrations and transmits all dynamic loads that appear in the wind turbine. That is why a dynamic analysis of the tower is worth executing. In this research such an analysis is made considering the tower of the generator as Euler-Bernoulli beam structure and considering it as a Love-Kirchhoff shell structure.

1 Introduction

The purpose of this study is to analyse the dynamic behaviour of a tower of a wind turbine generator of high power such as the NREL5MW wind turbine. Considering the aerodynamic interaction between the wind and the whole wind turbine, the station is subjected to dynamic loads. These loads are transmitted to the ground through the tower of the turbine. That is why a dynamic survey on the tower is of significant interest. The dynamic analysis can give the natural characteristics of the tower, can provide more knowledge on the forced vibrations and show a relation between wind speed and generated stresses.

In our previous studies [1] a modified BEM theory was used to obtain the aerodynamic forces acting on the turbine, using a non-homogenous, non-stationary wind field model. The data obtained from these studies will be used in the present one.

The following assumptions are made:

- The pitch angle of the blades is assumed to be constant and an active pitch control system is not considered;
- Any deformation of the blades as well as their mass and inertial characteristics are neglected;
- The wind field is considered non-homogenous, non-stationary.

2 A dynamic model of the tower of NREL5MW wind turbine

2.1. Geometrical and material properties of the tower

The tower of NREL5MW [2] is a truncated cone shown in Fig. 1. The geometric and material characteristics are given in Table 1.

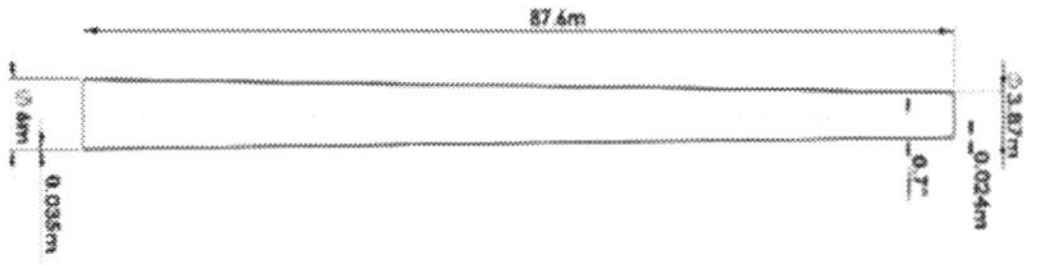

Fig. 1. NREL5MW Tower.

Table 1. Geometric and material characteristics of the tower of the NREL5MW wind turbine.

Geometric characteristics		Characteristics of the material	
Height, m	87,6	Density, kg.m^{-3}	8500
Cone angle, °	1,4	Mass, kg	347460
Outer diameter at the base, m	6	Young's module, GPa	210
Inner diameter at the base, m	5,93	Shear module, GPa	80,8
Outer diameter at the top, m	3,87	Poisson's ration	0,43
Inner diameter at the top, m	3,82	Damping ratio, %	1

2.2. External loads

The tower of the wind turbine transmits almost all forces that appear in the wind turbine generator structure. Three external forces are taken into consideration in the present study. These are the thrust force F_{th}, and the torque M_{wt}, both resulting from the aerodynamic interaction between the wind and the wind turbine [1] and a non-uniform distribution acting on the tower, resulting from the interaction between the wind and the tower itself - Fig. 2.

The determination of these forces was the subject of previous studies [3]. The purpose here is to derive an analytical form of the external loads. Using [1] one can

* Corresponding author: ivvoangelov@gmail.com

https://doi.org/10.1051/matecconf/201823404008

obtain a dependence between F_{th} and M_{wt} and the wind speed, defining them as a function of the wind speed V_∞ as shown in Fig. 3.

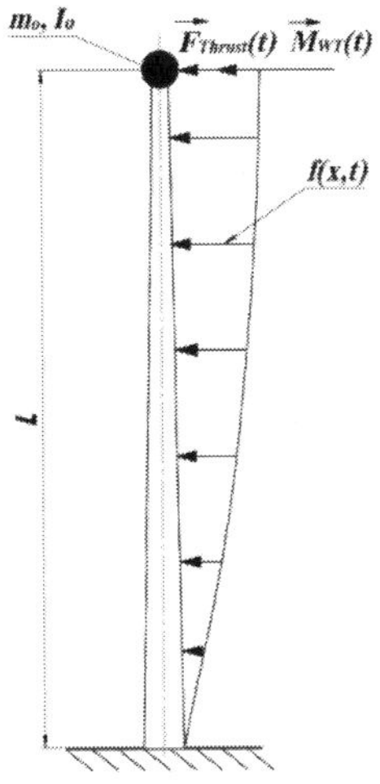

Fig. 2. Acting forces on the tower of NREL5MW.

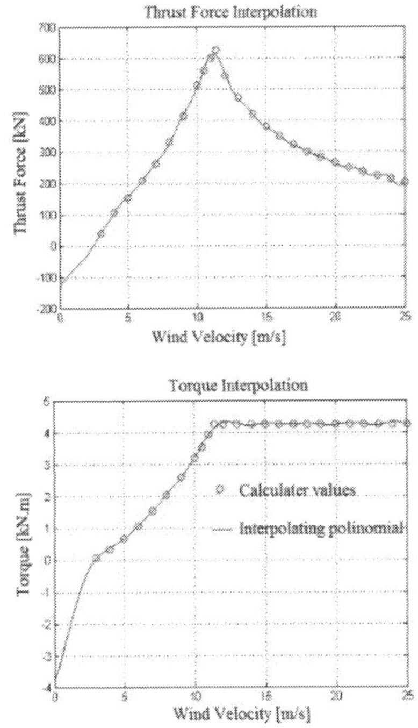

Fig. 3. Thrust force and torque as a function of the wind speed.

In order to obtain these dependencies into an analytical form a polynomial aproximation is made using the discrete values. The aproximation is in polynomial form:

$$F_{th}(V_\infty) = \sum_{m=1}^{M} a_m V_\infty^m$$
$$M_{WT}(V_\infty) = \sum_{m=1}^{M} b_m V_\infty^m, \qquad (1)$$

where:
a_m and b_m are coefficients to be defined;

V_∞ was derived in analytical form decomposing the power spectrum density function $S_{u'}$ in Shinozuka series [4]:

$$V_\infty = \sqrt{2}\sum_{i=0}^{N-1}\sqrt{S_{u'}(f)\Delta f}\cos(2\pi f_i t + f_i), \qquad (2)$$

where:
Δf –frequency summation step;
f_i – random phases.
Graphical visualization of (2) is shown in Fig.4.

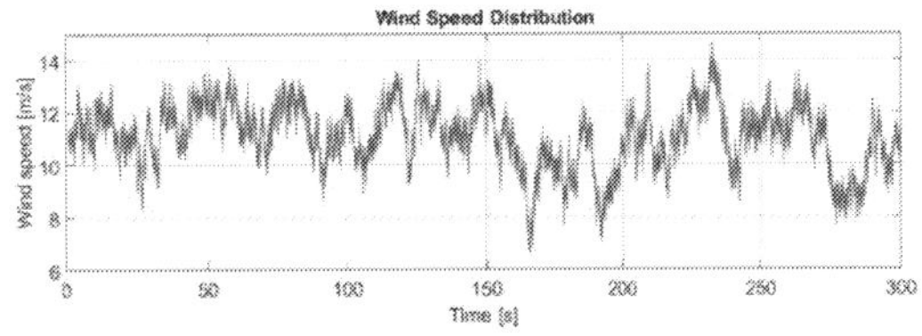

Fig. 4. Wind speed velocity distributed in time.

Substituting (2) in (1) gives an expresion of the thrust force $F_{th}(t)$ and the torque $M_{wt}(t)$ as a function of time. The obtained fuctions are shown in Fig. 5.

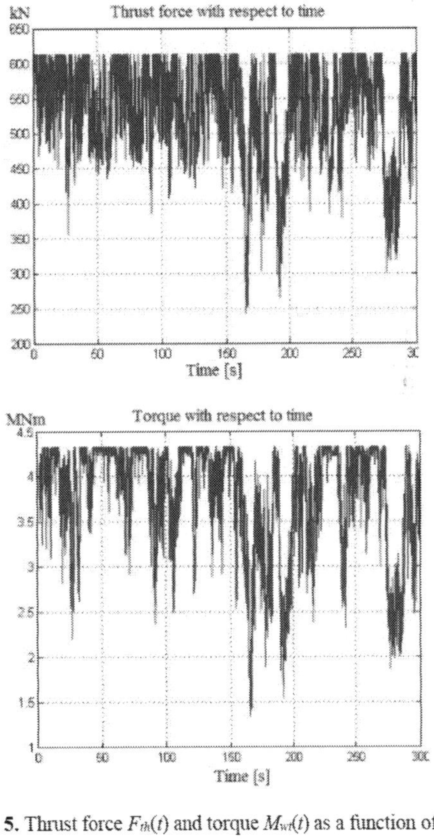

Fig. 5. Thrust force $F_{th}(t)$ and torque $M_{wt}(t)$ as a function of time.

The distributed load is defined similarly with the difference that one more function is required to represent the change in magnitude with respect to the height of the tower $f(x,t) = X(x)F(V_\infty(t))$, where $X(x)$ represents the vertical wind speed distribution [1]. Fig. 6 shows the distributed load $f(x,t)$.

 https://doi.org/10.1051/matecconf/201823404008

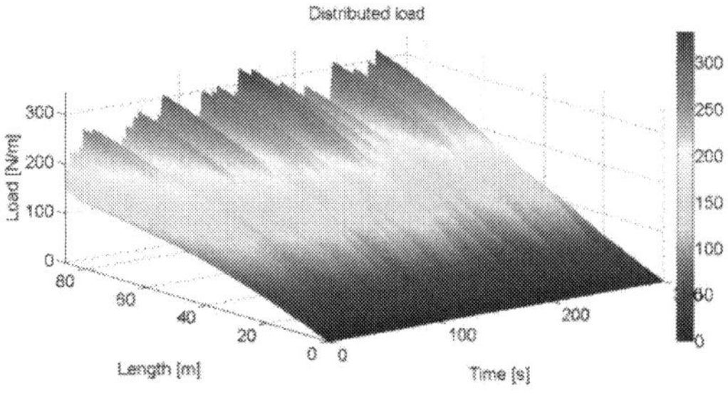

Fig. 6. Non-uniform distributed load as a function of time and height $f(x,t)$.

The boundary conditions of the model are as follows: a fixed end at the base of the tower and a point mass m_0 and inertia moment I_0 (with respect to the axis of the generator) at the top end - Fig.2. This point mass and inertia represent the mass characteristics of all bodies attached to the tower, such as wind wurbine, electric generator, hub, nacelle, shafts, bearings, etc.

2.3. Dynamic analysis of the tower in terms of the Euler-Bernoulli beam theory

In this section a dynamic analysis of the tower is executed using Euler-Bernoulli beam theory with equation in general form:

$$\frac{\partial^2}{\partial x^2}\left[EI(x)\frac{\partial^2 w(x,t)}{\partial x^2}\right] + \rho A(x)\frac{\partial^2 w(x,t)}{\partial t^2} = q(x,t)\,, \quad (3)$$

where:

- E – Young's modulus of the tower material;
- $I(x)$ – geometric moment of inertia of th tower;
- w – transverse displacement;
- ρ – density of the material of the tower's material;
- $q(x,t)$ – distributed load.

The implemented methodology is as follows. First, the natural frequencies ω_i are determined using Rayleigh-Ritz formula [5]:

$$\omega_i^2 = \frac{\int_0^L EI(x)\left[\frac{d^2W_i^*(x)}{dx^2}\right]^2 dx}{\int_0^L \rho A(x)[W_i^*(x)]^2 dx + m_0[W_i^*(L)]^2 + I_0\left[\frac{dW_i^*(L)}{dx}\right]^2}, \quad (4)$$

where $W^*_i(x)$ are aproximated functions, called Ansatzfunktion. For approximated functions are used the exact mode shapes of an uniform beam, with the same boundaries, the same length, the same material and mass properties, The mode shapes of this simplified beam model are shown in Fig. 7.

Substituting the aproximated mode shapes in (4) gives the corresponding frequencies, which are shown in Table 2.

Table 2. Natural frequencies of the tower of NREL5MW wind turbine with boundaries point mass and inertia at top end.

ω_1, Hz	0,3475
ω_2, Hz	3,3234
ω_3, Hz	11,0003
ω_4, Hz	22,2794
ω_5, Hz	37,0453

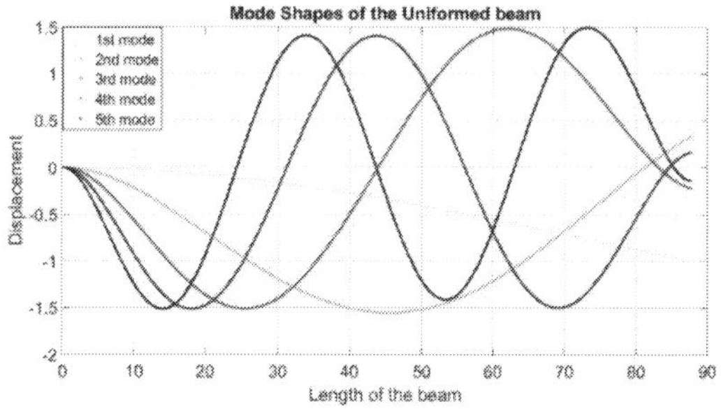

Fig. 7. First five mode shapes of the simplified beam.

Once the natural frequencies are obtained the partial differential Euler-Bernoulli equation (3) becomes a homogenous, ordinary differential equation ordinary with non-constant coefficions.

$$EI(x)\frac{d^4W_i(x)}{dx^4} + \rho A(x)\omega_i^2 W_i(x) = 0 \quad (5)$$

It will be solved in order to determine the mode shapes of the tower. Equation (5) does not have an exact solution, Here it is solved numerically in matlab using "bvp4c" solver [6]. The obtained solution is shown in Fig.8.

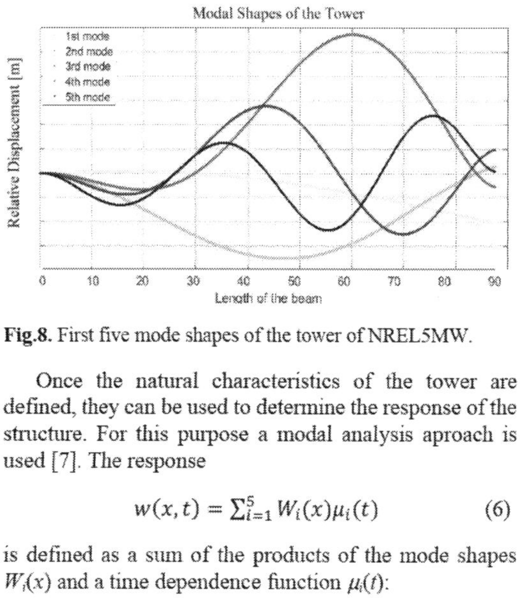

Fig.8. First five mode shapes of the tower of NREL5MW.

Once the natural characteristics of the tower are defined, they can be used to determine the response of the structure. For this purpose a modal analysis aproach is used [7]. The response

$$w(x,t) = \sum_{i=1}^{5} W_i(x)\mu_i(t) \quad (6)$$

is defined as a sum of the products of the mode shapes $W_i(x)$ and a time dependence function $\mu_i(t)$:

$$\mu_i(t) = \frac{1}{\omega_i}\int_0^t Q(\tau)\sin\left[\omega_i(t-\tau)\right]d\tau, \quad (7)$$

where

$$Q_i(t) = \int_0^L W_i(x) f(x,t) dx. \quad (8)$$

In order for the solution to show both qualitative and quantitative characteristics, the mode shapes must be normalized, using the following scaling coefficient C_i.

$$C_i \int_0^L \rho A(x) W_i^2(x) dx = 1 \quad (9)$$

The modal analysis solution shows the participation of each mode shape in the beam response. The described methodology is executed in Matlab and gives us the following results. Fig. 9 shows the response due to the distributed load $f(x,t)$ separated into the first four mode shapes. It is observed that the solution is dominated by the first mode. Every subsequent mode participates less as compared to the previous one. Fig. 10 shows the response due to the thrust force F_{th}. Fig. 11 shows the response due to the torque M_{wt}. Fig. 12 shows the total response of the tower due to all the external excitations. Fig. 13 shows the normal stresses, generated in the structure due to the external loads.

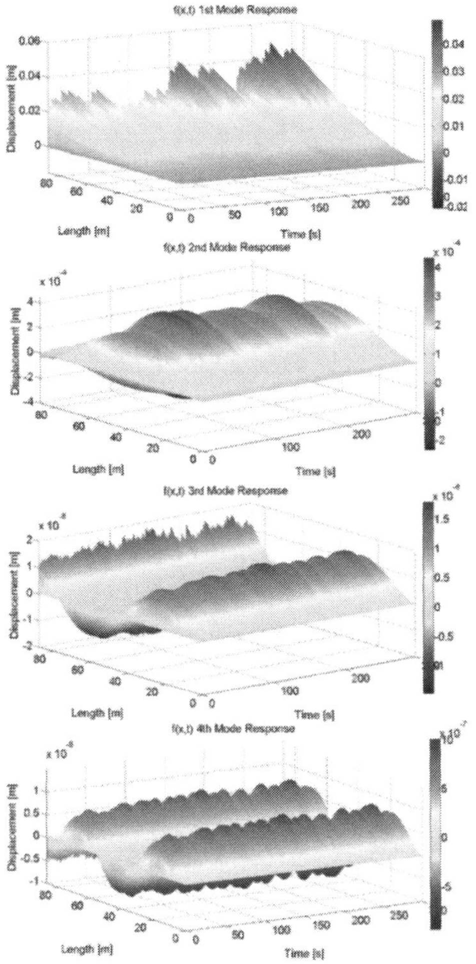

Fig. 9. Beam response due to the distributed load $f(x,t)$ with the exact participation of the first 4 mode shapes.

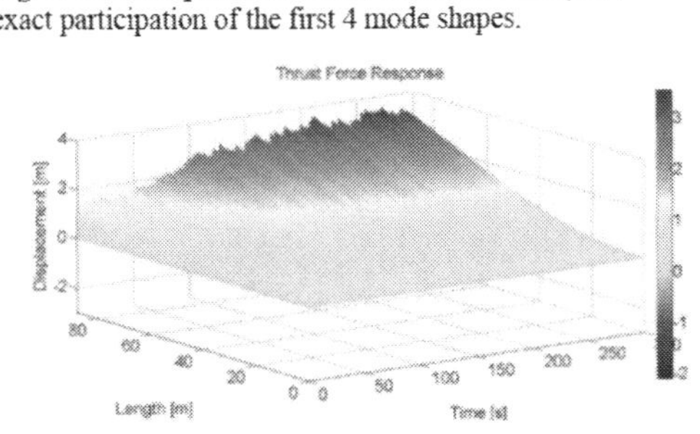

Fig. 10. Beam response due to the thrust force F_{th}.

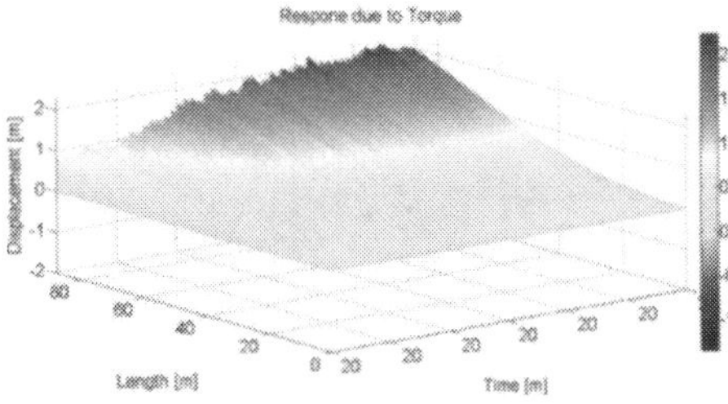

Fig. 11. Beam response due to the torque M_{wt}.

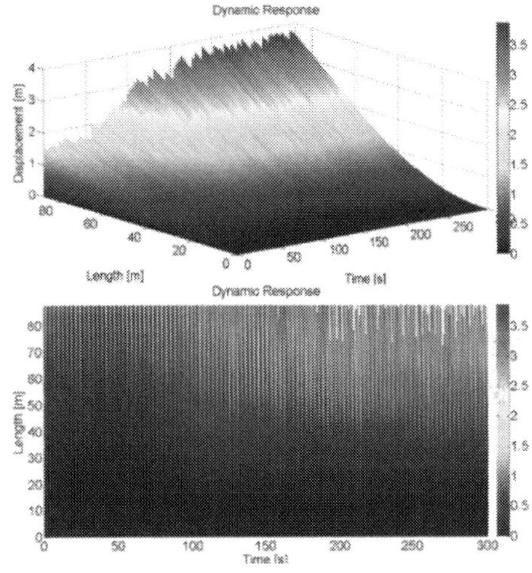

Fig. 12. Beam response due to all external loads.

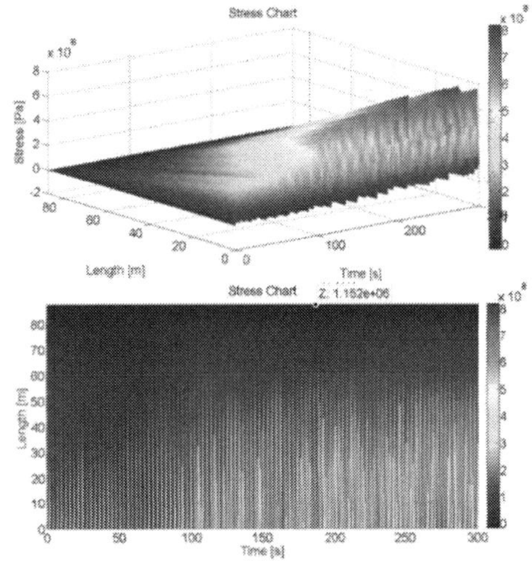

Fig. 13. Generated normal stresses due to all external loads.

2.4. Numerical analysis in ANSYS Workbench

The modal numerical survey made in ANSYS Workbech shows that the tower of NREL5MW has the behavior of shell structures. It shows the typical for a shell structure circumferential mode shapes. It also shows a difference in the natural frequencies corresponding to the shell bending mode shapes. The comparison between the frequencies is shown in Table 3.

Table 3. Natural frequencies of the structure, corresponding to the bending mode shapes.

Natural frequencies	Euler-Bernoulli beam analysis	ANSYS numerical results
ω_1, Hz	0,3475	0,331
ω_2, Hz	3,3234	1,967
ω_3, Hz	11,0003	4,61
ω_4, Hz	22,2794	10,531
ω_5, Hz	37,0453	18,874

Regardless of this difference, the performed analysis in section 2.3 is applicable for all kinds of continuous structures. Considering that the response is dominated by the first mode shape a significant difference in the solution will not appear. That is because there is only a slight difference in the frequency corresponding to the first mode.

2.5. Natural characteristics of a shell structure tower model

In order to confirm the shell behavior of the tower a shell model is analyzed in this section. For easier analysis and determination of the natural characteristics, the model of the tower is simplified as follows. It is considered uniform with a constant-circular cross section and constant thickness. The top end is set free without point mass and inertia. Fig. 14.

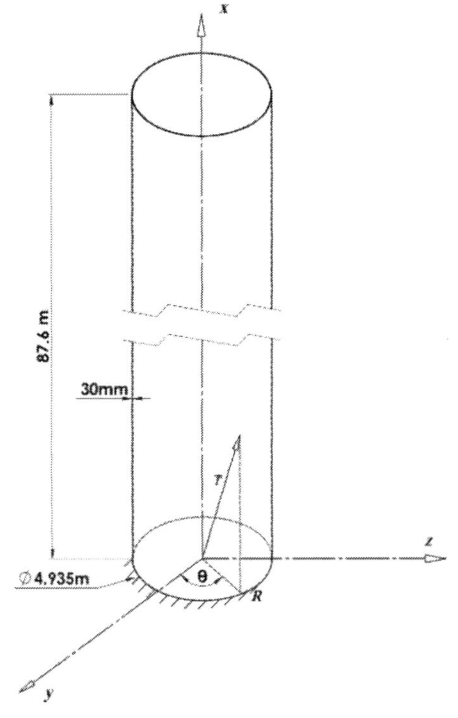

Fig. 14. Shell structure model. R and θ are polar coordinates.

The differential equations known as Love-Kirchhoff equations [6] describing the dynamic behaviour of a shell structure are:

$$\begin{cases} -\frac{\partial(N_{\alpha\alpha})}{\partial\alpha} - \frac{\partial(N_{\beta\alpha})}{\partial\beta} - N_{\alpha\beta}\frac{\partial A}{\partial\beta} + N_{\beta\beta}\frac{\partial B}{\partial\alpha} - AB\frac{Q_{\alpha z}}{R_\alpha} + AB\rho h\ddot{u} = ABq_\alpha \\ -\frac{\partial(N_{\alpha\beta})}{\partial\alpha} - \frac{\partial(N_{\beta\beta})}{\partial\beta} - N_{\beta\alpha}\frac{\partial A}{\partial\beta} + N_{\alpha\alpha}\frac{\partial B}{\partial\alpha} - AB\frac{Q_{\beta z}}{R_\beta} + AB\rho h\ddot{v} = ABq_\beta \\ -\frac{\partial(Q_{\alpha z}B)}{\partial\alpha} - \frac{\partial(Q_{\beta z}A)}{\partial\beta} + AB\left(\frac{N_{\alpha\alpha}}{R_\alpha} + \frac{N_{\beta\beta}}{R_\beta}\right) + AB\ddot{w} = ABq_z \\ \frac{\partial(M_{\alpha\alpha}B)}{\partial\alpha} + \frac{\partial(M_{\beta\alpha}A)}{\partial\beta} + M_{\alpha\beta}\frac{\partial A}{\partial\beta} - M_{\beta\beta}\frac{\partial B}{\partial\alpha} - Q_{\alpha z}AB = 0 \\ \frac{\partial(M_{\alpha\beta}B)}{\partial\alpha} + \frac{\partial(M_{\beta\beta}A)}{\partial\beta} + M_{\alpha\beta}\frac{\partial B}{\partial\alpha} - M_{\alpha\alpha}\frac{\partial A}{\partial\beta} - Q_{\beta z}AB = 0 \end{cases} \quad (10)$$

where:

- the α, β – curvilinear coordinates;
- A, B – lengths of the radius vector projection respectively on α, β coordinate;
- R_α, R_β – radiuses of the curvilinear coordinates
- z – linear coordinate;
- N_{ij}, Q_{ij} – normal and tangential forces per unit length
- M_{ij} – moments of the forces
- R_i – radiuses of curvature;

After a simplification for a cylindrical shell structure equations (10) take the form of:

$$\begin{cases} C\left(\frac{\partial^2 u}{\partial x^2} + \frac{1-v}{2R^2}\frac{\partial^2 u}{\partial\theta^2} + \frac{v}{R}\frac{\partial w}{\partial x} + \frac{1+v}{2R}\frac{\partial^2 v}{\partial x\partial\theta}\right) + q_x = \rho h\frac{\partial^2 u}{\partial t^2} \\ C\left(\frac{1-v}{2}\frac{\partial^2 v}{\partial x^2} + \frac{1}{R^2}\frac{\partial^2 v}{\partial\theta^2} + \frac{1}{R^2}\frac{\partial w}{\partial\theta} + \frac{1+v}{2R}\frac{\partial^2 u}{\partial x\partial\theta}\right) + \\ D\left(\frac{1-v}{2R^2}\frac{\partial^2 v}{\partial x^2} + \frac{1}{R^4}\frac{\partial^2 v}{\partial\theta^2} - \frac{1}{R^4}\frac{\partial^3 w}{\partial\theta^3} - \frac{1}{R^2}\frac{\partial^3 w}{\partial x^2\partial\theta}\right) + q_\theta = \rho h\frac{\partial^2 v}{\partial t^2} \\ D\left(\frac{\partial^4 w}{\partial x^4} + \frac{1}{R^2}\frac{\partial^3 v}{\partial x^2\partial\theta} - \frac{2}{R^2}\frac{\partial^4 w}{\partial x^2\partial\theta^2} - \frac{1}{R^4}\frac{\partial^4 w}{\partial\theta^4} + \frac{1}{R^4}\frac{\partial^3 v}{\partial\theta^3}\right) - \\ C\left(\frac{1}{R^2}\frac{\partial v}{\partial\theta} + \frac{w}{R^2} + \frac{v}{R}\frac{\partial u}{\partial x}\right) + q_z = \rho h\frac{\partial^2 w}{\partial t^2} \end{cases} \quad (11)$$

known as Donnell-Mushtari-Vlasov Theory [7, 8], where:

- w, v, u – displacements respectively in transverse, circumferential and axial direction;
- q_x, q_θ, q_z are projection of the distributed load;
- C and D - constants depending on material's properties;
- v – Poisson's ratio of the material;
- h – thickness of the shell.

The system of PDE (11) does not have an exact solution. The natural frequencies corresponding to the shell bending mode shapes are determined using a semi-analytical analysis based on Rayleigh-Ritz approach [9, 10]. The idea is that the solution of (11) is approximated in the following form:

$$\begin{cases} u(x,\theta,t) = \left[A_1\frac{dW_1(x)}{dx} + A_2\frac{dW_2(x)}{dx}\right]\cos(n\theta)\sin(\omega t); \\ v(x,\theta,t) = [B_1W_1(x) + B_2W_2(x)]\sin(n\theta)\sin(\omega t); \\ w(x,\theta,t) = [C_1W_1(x) + C_2W_2(x)]\cos(n\theta)\sin(\omega t), \end{cases} \quad (12)$$

where $W_1(x)$ are the beam bending modes for a clamped-free boundaries and $W_2(x)$ are the beam bending modes for a clamped-pinned boundaries. The described procedure is executed in Matlab. The bending modes of a shell structure are these for $n = 1$ and $m = 1, 2, 3...$ A comparison between the frequencies corresponding to the first five bending modes of the described structure considered as a beam, as a shell and results from ANSYS are shown in Table 4.

Table 4. Natural frequencies corresponding to the bending mode shapes of the model considered as a beam structure, as a shell structure and numerical results.

Natural frequencies	Beam structure	Shell structure	ANSYS numerical results
ω_1, Hz	0,7	0,64	0,63
ω_2, Hz	4,4	3,58	3,8
ω_3, Hz	12,32	9,82	10,18
ω_4, Hz	24,16	N/A	18,83
ω_5, Hz	39,96	N/A	29,19

The results confirm that the described structure from Fig. 14 has a behavior of a shell.

3 Conclusion

In the present paper several important conclusions were arrived at:

1. The tower of NREL5MW wind turbine shows a behavior of a shell structure, not a beam and should be analyzed and considered as a shell.
2. The dynamic response of the structure under a stochastic generated wind excitations is dominated by the first mode of the structure, which represents nearly 98% of the response. Every subsequent mode contributes less, as compared to the previous one. The second and the third mode respectively represent only 2% and 0,2% of the response.
3. Due to the stochastic characteristics of the wind, the absence of correlation between the frequencies of the excitation forces, a resonance state does not appear. The structure passes through resonanse very fast and amplitudes of the forced vibrations do not increase.

References

1. J. Genov, I. Angelov, Influence of the vertical distribution of wind on the dynamic load of high-class wind turbines (in Bulgarian), *Bultrans-2016 Proceedings*, 136-140 (2016)
2. J. Jonkman, S. Butterfield, W. Musial, G. Scott, Definition of a 5MW reference wind turbine for offshore system development, NREL Technical report NRELTP-500-38060, 15-16 (2009)
3. J. Genov, B. Gilev, Y. Slavchev, G. Venkov, Modeling and control of wind turbine tower vibrations, *Applications of Mathematics in Engineering and Economics – AMME-10,* American Institute of Physics, Conference Proceedings 1293, 30-38 (2010)
4. M. Shinozuka, G. Deodatis, Simulation of stochastic processes by spectral representation, Applied Mechanics Review, **44**, 191-204 (1991)
5. J. Wauer, *Kontinuumsschwingungen*, 92-97 (Springer Vieweg, 2008)
6. J. Tonchev, *Matlab 6-7, Transformations, calculations, visualisations* (in Bulgarian), 79-82 (Tehnika, Sofia, 2008)
7. S. Rao, *Vibrations of continuous systems*, 47-52; 556-590 (John Wiley & Sons Inc., 2007)
8. K. Forsberg, *A review of analytical methods used to determine the modal characteristics of cylindrical shells*, NASA Report Lockheed Aircraft Corporation, (1966)
9. G. Warburton, J. Higgs, Natural characteristics of thin cantilever cylindrical shells, Journal of Sound and Vibration, **11**(3), 335-338 (1970)
10. C. Sharma, Calculation of natural frequencies of fixed-free circular cylindrical shell, Journal of Sound and Vibration, **35**(l), 55-76 (1974)

Change in the maintenance strategy as a method of improving the efficiency of the process of operation of railway means of transport

Maciej Szkoda[1,*], and *Magdalena* Satora[2]

[1] Cracow University of Technology, Institute of Rail Vehicles, Jana Pawla II 37 Ave., 31-864 Cracow, Poland
[2] Cracow University of Technology, Faculty of Mechanical Engineering, Jana Pawla II 37 Ave., 31-864 Cracow, Poland

Abstract. The paper presents a description of methods of changing the maintenance strategy concerning measures of intervals between maintenance levels based on the example of a selected railway vehicle type. The process starts with monitoring the vehicle after it is put into operation and the operational database is built, through the statistical and RAMS analyses, to the approval of changes in the maintenance system documentation by qualified railway safety bodies. Following the applicable formal and legal requirements, one of the key problems involved in the introduction of changes in the railway vehicle maintenance system is an assessment of the proposed changes for safety.

1 Introduction

Projects intended to improve the efficiency of the process of operation (use and maintenance) of means of transport are amongst the fundamental areas of the strategies pursued by rail transport companies. In the group of maintenance projects, particularly noteworthy is the change in the maintenance strategy concerning the maintenance plan structure and the change in the measures of intervals between checks and scheduled repairs (so-called maintenance levels). Reduction of preventive maintenance costs and implementation of modern maintenance methods, such as Condition-Based Maintenance (CBM) is the subject of many scientific publications devoted to this issue [1-3]. The research and development work, which were done on the change in the maintenance strategy [4-10], prove that the costs of preventive maintenance of railway means of transport form up to 30,5% of the Life Cycle Costs (LCC) and are the second dominating cost category after the costs of power or fuel consumption. Currently, such projects are very often applied by Polish railway carriers. Following the model of leading foreign companies, in their strategic ventures, Polish carriers intend to introduce dynamic maintenance plans in which the criterion for referring a vehicle to a check /repair are measurements the vehicle actual operational work.

On the basis of an analysis of the writings and the experience gathered in conducting research and development work for Polish railway carriers, e.g.: [11-13], Table 1 presents a description of the praxeological sequence illustrating the change in the maintenance strategy concerning the measurements of intervals between checks and scheduled repairs with the fulfilment of the following criteria:

- ensuring a high level of safety in railway transport,
- ensuring effective outlays on maintenance.

The process starts with monitoring the vehicle after it is put into operation (purchased or modernised) and the development of an operational database, through a statistical and reliability analysis of the data, risk analysis of the proposed changes, to the approval of the changes in the maintenance system by the authorised railway safety bodies [14].

2 Example of method application

The method for changing the maintenance strategy is universal in its nature and may be applied to various types of rail vehicles (diesel and electric traction vehicles, cargo and passenger carriages). This paper presents selected aspects based on the example of a modernised diesel 6Dg type locomotive (Fig. 1).

Fig. 1. View of the 6Dg diesel locomotive (photo: M.Górowski).

* Corresponding author: maciej.szkoda@mech.pk.edu.pl

Table 1. Description of the method of change of the maintenance strategy concerning the measurements of intervals between checks and scheduled repairs [14].

No.	Name	Description of stage
1	Development of operational database	Operation monitored from the start of operation until the first repair at the P4 level. Collection of operational data concerning: kilometrages and time of operation (km/year, hr/year), time to failure, times of modernisation, causes for failures, costs of repairs. Archiving preventive maintenance documentation.
2	Formal requirements for the maintenance system	Analysis of the applicable law concerning the maintenance system and changes in rail vehicle maintenance systems.
3	Statistical analysis of the operational process	Analysis of statistical data on kilometrages and time of operation between maintenance levels and comparison with the assumptions in the maintenance system documentation.
4	Concept of changes in the maintenance process	Collection of practical knowledge from specialists from the Rolling Stock Maintenance Department about the possibility to change the maintenance plan for the analysed vehicle. Proposal of changes in the maintenance plan, comprising a change in the measurements of intervals between the maintenance levels and the scopes of activities within the Maintenance System Documentation (DSU).
5	RAMS analysis in the area of safety	Structure of failures. Identification of locomotive assemblies, subassemblies and elements (so-called relevant elements). Analysis of reliability data. Identification of models of distribution of probability of time of operation to failure and time of modernisation for vehicle assemblies and subassemblies. Identification of durability and reliability characteristics and RAMS indices.
6	Risk analysis for the proposed changes	Analysis of the risk involved in the proposed changes in accordance with Regulation (EU) No 402/2013 comprising: assessment of the relevance of changes, defining the system, identification of the interfaces, identification of hazards and risk assessment.
7	Assessment of the effectiveness of the changes in the maintenance process	Comparison of the costs of preventive maintenance before and after maintenance plan changes.
8	Approval of changes in the maintenance system documentation	Minor changes (according to Regulation 402/2013) are approved by the railway company management board whilst significant changes require the approval by the accredited body which supervises and verifies the correctness of the risk assessment process.

The analysis was the subject of a series of research and development papers [15-17]. Identification of the LCC cost structure of the 6Dg diesel locomotive demonstrated that scheduled checks done at the P2 level (P2/1 and P2/2 totalling 38,8%) and scheduled repairs done at the P4 maintenance level – 28,7% (Fig. 2) form the highest proportion of preventive maintenance costs.

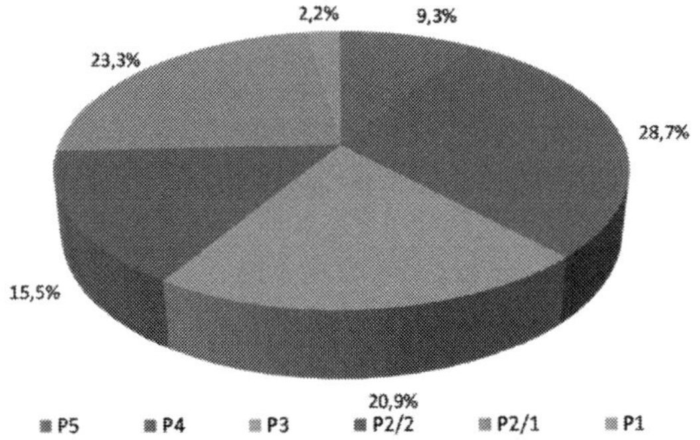

Fig. 2. Structure of the costs of preventive maintenance of a 6Dg diesel locomotive [14].

The cost verification, done in the operation stage, initiated the measures undertaken by the railway carrier (PKP Cargo S.A.) in cooperation with the research body (Institute of Rail Vehicles, Cracow University of Technology) intended to improve the efficiency of the locomotive maintenance process by changing the measurements of intervals between the maintenance levels and the scope of maintenance activities. It was assumed that the changes should correspond to the actual operating work and include advanced level of diagnostics and registration of the operating condition of modernised locomotives.

2.1 Development of operational database

During the period from the start of operation and the first repair at the P4 level, done after 24 480 mth or after 5 years, the modernised locomotives were monitored during their operation and operational data were gathered therein, concerning:

- their kilometrages in km/year and in km/day;
- time of operation in h/year and in h/day;
- time to failure;
- causes for the failure;
- duration of maintenance repairs;
- costs of maintenance repairs, checks and scheduled repairs.

The electronic information, which was gathered and processed, was formally recorded on a preliminary basis in Microsoft Excel. Universal failure codes were used comprising 7 locomotive systems and 30 subassemblies.

2.2 Formal requirements related to the maintenance system change

An assessment of the possibility to change the maintenance strategy should take account of formal and legal requirements under the law which applies in Poland, including:

1. Directive 2008/110/EC of the European Parliament and of the Council of 16 December 2008 amending Directive 2004/49/EC on safety on the Community railways) (OJEU L 345 of 23.12.2008);
2. Commission Implementing Regulation (EU) No 402/2013 of 30 April 2013 on the common safety method for risk evaluation and assessment and repealing Regulation (EC) No 352/2009 (OJEU L121 of 03.05.2013, as amended);
3. Announcement of the Speaker of the Sejm of the Republic of Poland of 13 October 2017 on the promulgation of the consolidated text of the Act on Rail Transport (Dz.U. 2017 item 2117);
4. Regulation of the Minister of Infrastructure and Construction of 28 July 2017 amending the regulation on the general technical conditions for the operation of rail vehicles (Dz. U. 2017, item 1525) and other.

Further, it is necessary to include in the analysis the railway carrier reference documents, such as:

1. Safety System Management Book with the necessary procedures, primarily regarding an assessment of the technical and operational risks;
2. Maintenance system and technical/operational documentation for the analysed locomotive.

The analysis of legal conditions has shown that the current regulations do not allow for performance of inspections or periodic repairs of railway vehicles according to a strategy based on the actual state of vehicle assemblies and subassemblies (CBM technique). The structure of the maintenance plan for railway vehicles must be defined with fixed distance measurements expressed in kilometers and/or units of time.

2.3 Statistical analysis of the process of operation

On the basis of the operational data gathered by the railway carrier for 2009÷2015, i.e. from the beginning of operation of a series of 121 vehicles, Table 2 comprises the average time of operation in mth/year and the average kilometrage in km/year of modernised 6Dg locomotives. Figures 3a and 3b present the histograms of these figures. The average annual kilometrage was found to be 23484,5 km/year, and the maximum figure does not exceed 35544,9 km/year. The average time of operation is 3916,5 mth/year or 78,3% of the figure assumed in the maintenance system documentation: 5000,0 mth/year.

2.4 Concept of changes in the locomotive maintenance plan

The concept of changes in the maintenance plan for the 6Dg type locomotive required, above all, the collection of practical knowledge from specialists engaged in the maintenance of rolling stock about the possibility to change the intervals between the maintenance levels and modify the scopes of activities in the Maintenance System Documentation (DSU). Additionally, the results of the theoretical analyses were used to identify the locomotive weakest elements from the viewpoint of the impact on the vehicle technical availability. The analysis used the Fault Tree Analysis (FTA) with Monte Carlo simulation which is presented in papers [14, 18, 19]. One of the indices serving the purpose of selecting the most relevant elements was the ReliaSoft's Downing Event Criticality Index (RS DECI) expressing the percentage of downtimes caused by a particular locomotive assembly or subassembly in the total downtime figure [20]:

$$RS\ DECI = \frac{C_{NSDE}}{N_{ALLdown}}\ 100\%, \quad (1)$$

where:
C_{NSDE} – number of downtimes caused by the particular locomotive assembly, subassembly;
$N_{ALLdown}$ – number of all locomotive downtime.

Table 2. Statistical data on time of operation and kilometrage of locomotives of SM42 series, 6Dg type.

Figure	Average	Total	Min	Max	Standard deviation
Time of operation, mth/year	3916,5	466064,9	2492,9	5297,4	579,9
Kilometrage, km/year	23484,5	2794650,5	3304,4	35544,9	6067,0

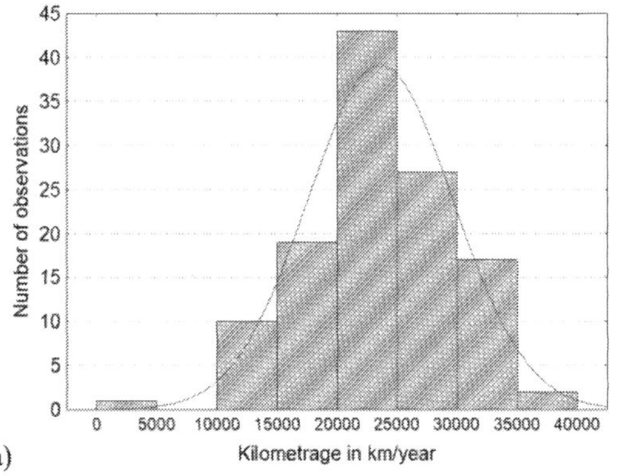

a)

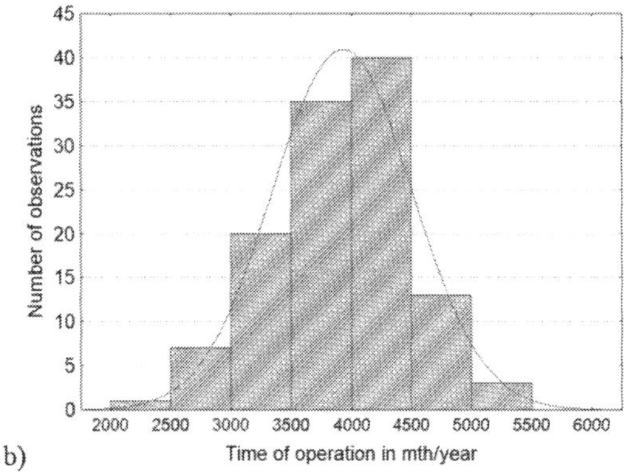

b)

Fig. 3. a) Histogram of locomotives of the SM42, 6Dg type a) kilometrage in [km/year], b) time of operation in [mth/year] [14].

On the basis of the analysis, changes in the current plan for the maintenance of the SM42 series 6Dg type locomotives were proposed (Table 3).

Table 3. Current and proposed maintenance plans for 6Dg locomotives.

No.	Maintenance level	Current	Proposed changes
1	P1	102 mth, 14 days, 1300 km	200 mth, 21 days, 2 500 km
2	P2/1	510 mth, 25 days, 2500 km	1000 mth, 90 days, 10500 km
3	P2/2	3060 mth, 5 months, 15000 km	3000 mth, 6 months, 21000 km
4	P3	12240 mth, 20 months, 60000 km	18000 mth, 4 years, 100 000 km
5	P4	24480 mth, 5 years, 200000 km	36000 mth, 8 years, 200000 km
6	P5	122400 mth, 30 years, 1000000 km	122400 mth, 30 years, 1000000 km

2.5 RAMS analysis in the area of safety

The RAMS analysis in the area of safety was based on the results of the operational tests of a selected sample of 75 locomotives of the 6Dg type, operated by a railway carrier over a period of 15 months. This enabled the observation of the course of operation of the locomotives in different conditions thus providing creditable data for the reliability analysis. The reliability data were gathered in the IT system supporting the management of the means of transport, enabling precise registration of the Time to Failure (TTF) and the Time to Restore (TTR).

The RAMS indices relating to safety were determined for relevant elements, i.e. assemblies and sub-assemblies of locomotives, whose failure results in hazards to the reliability and safety of railway operations (so-called hazardous failure). On the basis of the analysis of operation data for the locomotive sample tested over the 15-month period, a total of 151 failures were recorded which were qualified as hazardous ones. This is approximately 30,6% of the total number of failures. The structure of the hazardous failures is as follows [14]:

1. Running gear:
   1.1. Failures and wear of the wheel outer contour – 3,6%.
2. Brake system (pneumatic and mechanical parts):
   2.1. Failures of main or auxiliary compressor – 5,5%;
   2.2. Failures of pneumatic conduits – 2,4%;
   2.3. Failures of pneumatic valves (inter alia, main or auxiliary valve of the driver, reducing, end, safety valve) – 3,4%;
   2.4. Failures of elements of the brake (e.g. levers, couplers, coupling pins, bushings, couplings, brake blocks) – 1,2%;
   2.5. Failures of the actuator in the brake system – 0,4%;
   2.6. Failures of the engine driving the main or auxiliary compressor of rail vehicle – 0,6%;
   2.7. Failures of other elements of the pneumatic circuit – 2,0%.
3. Vehicle drive safety control devices:
   3.1. Failures ABP (SHP, CA, RS), metering device (speed meter, ammeter) or radiotelephone – 10,4%.

In the RAMS analysis of the 6Dg locomotive, the following safety ratios were determined:

a) FH(t) distribution function and the mean time to hazardous failure (MTTHF);
b) Mean time between hazardous failures, MTBHF and MDTHF;
c) Frequency of hazardous failures H for locomotive assemblies and subassemblies.

The definitions of the above ratios are available in specialist writings on reliability, availability, maintainability and safety, e.g.: [14, 21, 22] and in the applicable standards:

- PN-EN 50126:2002 Railway applications. The specification and demonstration of reliability, availability, maintainability and safety,
- PN-EN 61703:2016-12 Mathematical expressions for reliability, availability, maintainability and maintenance support terms.

2.6 Analysis of the risk involved in the proposed changes

The principles of the Common Safety Method (CSM) concerning risk analysis are described in Commission Implementing Regulation (EU) No. 402/2013 [23]. The detailed algorithm of the risk management process is presented in the Appendix to the Regulation entitled "Risk management process and independent assessment". The procedure of risk qualification where changes are made in the maintenance system requires:

- analysis of the relevance of the proposed changes,
- hazard identification and development of a register of hazards,
- assessment of risk acceptability.

In accordance with the Regulation, the acceptance of the risk concerning the assessed system is tested using one or several of the following risk acceptance principles:

- use of codes of practice,
- use of a reference system,
- explicit risk estimation.

In the analysed case, following the procedure for identification of hazards and risk assessment of the Safety Management System applied by the railway carrier operating 6Dg type locomotives, the risk was qualified using the (Failure Mode and Effects (FMEA) Analysis. The FMEA method is a quantitative one in which the risk of hazard occurrence is estimated using the Risk Priority Number (RPN) [22].

The analysis demonstrated a possibility to make changes in the plan for maintenance of 6Dg type locomotives without breaching the accepted safety level and introducing entries in accordance with Table 2. Nonetheless, the changes in the maintenance plan required the development of additional safety measures concerning the assemblies and subassemblies of major importance for rail traffic safety: wheel sets, brake system, bogie support and frame.

2.7 Assessment of the effectiveness of the changes in the maintenance process

The effectiveness of the changes in the 6Dg locomotive maintenance plan was assessed on the basis of a comparison of the costs of preventive maintenance (PMC_T), which was expressed through the formulae (2) and (3), taking into account the costs of checks and repairs

 https://doi.org/10.1051/matecconf/201823405001

of the 2017 level [11]. The calculations were based on the mean time of locomotive operation of 5 000,0 mth/year.

$$PMC_t = \sum_{i=1}^{n} PMC_i, \quad (2)$$

where PMC_i is costs of preventive maintenance in the year t for the maintenance level "i" in EUR:

$$PMC_i = PMCL_i + PMCM_i = NPMA_i(t)\left[\left(MMH_i \cdot CPH_p\right)\right] + ACM_i, \quad (3)$$

where:
$PMCL_i$ – the costs of labour for maintenance level "i";
$PMCM_i$ – the costs of materials and spare parts for maintenance level "i", EUR;
$NPMA_i(t)$ – number of preventive maintenance activities at maintenance level "i" in a year of operation;
MMH_i – mean labour intensity of preventive maintenance at maintenance level "i", man-hours;
CPH_P – cost of man-hour in preventive maintenance: CPH_P = 13,71 EUR/man-hour;
ACM_i – average cost of consumption of materials in preventive maintenance at maintenance level "i", EUR.

It was concluded on the basis of the calculations that as a result of the change in the maintenance strategy the costs of maintaining the modernised 6Dg type locomotive calculated in the full maintenance plan were reduced by 24,7% or EUR 232300,0 (Fig. 4a). For the entire series of 121 modernised vehicles owned by the carrier, the maintenance cost savings are EUR 28,1 million. In comparison with the unmodernised SM42 6D locomotive, the maintenance costs were reduced by 53,0% or EUR 799100,0 per vehicle (Fig. 4b).

a)
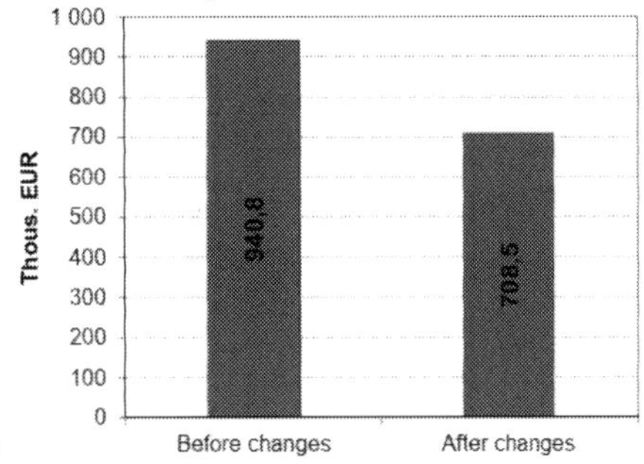

b)
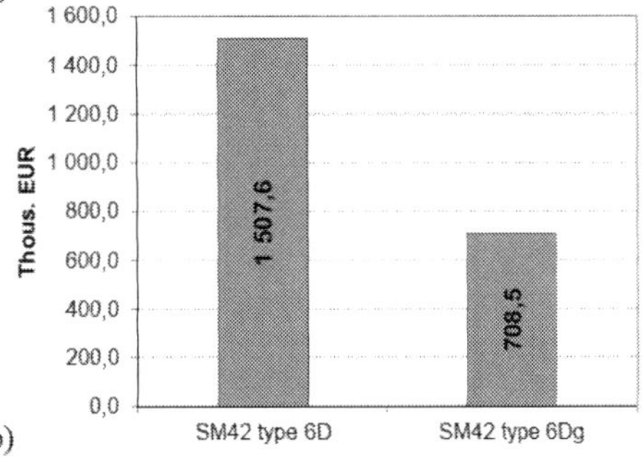

Fig. 4. a) Comparison of the costs of preventive maintenance before and after the change in the maintenance plan, b) Comparison of the costs of preventive maintenance of a non-modernised 6D locomotive and a 6Dg modernised one after the maintenance plan change [14].

3 Conclusion

The costs of maintenance are the second, following the costs of power or fuel consumption, dominating cost category in the Life Cycle Costs LCC of railway means of transport. Hence, projects undertaken by rail carriers to enhance the efficiency of the process of vehicle operation by changing the maintenance strategy are fully justified. The methods presented in the paper are based on many years of experience from research and development work related to the change in the strategy of maintenance of traction vehicles and rail carriages. They indicate the important role of the RAMS analysis in the area of safety thus enabling proper classification of hazards, quantification of the frequency of their occurrence and adoption of appropriate criteria for risk assessment.

The considerations presented in the paper are of general relevance. They may be applied not only to rail vehicles but to other technical objects, which are subject to a complex maintenance system.

References

1. F. Corman, S. Kraijema, M. Godjevac, F. Lodewijks, Optimizing preventive maintenance policy: A data-driven application for a light rail braking system, Journal of Risk and Reliability, **231**, 534-545 (2017)
2. J. Vatn, T. Aven, An approach to maintenance optimization where safety issues are important, Reliability Engineering and System Safety, **95**, 58-63 (2010)
3. M. Wolde, A. Ghobbar, Optimizing inspection intervals – Reliability and availability in terms of a cost model: A case study on railway carriers, Reliability Engineering and System Safety, **114**, 137-147 (2013)
4. M. Babeł, M. Szkoda, Diesel locomotive efficiency and reliability improvement as a result of power unit load control system modernization, Eksploatacja i Niezawodnosc – Maintenance and Reliability, **18**, 38-49 (2016)
5. Life Cycle Cost (LCC) analysis of a ST48 15D/A diesel locomotive. Paper No. M-8/494/2015/P, Cracow University of Technology, Institute of Rail Vehicles, Krakow (July 2015)
6. Life Cycle Cost (LCC) analysis of a 19WE rail set. Paper No. M-8/10/2009/1, Cracow University of Technology, Institute of Rail Vehicles, Krakow (March 2009)
7. Optimisation of the plan of maintenance of SM42 diesel locomotives operated by CTL Logistics Sp. z o.o. Paper No. M-8/160/2016/P, Cracow University of Technology, Institute of Rail Vehicles, Krakow (May 2016)
8. Optimisation of the plan of maintenance of 440Ra cistern wagons operated by CTL Logistics Sp. z o.o. Paper No. M-8/220/2016/P, Cracow University of Technology, Institute of Rail Vehicles, Krakow (May 2016)

9. Optimisation of the plan of maintenance of ET22 electric locomotives operated by CTL Logistics Sp. z o.o. Paper No. M-8/509/2016/P, Cracow University of Technology, Institute of Rail Vehicles, Krakow (October 2016)
10. Optimisation of the plan of maintenance of 418V self-discharging wagons operated by owned by CTL Maczki-Bór S.A. Paper No. M-8/543/2017/P, Cracow University of Technology, Institute of Rail Vehicles, Krakow (June 2017)
11. Analysis of the possibility of timely extension of the operating time of the P4 and P5 maintenance levels of the freight wagons of the Eas, Eaos and Eanos series. Paper No. M8/80/2014/P, Cracow University of Technology, Institute of Rail Vehicles, Krakow (2014)
12. Technical expertise on the possibility of introducing changes to the current inspection and repair cycle for repairs carried out at the fourth maintenance level of the ET22 modernised locomotive type 201Ek. Paper No. M-8/449/2016/P, Cracow University of Technology, Institute of Rail Vehicles, Krakow (2016)
13. Technical expertise to unify the existing maintenance plan in therms of P3, P4 and P5 levels of the modrnised traction vehicles ST45 type 301Dd and ST46 type 303 Da together with the appraisal and assessment of the risk. Paper No. M-8/629/2015/P, Cracow University of Technology, Institute of Rail Vehicles, Krakow (2015)
14. M. Szkoda, *Shaping the carrying potential of railway transport companies,* Wydawnictwo Politechniki Krakowskiej, Krakow (2017)
15. Strategy for the maintenance at the P1 maintenance level of SM42 modernised diesel locomotives (6Dg, 6Dg/A, 6Dg/B types). Feasibility study. Paper No. M-8/494/2014/P, Cracow University of Technology, Institute of Rail Vehicles, Krakow (May 2014)
16. Technical opinion on the possibility to change the maintenance plan as regards the P2 level of SM42 diesel locomotives of the 6Dg, 6Dg/A, 6Dg/B types together with the appraisal and assessment of the risk. Report of a research body on the assessment of safety. Paper No. M-8/410/2014/P, Cracow University of Technology, Institute of Rail Vehicles, Krakow (July 2014)
17. Technical opinion on the possibility to change the maintenance plan as regards P3 and P4 levels of SM42 diesel locomotives of 6Dg, 6Dg/A, 6Dg/B and 6Dk types together with the appraisal and assessment of the risk. Report of a research body on the assessment of risk permissibility. Paper No. M-8/236/2015/P, Cracow University of Technology, Institute of Rail Vehicles, Krakow (2015)
18. G. Kaczor, M. Szkoda, Reliability and availability assessment of diesel locomotive using Fault Tree Analysis, Archives of Transport, **40**, 79-89 (2016)
19. M. Szkoda, G. Kaczor, Application of fault tree analysis for reliability and availability assessment of the rolling stock, *The 11th World Congress on Railway Research WCRR 2016*, Milano, Italy (2016)
20. *ReliaSoft Corporation: System analysis reference: reliability, availability & optimization*, ReliaSoft Publishing, Tucson, AZ (2009)
21. A. Kadziński, *Study on selected dependability aspects of systems and rail vehicles objects*, Wydawnictwo Politechniki Poznańskiej, Poznan (2013)
22. R. Manzini, A. Regattieri, H. Pham, E. Ferrari, *Maintenance for Industrial Systems*, (Springer, 2010)
23. Commission implementing regulation (EU) No 402/2013 of 30 April 2013 on the common safety method for risk evaluation and assessment and repealing Regulation (EC) No 352/2009 (L 121/8, 3 May 2013)

Dynamic loading of the tank container on a flat wagon considering fittings displacement relating to the stops

Oleksij Fomin[1], *Juraj* Gerlici[2], *Alyona* Lovskaya[3], *Mykola* Gorbunov[4], *Kateryna* Kravchenko[2,*], *Pavlo* Prokopenko[5], and *Tomas* Lack[2]

[1]State University of Infrastructure and Technology, Department of Cars and Carriage Facilities, 9 Kyrylivska str., 04071 Kyiv, Ukraine
[2]University of Zilina, Department of transport and handling machines, Univerzitná 8215/1, 01026 Zilina, Slovak Republic
[3]Ukrainian State University of Railway Transport, Department of Wagons, 7 Feuerbach Square 7, 61050 Kharkov, Ukraine
[4]Volodymyr Dahl East Ukrainian National University, Department of railway, road transport and handling machines, 59a Central Avenue, 93400 Sewerodonetsk, Ukraine
[5]Branch "Research and Design and Technological Institute of Railway Transport", PJSC "Ukrzaliznytsya", 39 Ivana Fedorova, 03038 Kyiv, Ukraine

Abstract. The mathematical modelling of the dynamic loading of the container-tank during maneuvering collision of the flat wagon has made it possible to obtain an accurate value of the accelerations acting on its supporting structure. At the same time, possible movements of the tank-container relative to the frame of the flat wagons due to the presence of a technological gap in the pair "fitting stop-fitting" are taken into account. It is established that this acceleration, which is a component of the dynamic load, exceeds the one permitted by the regulatory documents. A computer simulation of the dynamic loading of the tank container in the CosmosWorks software environment was also carried out. The calculation was carried out using the finite element method. The results of the simulation made it possible to determine the distribution fields of the accelerations acting on the supporting structure of the tank-container during the shunting of the flat wagon, as well as their numerical values. Checking the adequacy of the mathematical and computer models using the Fisher criterion allowed us to conclude that the hypothesis of adequacy is not rejected.

1 Introduction

Ensuring the effectiveness of the developing economic ties between the states of Europe and Asia calls for the introduction of combined transport operations. To date, one of the most promising vehicles that have found application in combined transport is a container, and for the carriage of fluid cargo – a container-tank.

The increase in the volume of fluid cargo transportations towards the international transport corridors foreshows an increase in the level of demand for container-cisterns. In connection with this, there is a need to develop and put into operation new tank-containers with improved technical and economic indicators [1-5].

2 Analysis of recent research and publications

A study of the dynamic loads acting on the tank-container placed on the flat wagon during the manoeuvring collision is presented in [6].

The determination of the accelerations acting on the tank-container was carried out taking into account the gaps between the fitting stops of the flat wagon and the fittings of the tank-container. The longitudinal force acting on the flat wagon from the wagon-striker was about 2200-2800 kN, depending on the capacity of the tank-container.

The studies were carried out on tank containers for transportation of gasoline and nitric acid with a gross mass of 21.9 tons and 24.0 tons, respectively.

It is important to note that the maximum value of the longitudinal impact force, which can act on the flat wagon with the cargo placed on it, including the tank containers, with a shunting collision, is 3.5 MN [7, 8]. Therefore, in order to obtain an accurate value of the accelerations acting on the container-tank in operation, additional studies are needed.

A study of the strength of tank-container model TK25 and the optimization of its supporting structure are given in [9, 10]. When compiling the model of the strength of the tank-container, the normative values of the loads given in [11] are taken into account.

A study of the strength of the load-bearing structure of the tank-container, taking into account the specified operational values of the loads (the compliance of the fluid cargo, the movement of fittings relative to the fitting stops, etc.) is not carried out.

In works [12, 13], the feasibility of designing and commissioning tank containers as vehicles has been justified, the results of optimizing the design of tank containers are given. Advanced designs of tank containers for transportation of oil products have been developed.

* Corresponding author: kkatherina@ukr.net

The structural features of tank-containers for transportation of liquefied gases are discussed in [14]. The material of the supporting structure of the container is stainless steel. To limit the thermal impact of the environment on the container, it is proposed to encapsulate it with an insulating system.

Features of computer modelling of the supporting structure of a tank-container are given in [15]. The finite element method is used as the calculation method.

A study of the transfer of heat flow through the internal supports of cylindrical vessels using the example of a tank-container was considered in [16]. The paper simulates the heat flow through a multilayer support made of plastic.

Tests of metal and composite containers under the impact of low temperatures are given in [17]. It determines what types of containers it is expedient to use for transportation of the specified nomenclature of cargoes taking into account low ambient temperature.

Questions about the creation of ideal load-bearing structures of railway vehicles and the requirements that they must meet at the current stage of development are covered in [18-20].

The aforementioned works, studies of the dynamic loading of tank containers by methods of mathematical and computer modelling, taking into account possible displacements relative to the flat wagon frame, did not pay attention to the shunting collision.

3 Exposition of the object of research

An analysis of the regulatory documents on the issues of ensuring the strength of tank containers in operation allowed us to conclude that the greatest values of the dynamic loads acting on their load-bearing structures and fastening devices are indicated in [11].

It is agreed that the construction of the tank-container must withstand the inertia forces arising during the movement of the vehicle, as well as during the shunting of the wagons, including during the dissolution from the hills, emergency braking in trains at low speeds, with the following accelerations: in the longitudinal direction P_l – 2g; in the transverse direction P_t – 1g; in the vertical direction P_v – 2g; at impact: for the loaded container – 4g; for empty container (for the purpose of checking the fittings) – 5g.

To determine the operational values of the dynamic loads acting on the supporting structure of the tank-container placed on the flat wagon during manoeuvring collision, the mathematical model given in [6] was used. This model takes into account that three tank containers are placed on a long-based flat wagon, and the connection between them was simulated as an elastic-friction one, that is, each tank-container has its own degree of freedom in the vertical plane (formulas: 1-7).

The components of a mathematical model are determined by the formulas (8-11).

It is important to note that in the transportation of tank-containers, flat wagons on the frame of which two containers can be accommodated are also used. Such wagons have a smaller base, and accordingly less than the amount of movement of the supporting structure under the influence of the vertical load from the containers [21].

Therefore, in this study, research was conducted on the 13-4085flat wagon model, the construction of JSC Dniprovagonmash and the TK25tank-container model, built JSC Zarechensky Chemical Mechanical Engineering Factory. The tank-container is of ISO-1CC standard size and is intended for transportation of fuels and lubricants, gasoline, diesel fuel, engine oil, oil, oil solvents, nefras, foaming agents.

The tank-container was considered as an attached mass in relation to the flat wagon frame, having long-distance compliance due to the presence of gaps between the fitting stops of the flat wagon and the container-tank fittings. That is, the tank-container has its own degree of freedom up to the moment of fitting in the fitting stop, after which the container-tank repeats the trajectory of moving the flat wagon (Fig. 1).

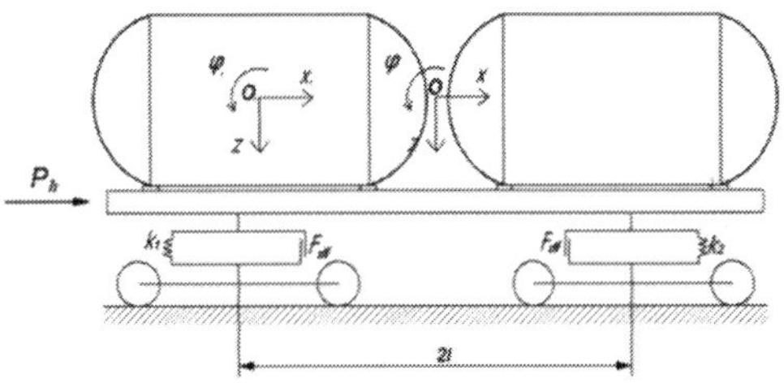

Fig. 1. Scheme of action of the longitudinal force on the flat wagon with the tank-containers placed on it.

The connection between the frame of the flat wagon and the container-tank fittings was simulated as frictional. At the same time, it is taken into account that the tank-containers placed on the flat wagon have the same boiler workload of the fluid cargo.

$$M'_{fw}\,\ddot{x}_{fw} + M_{fw}\,h\,\ddot{\varphi}_{fw} = S_c - \textstyle\sum_{i=1}^{2} S_i, \tag{1}$$

$$I_{fw}\,\ddot{\varphi}_{fw} + M_{fw}\,h\,\ddot{x}_{fw} - g\,\varphi_{fw}\,M_{fw}\,h = l\,F_{df}\left(sng\dot{\Delta}_1 - sng\dot{\Delta}_2\right) + l(k_1\,\Delta_1 - k_2\,\Delta_2), \tag{2}$$

$$M_{fw}\,\ddot{z}_{fw} = k_1\,\Delta_1 + k_2\,\Delta_2 - F_{df}\left(sng\dot{\Delta}_1 - sng\dot{\Delta}_2\right), \tag{3}$$

$$\left(m_i + \textstyle\sum_{j=1}^{k} m_{ij}\right)\left(\ddot{x}_{fw} + \ddot{x}_i\right) + \left(m_i\,z_{ci} + \textstyle\sum_{j=1}^{k} m_{ij}\,c_{ij}\right) \times \left(\ddot{\varphi}_{fw} - \ddot{\varphi}_i\right) - \textstyle\sum_{j=1}^{k} m_{ij}\,l_{ij}\,\ddot{\xi}_{ij} = S_i \tag{4}$$

$$\left(I_{\theta i} + \textstyle\sum_{j=1}^{k} m_{ij}\,c_{ij}^2\right)\left(\ddot{\varphi}_{fw} - \ddot{\varphi}_i\right) + \left(m_i\,z_i + \textstyle\sum_{j=1}^{k} m_{ij}\,c_{ij}\right)\left(\ddot{x}_{fw} + \ddot{x}_i\right) + \textstyle\sum_{j=1}^{k} m_{ij}\,c_{ij}\,l_{ij}\,\ddot{\xi}_{ij} - g\left(m_i\,z_i + \textstyle\sum_{j=1}^{k} m_{ij}\,c_{ij}\right)\left(\varphi_{fw} - \varphi_i\right) = 0 \tag{5}$$

$$\left(m_i + \textstyle\sum_{j=1}^{k} m_{ij}\right)\ddot{z}_{fw} = 0, \tag{6}$$

$$I_{ij}\,\ddot{\xi}_{fw} - m_{ij}\,l_{ij}\,\ddot{x}_{ij} - m_{ij}\,c_{ij}\,l_{ij}\,\ddot{\varphi}_{ij} + g\,m_{ij}\,l_{ij}\,\ddot{\xi}_{ij} = 0, \tag{7}$$

where

$$M'_{fw} = M_{fw} + 2m_b + \frac{n\,I}{r^2}; \tag{8}$$

$$\Delta_1 = z_{fw} - l\,\varphi_{fw}; \tag{9}$$

$$\Delta_2 = z_{fw} + l\,\varphi_{fw}; \tag{10}$$

$$S_i = f_{df}\,sng\left(x_{fw} - x_i\right)', \tag{11}$$

M_{fw} – weight of the supporting structure of the flat wagon, t; I_{fw}– moment of inertia of the flat wagon with respect to the longitudinal axis, t·m^2; S_c – magnitude of the longitudinal impact force in the coupler, kN; F_{df} – amplitude value of the dry friction force, kN; m_b – weight of the bogie, t; I – moment of inertia of the wheel pair, t·m^2; r – radius of the medium-worn wheel, m; n – number of axles of the bogie; l – half the base of the flat wagon, m; F_{df} – the absolute value of the dry friction force in the spring set, kN; k_1, k_2 – rigidity of springs spring suspension of carriages of the flat wagon, kN/m; k – number of tones of fluctuations in fluid cargo; m_i – mass of the body equivalent to the i-th tank-container with a part of the fluid cargo that does not participate in the movement relative to the boiler, t; m_{ij} – mass of the j-th pendulum in the i-th container-cistern, t; z_{ci} – height of the center of gravity of the tank-container, m; c_{ij} – distance from the plane to the point of fixation of the j-th pendulum in the i-th container-tank; l_{ij} – length of the j-th pendulum, m; I_θ– the given moment of inertia of the i-th tank-container and the fluid cargo, not taking part in the movement relative to the boiler, t·m^2; I_{ij} – moment of inertia of the pendulum, t·m^2; x, y, z – coordinates corresponding, respectively, longitudinal, angular around the longitudinal axis and vertical displacement of the flat wagon; x_i, φ_i – coordinates corresponding, respectively, longitudinal and angular around the longitudinal axis to the displacement of the tank-container; $\ddot{\xi}_{ij}$ – angle of deviation of the j-th pendulum from the vertical.

In this case x_i < 30 mm [6, 22], if $x_i \geq 30$ mm, then $x_i = x$.

Vertical movements of the tank-container relative to the flat wagon frame were not taken into account. The compliance of the fluid cargo with respect to the walls of the tank-container cistern was also taken into account. The movement of fluid cargo was described by a set of mathematical pendulums [6]. The force of the longitudinal impact acting on the wagon frame is taken to be 3.5 MN.

Determination of the hydrodynamic characteristics of fluid cargo was carried out according to the procedure given in [23]. Petrol was assumed to be the bulk cargo, On the basis of the calculations carried out in the case of the maximum permissible workload of the tank-container boiler, in accordance with [24], the value of $m_{ij} \approx 6.8$ t, $I_{ij} \approx 250$ t·m^2, was obtained.

The solution of the differential equations was carried out using the Runge-Kutta method in MathCad software environment [25, 26].

The results of the studies made it possible to conclude that the maximum accelerations that act on the tank-container are about 300 m/s^2.

Also, a study of the dynamic loading of a tank-container was carried out using computer simulation in CosmosWorks software environment, version 2015 [27]. A spatial model of the load-bearing structure of a flat wagon with containers is shown in Fig. 2.

The calculation was carried out using the finite element method. When compiling the finite element model, spatial isoparametric tetrahedra are used. The number of grid nodes was 707359, the elements – 2150500. The maximum element size was 50 mm, the minimum – 10 mm. The minimum number of elements in the circle was 9, the ratio of the increase in the dimensions of the elements in the grid is 1.7. The maximum aspect ratio is 689.01, the percentage of elements with aspect ratio of less than 3 – 52.1, more than 10 – 4.75.

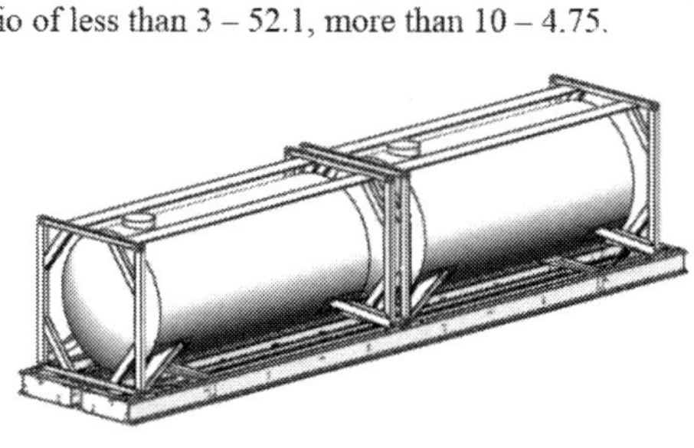

Fig. 2. Spatial model of the flat wagon with tank containers placed on it.

When compiling the strength model (Fig. 3), it is taken into account that, apart from the longitudinal force P_h, kN, there are vertical forces acting on the flat wagon in the container support areas on the fitting stops. The container has a vertical response in the fitting area of the fitting to the fitting stop, the internal pressure of the fluid cargo P_{fc}, kPa on the boiler, the longitudinal load P_{ll}, kN on the bottom of the boiler when the fluid cargo moves, due to the action of the longitudinal force P_h on the rear stop of the coupler and the movement of the fittings relative to the fitting stops by the amount of the technological gap. To account for the frictional force in the areas of interaction between the flat wagon and the containers, the modal damping coefficient was used.

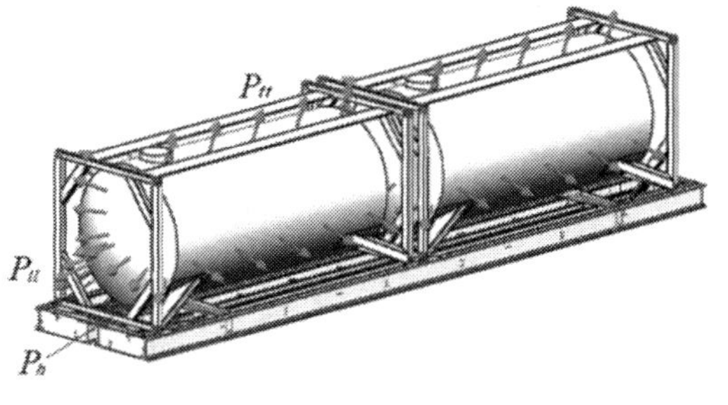

Fig. 3. The model of the strength of the flat wagon with tank containers placed on it under the action of longitudinal force on the rear stop of the automatic coupler.

The fixing of the model was carried out in the areas of supporting the supporting structure on the running gear. As the material for the load-carrying structure of the flat wagons and containers, steel 09G2C was used with a tensile strength value of 490 MPa and a yield strength of 345 MPa [7, 8].

The results of the computer simulation are shown in Fig. 4.

From the studies, which were carried out, it can be concluded that the maximum acceleration acting on the tank-container was about 320 m/s^2.

To check the adequacy of the developed model, Fisher's criterion was used.

 https://doi.org/10.1051/matecconf/201823405002

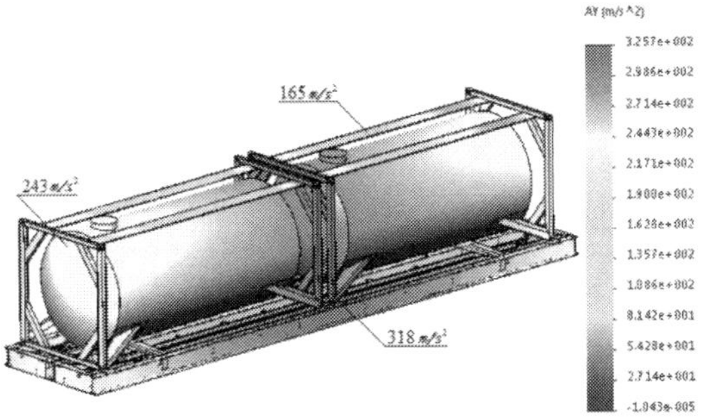

Fig. 4. Acceleration of the flat wagon with the containers placed on it when the longitudinal force acts on the rear stop of the automatic coupler.

It was established that the considered model is linear and characterizes the change in the acceleration of the tank-container placed on the flat wagon from the longitudinal force acting on the rear fence of the automatic coupler. In this case, the number of degrees of freedom for $N = 6$ will be $f = 4$.

On the basis of the calculations, it has been established that the actual value of criterion $F_{cl} = 1.6$ is less than the table value of criterion $F_t = 6.16$. Consequently, the hypothesis of adequacy is not rejected.

The obtained results for the accelerations, as components of the dynamic load, are taken into account in the study of the strength of the load-bearing structure of the tank-container. The results of the calculations made it possible to conclude that the maximum equivalent stresses in the construction of a tank container arise in the zone of interaction between the boiler support and the fitting stop and are about 1015 MPa (Fig. 5), the maximum displacements in the structural units were 13.6 mm, the deformations – 6.08×10^3 [21].

The distribution of stresses along the length of the lower transverse belt of the tank-container frame is shown in Fig. 6. It can be seen from the figure that the maximum equivalent stresses are concentrated in the areas of placement of the container-tank fittings, and they decrease towards the central part of the belt.

4 Conclusion

The conducted research allowed us to define the specified value of acceleration, which acts on the supporting structure of the tank-container taking into account the movements of the fittings relative to the fitting stops of the flat wagon. Calculation of the strength showed that with possible movements of the tank container relative to the platform car frame in the conditions of the shunting collision and the underload of the boiler with the fluid load, the stresses in the load-bearing structure exceed the permissible ones. The results of the research will contribute to the creation of a new-generation tank-containers with improved technical and economic parameters that will improve the efficiency of the transport process towards international transport corridors.

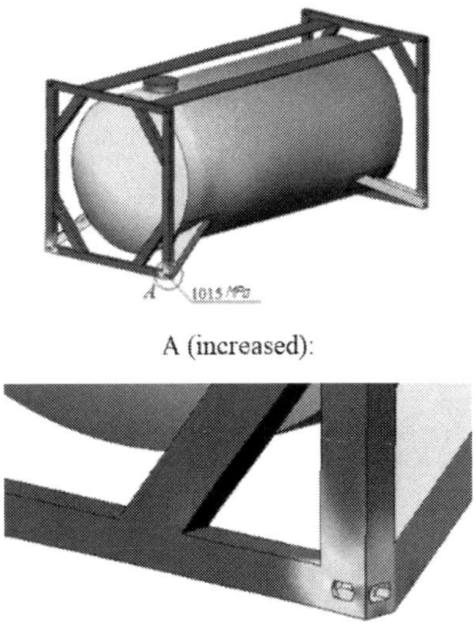

Fig. 5. Results of the calculation of the strength of the load-bearing structure of the tank-container TK25.

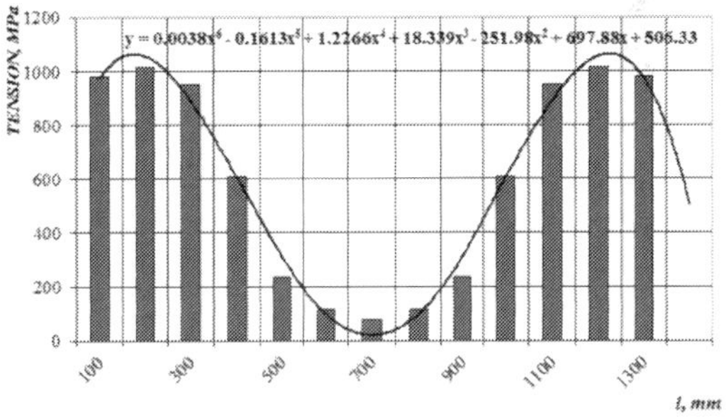

Fig. 6. Distribution of stresses along the length of the lower transverse belt of the tank-container frame.

References

1. P. Fabian, J. Gerlici, J. Masek, P. Marton, Versatile, efficient and long wagon for intermodal transport in Europe, Komunikacie, **15**(2), 118-123 (2013)
2. J. Dižo, J. Harušinec, M. Blatnický, Computation of modal properties of two types of freight wagon bogie frames using the finite element method, Manufacturing Technology, **18**(2), 208-214 (2018)
3. I. Gritsuk, Y. Gutarevych, V. Mateichyk, V. Volkov, Improving the processes of preheating and heating after the vehicular engine start by using heating system with phase-transitional thermal accumulator, *SAE Technical Paper* 2016-01-0204 (2016), https://doi.org/10.4271/2016-01-0204
4. V. Hauser, O. Nozhenko, K. Kravchenko, M. Loulová, J. Gerlici, T. Lack, Car body and bogie connection modification for track curves passability improvement, *22nd Slovak-Polish Scientific Conference on Machine Modelling and Simulations, MMS 2017*, Sklene Teplice, Slovakia, MATEC Web of Conferences, **157** (5-8 September 2017)

5. P. Štastniak, M. Moravčík, Freight bogie prototype properties analysis by means of simulation computations, Manufacturing Technology, **17**(3), 381-388 (2017)
6. G. Bogomaz, D. Mehov, O. Pilipchenko, Y. Chernomashentseva, Load of tank containers located on a railway platform, when striking a coupler (in Russian), *Collection of scientific works "Dynamics and control of movement of mechanical systems"*, Kiev, 87-95 (1992)
7. Norms for the calculation and design of railroad wagons of the Ministry of Railways of gauge 1520 mm (non-self-propelled) (in Russian), *Moscow: GosNIIV, VNIIZhT*, 319 (1996)
8. GOST 33211. Freight wagons. Requirements for strength and dynamic qualities (in Russian). *Moscow: Standartinform*, 54 (2016)
9. O. Fomin, O. Burlutsky, Y. Fomina, Development and application of cataloging in structural design of freight car building, Scientific and technical journal "Metallurgical and Mining Industry", **2**, 250-256 (2015)
10. O. Fomin, A. Lovskaya, A. Plakhtiy, V. Nerubatsky, The influence of implementation of circular pipes in load-bearing structures of bodies of freight cars on their physico-mechanical properties, Scientific Bulletin of National Mining University, **6**(162), 89-96 (2017)
11. GOST 31232. Containers for the transport of dangerous goods. Requirements for operational safety (in Russian), Belarusian State Institute for Standardization and Certification, Minsk, 6 (2005)
12. S. Myamlin, Y. Kebal, S. Kondratyuk, Perspective constructions of tank-containers for transportation of light oil products, ammonia and hydrocarbon gases (in Russian), Journal "Rail transport Ukraine", **2**, 44-46 (2012)
13. E. Lisowski, W. Czyzycki, Transport and storage of lng in container tanks, Journal of KONES Powertrain and Transport, **18**(3), 193-201 (2011)
14. E. Lisowski, W. Czyżycki, K. Łazarczyk, Wdrożenie produkcji cystern kontenerowych do transportu i przechowywania gazów skroplonych, Biuletyn Instytutu Spawalnictwa w Gliwicach, **56**(2), 25-31 (2012)
15. M. Ghiţă, C. Micu, M. Ţălu, Ş. Ţălu, 3D modeling of a shrink fitted concave ended cylindrical tank for automotive industry. Acta Technica Corviniensis: Bulletin of Engineering, **6**(4), 87-92 (2013)
16. Wojciech czyżycki modeling of heat flow through multilayer internal supports of cryogenic vessels, *Technical transportations*, **2**, 27-34 (2015)
17. J. Föhr, K. Karttunen, J. Enström, T. Johannesson, T. Ranta, Metal and composite intermodal containers in comparative cold tests with wood chips, Journal of Sustainable Bioenergy Systems, **5**(1), 32-39 (2015)
18. O. Fomin, Increase of the freight wagons ideality degree and prognostication of their evolution stages, Scientific Bulletin of National Mining State University, **2**, 68-76 (2015)
19. O. Fomin, Improvement of upper bundling of side wall of gondola cars of 12-9745 model, Metallurgical and Mining Industry, **1**, 45-48 (2015).
20. A. Lovska, Simulation of loads on the carrying structure of an articulated flat car in combined transportation. International Journal of Engineering & Technology, **7**(4.3), 140-146 (2018)
21. O. Lovska, Investigation of the durability of the bearing structure of the tank-container placed on the carriage-platform during the maneuvering co-operation (in Ukrainian), Collection of scientific works of State University of Infrastructure and Technology, **28**, 90-98 (2016)
22. Rules for the placement and fastening of goods in wagons and containers when transporting them by railways of the gauge of 1520 mm of the SMGS participating countries (in Russian). Part 1. General provisions. OSJD, 681 (2012)
23. Y. Krivovyazyuk, Estimating the equivalent loading of four-axle rail tank wagons with liquid cargo of different density for longitudinal impacts (in Russian), *Tesis of PhD Dnepropetrovsk*, 157 (1986)
24. Rules for the transport of dangerous goods, To the agreement on international railway cargo communication, **3**, OSJD, 531 (2011)
25. V. Dyakonov, *MATHCAD 8/2000: Special directory* (in Russian), 592 (St. Petersburg, 2000)
26. D. Kiryanov, *Mathcad 13* (in Russian), 608 (St. Petersburg, 2006).
27. A. Alyamovskiy, *SolidWorks/COSMOSWorks 2006-2007. Engineering analysis by the finite element method* (in Russian), 784 (Moscow, 2007).

MATEC Web of Conferences **234**, 05003 (2018) https://doi.org/10.1051/matecconf/201823405003
BulTrans-2018

Research of safety indicators of diesel train movement with two-stage spring suspension

Andriy Kuzyshyn[1,*], *Andriy* Batig[2], *Sergei* Kostritsa[1], *Julia* Sobolevska[3], *Vitalii* Kovalchuk[3], *Stepan* Dovhanyuk[1], and *Oleg* Voznyak[3]

[1]Dnipropetrovsk National University of Railway Transport named after Academician V. Lazaryan, 49000 Dnieper, Ukraine
[2]Lviv Research Institute of Forensic Expertise, 79000 Lviv, Ukraine
[3]Lviv branch of Dnipropetrovsk National University of Railway Transport named after Academician V. Lazaryan, 79052 Lviv, Ukraine

Abstract. The problem of the interaction of rolling stock with the rail track has been analyzed in the present paper. It has been established that in the process of transport science development a number of methods for determining the causes of wheel pairs derailment are developed, which, in a varying degree, take into account the peculiarities of their interaction. The problem of choosing the most accurate method for estimating the causes of rolling stock derailment becomes more complicated because of the lack of sufficient experimental data that would allow us to verify the adequacy of the models. The indicators of stability of the wheel against derailment, which are used on the railways of Ukraine and Europe, have been examined. Their dependences on the speed of movement were derived. It has been established that the increase of the speed of motion leads to the increase of the interaction power of the rolling stock with the rail track, which may, under certain operational parameters, provoke its derailment. As a result of the calculations, it has been shown that the use of norms for car design and calculation used on Ukrainian railways can lead to an inadequate estimation of traffic safety parameters, since it does not take into account the unevenness of the railway track. It has been established that the requirements of BS EN 14363: 2005 European norms are stricter in comparison with the norms for calculation and evaluation of the bearing elements strength and dynamic qualities of motor-vehicle rolling stock used on Ukrainian railways. A comparison of the experimentally and theoretically calculated values of the stability margin coefficient against wheel derailment of the first wheel pair of the diesel train car was carried out.

1 Introduction

The most important problem on rail transport is to ensure the safety of train movement during operation. At the same time, the primary task is to prevent accidents and derailments. Rolling stock derailment is due to the loss of the interaction of at least one wheel of the rolling stock of the railway transport with a rail as a result of the displacement of the wheel from its normal position relatively to the rail head [1].

During movement of the wheel along the rail, vertical and horizontal forces occur, which cause the wheel flange to press on the head of the rail. Taking into account the geometry of the wheel and rail contact, such a process can lead to the displacement of the contact point on the wheel flange, and in the worst case, can lead to the wheel to roll onto the head of the rail (Fig. 1).

As a rule, the action of a considerable vertical force on the wheel is enough to make such a process impossible. However, under certain operating conditions, there can be substantial increase in horizontal force or vertical force reduction, which may cause wheel derailment.

The level of safe operation of the rolling stock on the railways is determined, mainly, by the availability of stability margin of the train. Therefore, the issue of derailment is given great importance both in theoretical and experimental studies.

The object of our research is the DPKr-2 modern passenger diesel train [2] of Kryukiv, Ukraine, carriage plant production (Fig. 1).

It should be noted that at this stage only diesel train operational tests were carried out. Theoretical studies of the power interaction of this diesel train with the rail track, as well as theoretical determination of the level of safe operation with the help of the coefficient of stability margin of the wheel against derailment have not been carried out.

Fig. 1. The wheel rolling onto the head of the rail.

* Corresponding author: kuzyshyn1993@gmail.com

2 Analysis of literary data and problem statement

Over a hundred years ago Nadal in his paper [3] proposed a method for determining the ratio of forces acting on the wheel, under which wheel derailment does not occur. But in this method a separate wheel is considered, and the criterion itself is based on the quasi-static conditions of the vehicle movement. Under real conditions of operation, the process of wheel pair derailment is dynamic and is the subject of further research [4, 5].

For dynamic simulation of wheel pair derailment, it is suggested to use modern software packages [6]. This makes it possible to simulate the movement of the whole train rather than a separate wheel pair.

In paper [7] a technique using acceleration sensors to detect early signs of wheel pair derailment is presented. The algorithm for detecting wheel pair derailment is tested by means of simulation and experimental research on a scale model of a train.

In paper [8], two safety criteria against wheel derailment are compared. The first one is based on Nadal's formulas, and the second criterion is a power one. It was noted that the power criterion is more convenient for railway transport state monitoring, since it does not require the measurement of contact forces.

In paper [9], the kinematic and geometric safety criteria against wheel derailment were researched and compared.

Applying the concept of dynamic system stability in paper [10], the mechanism of rolling stock derailment on a railway bridge was investigated. The theory of random power analysis is proposed.

In papers [11, 12], the studies were carried out on the modification of derailment criteria, which took into account the angle of wheel rolling on the head of the rail. It was established that taking into account the angle of wheel rolling onto the rail, when determining the possibility of wheel pair derailment, the accuracy of calculations increases.

In papers [13, 14] the method for assessing the stability of motion against wheel derailment according to the specified formula was proposed. It determines the critical magnitude of the ratio of horizontal frame strength and vertical force under given vertical load distribution. The refined formula has found wide application in experimental works on determining the maximum permissible speeds of movement of new and modernized units of rolling stock.

In current regulations of railways in Ukraine [15, 16] that are used to estimate the probability of wheel derailment, the coefficient of stability margin against wheel rolling onto the head of the rail is suggested for usage. It is believed that in the case when the value of this coefficient is less than one, the wheel pair can derail.

In paper [17], to determine the safety conditions against wheel pair derailment, the author suggests not to consider a single wheel derailment, but a wheel pair as a whole regarding it as a solid body. In this case, it is necessary to consider the possible movement of the wheel pair when wheel flange is sliding in the vertical transverse plane and runs on the head of the rail. The disadvantages of the proposed method are the failure to take into account the impact of the other wheel pairs of rolling stock, which leads to inaccuracies in the calculations.

Paper [18] formulates the power interpretation of the wheel rolling onto the head of the rail. According to the authors, this is possible when the kinetic energy of motion, which falls on a wheel pair, exceeds the work of forces that arise in the contact between the surfaces of the flange and the wheel.

Article [19] presents a brief overview of selected problems of railway vehicle dynamics that are related to traffic safety. The study of the corresponding dynamic phenomena, including the mechanism of the wheel pair derailment, is also discussed.

Works [20, 21] outline the approaches that are used in conducting forensic rail-transport expertise.

The analysis of publications and normative documents has shown that most of the considered methods estimate the probability of derailment for one wheel pair, without taking into account the impact of other wheel pairs, which may lead to incorrect identification of the reason for wheel pair derailment. In addition, horizontal and vertical unevenness of the track is not taken into account, although, as experience shows, the main causes of accidents and derailments on the railways of Ukraine are, first of all, the deviation in the condition of the railway track and running gears of the rolling stock from the standards.

Therefore, to approximate the results of mathematical modeling to the real conditions of operation, it is proposed to calculate wheel pair derailment because of the wheel rolling onto the head of the rail, using the dynamic equations of spatial oscillations of the car arising from the geometrical unevenness of the track in the horizontal and vertical directions.

At present the theoretical studies of power interaction of the diesel train DPKr-2 with a rail track have not been conducted. Therefore, the problem of determining the conditions for its safe operation with the help of the coefficient of stability margin against wheel derailment and the theoretical prediction of the conditions for accidents and derailments prevention is urgent.

3 Purpose and objectives of the study

The purpose of the work is to determine the traffic safety index, namely the stability of the wheel against rolling on the head of the rail for DPKr-2diesel train.

To achieve this goal, the following tasks were performed:

- wheel stability indicators against derailment used on the railways of Ukraine and Europe were calculated;
- operation conditions of DPKr-2 diesel train moving on a straight section of the track, which are acceptable according to Ukrainian and European norms, were studied;
- adequacy of the mathematical model was studied by comparing the obtained results during theoretical calculations with the experimental ones.

4 Theoretical determination of traffic safety indicators

To obtain a spatial mathematical model that will be used to determine the dynamic characteristics of the criteria for the probability of wheel pair derailment, an analysis of the design of the diesel train car was made and its spatial mechanical model was adopted [22].

The chosen mechanical model consists of 7 solid bodies (body, two bogies, four wheel pairs), connected by elastic-dissipative joints. The body and bogies have 6 degrees of freedom, and wheel pairs have 5, because there is no galloping for them. The possibility of taking into account the unevenness of the track in the horizontal and vertical directions is foreseen.

To describe the oscillations of the diesel train car, the mathematical model consisting of 38 second-order differential equations is obtained [23].

At present, there are a number of standard methods for determining the possibility of wheel pair derailment, which, in varying degrees, take into account the features of a wheel pair and rail track interaction. Thus, the railways of Ukraine are governed by the norms for calculating and designing motor-vehicle rolling stock [15] or carriages [16], while European railways are governed by BS EN 14363: 2005 norms.

On the basis of the performed dynamic calculations of the car of the DPKr-2 diesel train, we determine its coefficient of stability margin in motion on the straight section of the track according to different methods.

In accordance with norms [15], the coefficient of stability margin of the wheel against derailment is used as a traffic safety indicator. This criterion is the ratio of the permissible theoretical value of the driving force to the vertical load up to its experimental value. The minimum permissible value of the stability margin coefficient against wheel derailment is 1,4.

The expression for determining the coefficient of stability margin has the form [15]:

$$K_{\mathrm{st}} = [\left(\frac{Y_{\mathrm{d}}}{N_{\mathrm{n}}}\right)_{\mathrm{T}} / \left(\frac{Y_{\mathrm{d}}}{N_{\mathrm{n}}}\right)_{\mathrm{E}}] = \langle \frac{Q[2(b-a_2)-K_{\mathrm{D1}}(l_1+a_1)+a_2K_{\mathrm{D2}}]+Y_f r+q(b-a_2)}{Y_f l_1} + \mu'\{Q[2(b-a_1)-K_{\mathrm{D2}}(l_1+a_2)+a_1K_{\mathrm{D1}}]-Y_f r + q(b-a_1)\}\frac{1+\mu\sin(2\beta)}{\beta} + \{Q[2(b-a_2)-K_{\mathrm{D1}}(l_1+a_1) + a_2K_{\mathrm{D2}}]+Y_f r+q(b-a_2)\}\mu\cos^2\beta\rangle\frac{1}{\cot\beta+\mu}, \quad (1)$$

where Y_d – the driving force; N_n – vertical force from the wheel, which is rolling onto the head of the rail; μ – the coefficient of friction between the flange of the rolling wheel and the rail; β – angle of inclination of the wheel flange generatrix to the horizontal plane; Y_f – frame force acting on a wheel pair (modulo); Q – weight of the suspension part of the car, which falls on the neck of the axle of the wheel pair; K_{D1}, K_{D2} – coefficients of vertical dynamics in the 1st degree of suspension; μ' – the coefficient of friction between the rolling surface of the non-rolling wheel and the rail; q – the weight of non-suspended parts that falls on the wheel pair; b – half distance between the points of application of vertical loads and the necks of the axle of the wheel pair; a_1 – the distance between the point of application of the vertical load on the axle neck of the rolling wheel and the contact point on the flange; a_2 – the distance between the point of vertical load application on the axle neck of the non-rolling wheel and the point of contact on the surface of its rolling; r – the radius of the wheel in the circle of rolling.

Analyzing formula (1), we see that the change in the stability margin coefficient is due to the change in the frame strength and coefficients of vertical dynamics. The other parameters are constant values that do not result in the change of the stability margin coefficient against wheel's derailment during diesel train movement.

To determine the frame strength and coefficients of vertical dynamics, we use the adopted spatial mathematical model.

Since the frame force is a horizontal force acting on the axle of the wheel pair of the bogie on the body side in the transverse direction to the axis of the track, then it can be calculated by the formula:

$$Y_f = 2S_{ts}\left[\left(y_{bj} \pm a_1\psi_{bj} + a_7\theta_{bj}\right) - \left(y_{wpi}\right)\right], \quad (2)$$

where S_{ts} – the transverse stiffness of the spring of an axlebox suspension; y_{bj}, y_{wpi} – horizontal transverse displacements of the $j^{\text{-th}}$ bogie of the diesel train car and its $i^{\text{-th}}$ wheel pair, respectively, ψ_{bj}, θ_{bj} – angles of wobbling and lateral oscillation of the bogie, obtained during the numerical integration of the equations of motion; a_1, a_7 – corresponding lengths.

In formula (2), the upper "+" sign is used for the first wheel pair of the bogie, the "–" sign is used for the second wheel pair of the bogie.

The coefficients of vertical dynamics are determined by deflections Δ or forces F in separate sets of spring suspension:

$$K_{\mathrm{D}} = \frac{\Delta_{\mathrm{D}}}{\Delta_{\mathrm{st}}} = \frac{(z_{bj} \pm a_1\varphi_{ybj} \mp a_4\theta_{bj}) - (z_{wpi} \mp a_4\theta_{wpi})}{\Delta_{\mathrm{st}}}, \quad (3)$$

where z_{bj}, φ_{ybj}, θ_{bj}, z_{wpi}, θ_{wpi} – generalized coordinates of car movement of a diesel train, obtained during calculations of the spatial mathematical model; a_1, a_4 – corresponding lengths.

After calculating the given formulas at different speeds of the diesel train movement, we obtain the dependence of the coefficient of stability margin against wheel derailment on the speed of the first and second wheel pairs (Fig. 2).

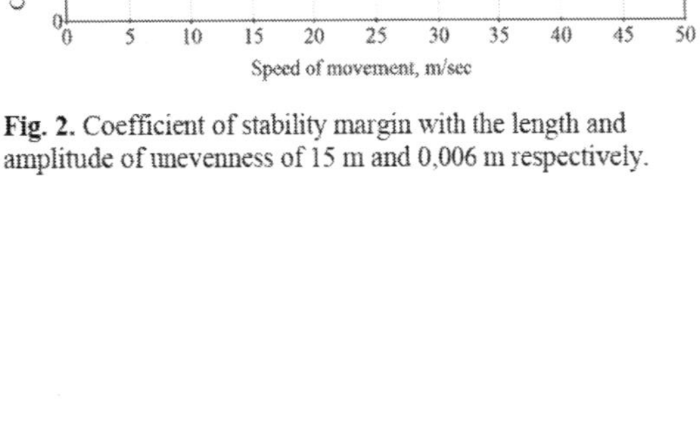

Fig. 2. Coefficient of stability margin with the length and amplitude of unevenness of 15 m and 0,006 m respectively.

The analysis of the obtained dependencies shows that when the speed of movement increases from 0 m/s to 50 m/s, the stability coefficient decreases in the range: I wheel pair – from 3,73 to 2,84; II wheel pair – from 3,73 to 2,14.

The reduction of the stability margin coefficient against wheel derailment shows the increased power effect of the rolling stock on the railway track. Under certain operating conditions, it can lead to the loss of stability of the rolling stock and its derailment.

The second technique is based on the use of norms [16] for cars. According to them, the coefficient of stability of the wheel pair against derailment (under the condition of rolling onto) is determined by the formula:

$$K_{\mathrm{st}} = \frac{\tan\beta-\mu}{1+\mu\cdot\tan\beta} \times \frac{2Q_n\left[\frac{b-a_2}{l}(1-C_{vd1})-\frac{b}{l}C_{dlw}\right]+H_f\frac{r}{l}+q_{wp}\frac{b-a_2}{l}}{H_f+\mu\left\{2Q_n\left[\frac{b-a_1}{l}(1-C_{vd})+\frac{b}{l}C_{dlw}\right]-H_f\frac{r}{l}+q_{wp}\frac{b-a_1}{l}\right\}} \geq [K_{\mathrm{st}}], \quad (4)$$

where β – the angle of inclination of the wheel flange generatrix to the horizontal plane; μ – the coefficient of friction of surfaces of wheels and rails; Q_n – the force of the weight of the suspended parts of a car, acting on the neck of the axle of the wheel pair; q_{wp} – the weight of non-suspended parts that falls on the wheel pair; C_{vd1} – the estimated value of the coefficient of vertical dynamics of a car; C_{dlw} – the estimated value of the coefficient of lateral wobbling dynamics; H_f – the calculated mean value of the frame force; b – half distance between the middle parts of the axle necks; l – an average distance between contact points of wheels with rails; $a_{1,2}$ – the calculated distance from the contact points to the middle of the necks of the wheel pair axle; r – the radius of the wheel.

According to the results of the calculations with formula (4), the dependence of the stability coefficient of the wheel pair against derailment (under condition of rolling onto) on the speed of the diesel train car (Fig. 3) was obtained.

The analysis of the calculated data shows that when the speed of movement increases from 10 to 50 m /s, the stability coefficient of the wheel pair decreases from 7,27 to 2,01, but remains permissible.

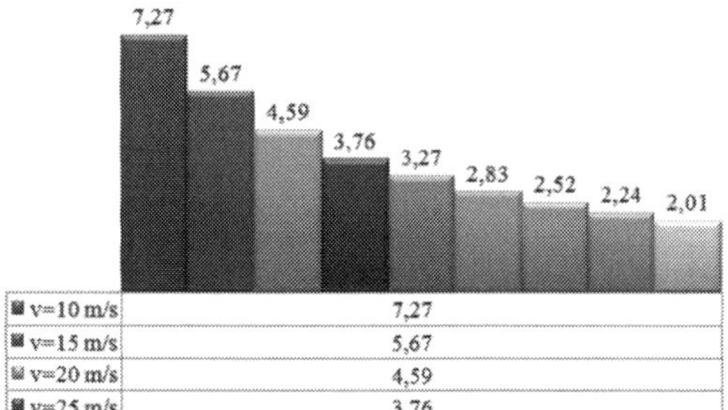

v=10 m/s	7,27
v=15 m/s	5,67
v=20 m/s	4,59
v=25 m/s	3,76
v=30 m/s	3,27
v=35 m/s	2,83
v=40 m/s	2,52
v=45 m/s	2,24
v=50 m/s	2,01

Fig. 3. Dependence of the stability coefficient of the wheel pair on the speed of movement.

It should be noted that formula (4) does not allow us to take into account the real conditions of operation, namely, vertical and horizontal unevenness of the rail track, as well as the peculiarities of the interaction of separate elements of the diesel train car design. The frame strength and coefficients of dynamics are considered constant values. All this makes it impossible to use the data obtained during the calculations of the mathematical model.

Let us turn to the analysis of European norms BS EN 14363:2005, according to which the rolling stock is considered secure against derailment if the condition (5) is performed at each stage of the test:

$$\left(\frac{Y}{Q}\right)_{\max} \leq \left(\frac{Y}{Q}\right)_{\lim}, \quad (5)$$

where Y – horizontal guiding force; Q – vertical force from the wheel on the rail.

Using the model of the diesel train car DPKr-2, the guiding force was found to be the resultant force of the creep force and the frame strength and is determined by the formula:

$$N_{\mathrm{n}} = \frac{Y_f-Y_{fr1}-Y_{fr2}}{\sin\gamma_1}, \quad (6)$$

where Y_f – frame strength; $Y_{fr1,2}$ – the forces of the creep on the first and second wheels of the wheel pair respectively; γ_1 – the conicality of the wheel flange of the wheel pair.

It should be noted that the guiding force occurs only when the gap between the wheel flange and the rail head is exhausted. In the presence of a gap, the guide force is zero. Since the mutual horizontal movement of the wheel and the rails is calculated at each step of the integration, the proposed model allows for tracking the instant when the clearance between the flange and the rail head becomes zero and for calculating the guide force.

It should be noted that the use of European norms BS EN 14363: 2005 to determine the possibility of a wheel derailment is practically impossible without the use of computer simulations, since it is necessary to calculate a guiding force, which requires tracking the instant when the gap between the flange and the rail head becomes equal to zero.

By calculation, we check the fulfillment of condition (5) at different speeds of the diesel train movement (Fig. 4). The calculation is performed with a horizontal unevenness for 15 m and its amplitude for 0,006 m, which corresponds to the good condition of the railway track.

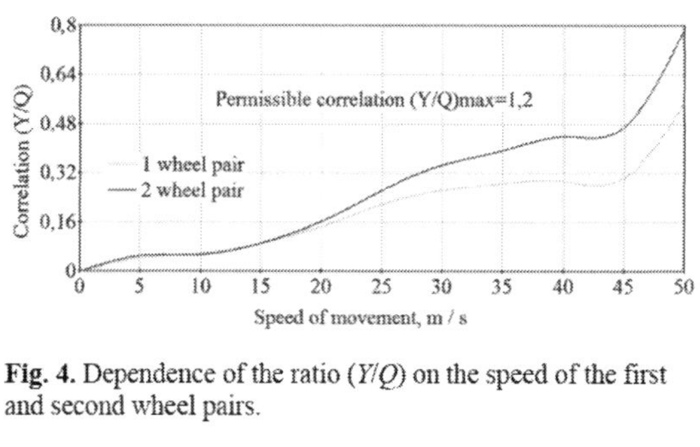

Fig. 4. Dependence of the ratio (Y/Q) on the speed of the first and second wheel pairs.

https://doi.org/10.1051/matecconf/201823405003

From the obtained dependences, it follows that an increase in the speed from 0 m/s to 50 m/s leads to the increase in the numerical value of the ratio (*Y/Q*): for the first wheel pair – from 0 to 0,55; for the second one – from 0 to 0,79, which is close to the permissible value and can, under certain operating conditions, cause accidents and derailments.

Using the proposed approach for determining the safety indicators, the maximum permissible speed of the DPKr-2 diesel train car was determined at a given horizontal unevenness of the track (Table 1).

The sign "–" means the realization of the condition of derailment and the danger for the diesel train car movement.

The norms [15] used in Ukraine allow for movement of a diesel train car at a speed of 115-120 km/h, but BS EN 14363: 2005 European norms do not allow it.

Table 1. Permissible speed of the diesel train car movement.

Speed, km/h	Traffic safety indices with an unevenness of 20 m and an amplitude of 10 mm		
	European norms on the condition of derailment	Norms [15] on the condition of derailment	Car norms [16]
150	–	–	+
145	–	–	+
140	–	–	+
135	–	–	+
130	–	–	+
125	–	–	+
120	–	+	+
115	–	+	+
110	+	+	+
105	+	+	+

It was found out that the norms for cars [16], which do not take into account rail track unevenness, permit the diesel train to drive at the maximum permissible speed. However, according to the norms [15], with a given horizontal unevenness, the speed of up to 120 km/h is permissible. Thus, consideration of the track state significantly affects the traffic safety indicators. Using the coefficient of stability margin against the wheel derailment, which is calculated according to the norms [16], may lead to an unreliable estimation of traffic safety parameters, because the failure to take into account actual operating conditions can cause serious accidents.

In general, it can be stated that the requirements of the European norms for the DPKr-2 diesel train are more stringent than the requirements of norms used in Ukraine.

The obtained results can be used to establish safe speeds of diesel train movement on straight sections of the track in the presence of vertical and horizontal unevenness.

5 Experimental determination of the stability margin coefficient against wheel derailment

To verify the adequacy of the mathematical model of the diesel train car, we will compare the values of the stability margin coefficient against wheel derailment obtained by calculation with the data of experimental studies (Fig. 5).

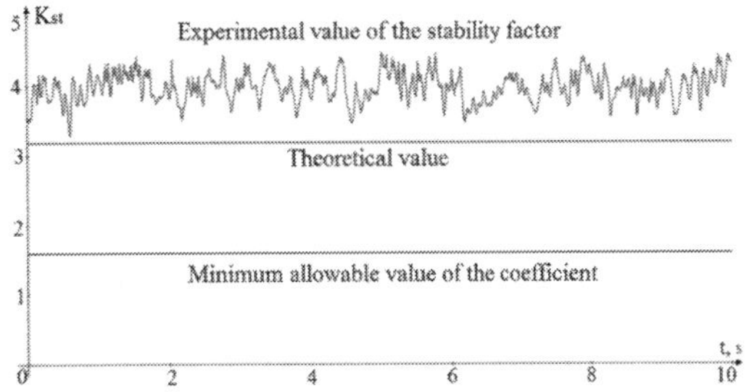

Fig. 5. Comparison of the experimental values of the stability margin coefficient against wheel derailment of the first wheel pair of a diesel train car at the speed of 30 m/s on a straight section of the railway with theoretically calculated ones.

Experimental study of the DPKr-2 diesel train was carried out by the "Branch research laboratory of dynamics and rolling stock strength" of Dnipropetrovsk National University of Railway Transport named after Academician V. Lazaryan.

The determination of the dynamic characteristics included in the criteria for the probability of derailment was carried out using strain gauges of the type KF5-P1-10-200 installed on the axle boxes, as well as sensors of linear displacements WS31-250 with a maximum stroke of 250 mm and a relative error of ± 1%.

All processes using a tensor amplifier of the type TMA-32 with a block of filters and ADC (analog-to-digital converter) were transferred to the corresponding software modules.

From Fig. 5 it is evident that the minimum experimental value of the stability margin coefficient against wheel derailment at a speed of the diesel train car of 30 m/s is 3,4. In this case, the minimum theoretical value of the stability margin coefficient against wheel derailment on the first wheel pair at the speed of 30 m/s and the average amplitude of horizontal unevenness of 0,006 m is 3,2. The difference in the results does not exceed 6%.

6 Conclusions

It has been shown that the increase in the speed of the diesel train car leads to deterioration of traffic safety indicators determined by all methods of study. However, the values of the maximum permissible speeds under the same conditions are different. This is conditioned, first of all, by the fact that the coefficient of stability against wheel pair derailment calculated according to car norms [16] does not take into account the influence of rail track unevenness and, therefore, may give a false assessment of the traffic safety parameters.

In assessing the possibility of wheel derailment according to the norms valid on the railways of Ukraine, the values of the frame strength and the coefficient of vertical dynamics are analyzed. At the same time BS EN 14363: 2005 European norms are guided by horizontal guiding force. Therefore, the results obtained by these techniques are different. It has been shown that the requirements of European norms are more stringent than those used in Ukraine. So under the same conditions of diesel train movement (straight track section, horizontal unevenness of 20 m long and amplitude of 10 mm) European standards allow for speeds of up to 110 km/h, norms of a motorized rolling stock [15] allow speeds up to 120 km/h. and car norms [16] allow for speeds of up to 150 km/h.

The comparison of the results of the theoretical studies with those from the experimental ones proves the adequacy of the model and the possibility to use it in solving the problems of wheel derailment, since the difference in the results is 6%.

References

1. Instructions for organization the rebuilding work in the aftermath of transport accidents on the railways of Ukraine. Order of the Ministry of Transport of Ukraine from April 27, 2001 № 258
2. Passenger diesel train DTKr-2. Operating manual DTKr-2.000.000 RE. Part 1, 563 (2014)
3. M. Nadal, Locomotives a Vapeur, Collection encyclopedie scientifique, biblioteque de mecanique appliquee et genie, **186** (1908).
4. A. Chudzikiewicz (eds.), *Monitorowanie stanu układu dynamicznego pojazd szynowy – tor*, Oficyna Wydawnicza PW, 233 (2012)
5. J. Droździel, E. Kardas-Cinal, B. Sowiński, Railway vehicle safety assessment affected by wheel and rail wear, *VSDIA 2010 – 12th International Conference on Vehicle System Dynamics, Identification and Anomalies*, Budapest, 8-10 Nov., 81-88 (2010)
6. J. Evans, M. Berg, Challenges in simulation of rail vehicle dynamics, Vehicle System Dynamics: International Journal of Vehicle Mechanics and Mobility, **47**, 1023-1048 (2009)
7. C. Hung, Y. Suda, M. Aki, T. Tsuji, M. Morikawa, T. Yamashita, T. Kawanabe, T. Kunimi, Study on detection of the early signs of derailment for railway vehicles, Vehicle System Dynamics, **48**, 451-466 (2010)
8. M. Opala, Analysis of experimental data in the context of safety against derailment of a railway vehicle, Key Engineering Materials, **518**, 16-23 (2012)
9. A. Shabana, Nadal's formula and high speed rail derailments, J. Comput. Nonlinear Dynam., **7** (2012)
10. J. Xiang, Q. Zeng, Mechanism and energy random analysis of train derailment on railway bridges, International Journal of Structural Stability and Dynamics, **9**, 585-605 (2009)
11. J. Elkins, H. Wu, New criteria for flange climb derailment, Railroad Conference, **1-7** (2000)
12. H. Wu, X. Shu, N. Wilson, Flange climb derailment criteria and wheel/rail profile management and maintenance guidelines for transit operations, TCRP Report, **5**, 71 (2005)
13. M. Verigo, *Interaction of the track and rolling stock*, 559 (Moscow, Transport, 1986)
14. M. Verigo, Questions of the interaction of the railway track and rolling stock and the problems of calculating the path, *Proceedings of the Central Research Institute of the Ministry of Railways*, **268**, 125 (1963)
15. Norms for calculating and evaluating the strength of load-bearing elements and the dynamic qualities of a motorized rolling stock. etc. The Ministry of Railways of the Russian Federation is 1520 mm gauge, VNIIZhT, 147 (1997)
16. Norms for the calculation and design of railroad cars of the Ministry of Railways of gauge 1520 mm (non-self-propelled), GosNIV-VNIIZhT, 319 (1996)
17. I. Klimenko, Development of theoretical foundations and methods for assessing and improving the safety of rolling stock railroad traffic [Manuscript]: dis. ... dr. techn. sciences, DNUZT, 284 (2015)
18. A. Azovsky, V. Koturanov, M. Ovechnikov, I. Plotnikov, On the estimation of safety from the wheels derailment, *Collection of articles of the international conference "Traffic safety of trains"*, MIIT, (2007)
19. E. Kardas-Cinal, Selected problems in railway vehicle dynamics related to running safety, The archives of transport, **31**, 37-45 (2014)
20. E. Sokol, Derailment and collisions of rolling stock (Forensic Examination. Elements of the theory and practice), Transport of Ukraine, 386 (2004)
21. E. Sokol, *Train crash (Forensic Expertise, Elements of Theory and Practice)*, 355 (Phoenix, 2007)
22. A. Kuzyshyn, A. Batig, Construction of the mechanical model of the diesel train DTKr-2 and its features, Bulletin of the DNUZT, **6 (72)**, 20-30 (2017)
23. S. Kostritsa, Yu. Sobolevskaya, A. Kuzyshyn, A. Batig, Mathematical model of diesel train wagon DPKr-2, Bulletin of the DNUZT, **1 (73)**, 56-65 (2018)

Approach for siting a support facility for transporting supplies in emergency cases in the Republic of Bulgaria

Emre Demir[1,*]

[1]Antalya Bilim University, College of Engineering, Department of Civil Engineering, Antalya, Turkey

Abstract. In this paper, the author determines the most suitable transportation location for intervention in a large scale disaster in the Republic of Bulgaria, by means of the Weber Problem and the Weiszfeld method. The objective is to minimise the cost of transporting emergency supplies across the country by locating a support facility, and thus reaching the area of event at the lowest possible cost. A brief description of the Weiszfeld method is provided in the paper. Further, using recent population data of both provinces and municipalities, the method is applied respectively to obtain the results. They were compared in terms of spatial correspondence and the final facility location was fixed. Erecting the supply site is recommended to help decrease the losses.

1 Introduction

Providing reliable and viable transportation between populated areas is one of the main issues in many countries [1, 2]. In places where the population density is relatively high, control of the traffic within the area is difficult but important [3, 4]. Especially in the case of a disaster, access to the area is really crucial and vital. Not only natural disasters such as forest fires, landslide, flood, and earthquake, but also large scale man-made disasters can be more effective and harsh as the population grows [5, 6]. The Republic of Bulgaria does have many facilities in order to intervene under the European Union Civil Protection Mechanism in case of major emergencies and actions requiring urgent response [7]. However, if the event is on a larger scale than expected, the utilities and the equipment may not be enough, and extra support may be needed. Surely, help can come from the neighbourhood of the impact cities or towns, nevertheless the equipment and staff may not be adequate to satisfy the needs in the case of a large scale disaster. For instance, recent forest fires in Greece required extra help and supplies from out of the impact regions and towns. Thus, the concern here is the necessity to supply additional support to the response area in unforeseen circumstances. The necessary service must be provided as soon as possible so that the loss of life and financial loss can be minimised. In this direction, for transporting supplies within the country, a comprehensive support facility that stores the emergency necessities (e.g. first aid equipment deliveries), and includes healthcare professionals is notably needed. Therefore, the motivation to study this problem comes from the need for one large facility with a lot of emergency equipment and crew ready for departure to any part of the country rather than making investment in all towns and cities. Hence, this paper focuses on determining the most suitable transportation location for an intervention. For this reason, a single point facility location is established; therefore techniques in the literature can be applied to solve the case. For example, central median location problem which was used many times in the literature can be functional. In this application, because the sum of the geometric planar distances to the facility location from the demand points is a minimum, the facility location can be simply found. Although, this technique seems simple and viable, it does not include the weights of demand points. Another technique considers the weights of demand points by minimising the weighted Euclidean distances to a single facility location in two-dimensional space, which is introduced by Alfred Weber in 1909 [8], and named as the Weber Problem (WP).

In this research, since the weights of demand points will be taken into account, the most suitable approach to solve our case is by applying WP. In order to solve WP, Weiszfeld has proven in his research that an iterative procedure can be used [9]. This iterative fashion, called Weiszfeld method or algorithm (WM), is accepted as a very powerful method in practice to identify the optimal facility location of WP [10].

In this study, transportation data including costs such as distances are evaluated by the application of WP and WM, and the detected locations are compared. According to the findings, this research is surely instructive to decrease the losses in case of a large scale disaster. The supply site, which is most appropriate in terms of location, to provide service to the regions throughout the Republic of Bulgaria is determined according to the analysis. The following sections of the paper describe the methodology and then explain the use of input data. Finally, the evaluation of the results is presented, and then concluding remarks and further research are exhibited at the end of the paper.

* Corresponding author: emre.demir@antalya.edu.tr

2 Methodology

In this paper, the methodology includes the solution of WP which is an operational research that was made by Alfred Weber in 1909, as mentioned in [8]. In this operation research problem, the optimization of location is the main purpose. WP is mainly solved by minimising the weighted Euclidean distances in order to find an optimal location serving the demand locations [11-14]. A brief description of WP can be seen as follows.

$$\min_{\bar{x}} \sum_{j=1}^{n} w_j \|\bar{x} - \bar{x}_j\| \qquad (1)$$

The description of (1) is subject to, $\bar{x} = (x, y)$ which is the coordinate of optimum location that is desired, and $\bar{x}_j = (x_j, y_j)$ which denotes the particular demand point location of j where $j = 1, ..., n$. Therefore, by subtracting $\bar{x}$ from the location data of $\bar{x}_j$, the optimisation process can identify the amount of the distance in between two locations. Moreover in the formula (1), the weight of a specific demand point is introduced by w_j. In other words, w_j stands for the importance of the particular location point of $\bar{x}_j$ compared to the other location points. Additionally, all of the location points (i.e. $\bar{x}$ and $\bar{x}_j$) in the description (1) are the points of the surface of E^2 which is the Euclidean plane called in older literature. It is the two dimensional space of real numbers (i.e. commonly denoted by R^2). Finally, the minimisation of the multiplication of the particular distances and the related weights is completed. Thus the optimization of (1) discovers the most suitable location of $\bar{x}$.

WP provides a useful mathematical problem to apply in many regional location assessments. Most importantly, WP can be solved by WM, which was built by Enrich Weiszfeld [9, 15, 16]. In the market, there are computer programs which are used for running and achieving solutions from WM. Because the computer program of LINGO can achieve the solution of WP by applying WM [10, 17], it is used as a tool for analysing the location and thus optimizing this research. Moreover, it analyses and detects the optimum location of a solution point with regards to the weighted Euclidean distances of demand points in a predetermined surface of a topography [18]. The program does not stop running the iterations, until all the demand points in the analysis are satisfied in terms of their weights [19]. Whenever the solution location does not change according to the iterations running repeatedly, the iterations are stopped and the final decision of the optimization is reported by the program.

As discussed previously, WP which is a mathematical problem for solving transportation related problems, one way of assessing the solution is to apply WM. Since the locations of cities and their weighted importance (i.e. including the population of the provinces, the population in the territories of municipalities, etc.) are the inputs, the methodology discussed in this section can be applicable for the problem of this research. Related formulation of the aim can be found as follows:

$$min\ [a_{r1}\sqrt{(X_{loc} - x_{r1})^2 + (Y_{loc} - y_{r1})^2} + a_{r2}\sqrt{(X_{loc} - x_{r2})^2 + (Y_{loc} - y_{r2})^2} + \dots + a_{rn}\sqrt{(X_{loc} - x_{rn})^2 + (Y_{loc} - y_{rn})^2}]. \qquad (2)$$

The formulation of (2) is subject to some variables which are described as follows:
(X_{loc}, Y_{loc}) the most suitable location of the facility;
n the number of demand points;
$i = 1, 2, ..., n$ the demand points from *1* to *n*;
(x_{ri}, y_{ri}) the spatial coordinates of the demand points from *1* to *n*;
a_{ri} the total demand of the demand points from *1* to *n*;
D_i Euclidean distance to (X_{loc}, Y_{loc}):

$$D_i = \sqrt{(X_{loc} - x_{ri})^2 + (Y_{loc} - y_{ri})^2}; \qquad (3)$$

$a_{ri}D_i$ weighted Euclidean distance to (X_{loc}, Y_{loc}):

$$a_{ri}D_i = a_{ri}\sqrt{(X_{loc} - x_{ri})^2 + (Y_{loc} - y_{ri})^2}. \qquad (4)$$

As the aim function (2) is regulated in the following with a constraint of $(X_{loc}, Y_{loc}) \geq 0$:

$$min\ \sum_{i=1}^{n} a_{ri}D_i. \qquad (5)$$

Basically, as can be seen in the formulation of (5), the purpose is to minimise the cost which is the distance for all demand points within the scope of the research. As the formulation of (5) is revealed via WM, the following section describes the scope of the input data as well as the use of data including the conversions and the necessary processes.

3 Input data

The input data is provided by inspecting the report titled "Cities and their urbanised areas in the Republic of Bulgaria" which is prepared by the National Statistical Institute [20]. The number of people (population), in this research, is used to represent the amount of demand at each city in an emergency case. Furthermore, the transportation costs such as distances between the attraction areas (i.e. populated cities) are used in the evaluation. In this study, the cost of the Euclidean distance is considered only. The main reason of using Euclidean distances is that immediate help or emergency support can be provided faster by air transportation such as the transportation by means of helicopter. For discovering the cost, the Euclidean distances between the attraction areas are computed by a Mercator projection conversion, because of the geoid shape of the Earth. Therefore, the coordinate of a location can be used as it is on a rectangular (*x*, *y*) projection [21]. The coordinates of the locations can be entered the equations of (6) and (7) in order to convert them on a rectangular projection:

$$x = \pi\ R\ (\lambda° - \lambda_0°)/180°; \qquad (6)$$

$$y = R\ \ln \tan(45° + \Phi°/2). \qquad (7)$$

 https://doi.org/10.1051/matecconf/201823406001

For this computation, x (the rectangular coordinate on x-axis), y (the rectangular coordinate on y-axis) can be found by using R (the radius of the sphere), $\lambda°$ (the latitude in degrees), $\lambda_0°$ (the central latitude in degrees), and $\Phi°$ (the longitude in degrees). With the help of the conversion, WM can be applied for the location analysis on a rectangular projection.

3.1 Data of provinces

There are 17 provinces in total as described in the documentation of "Cities and their urbanised areas in the Republic of Bulgaria" [20], and this information was used as the inputs in this paper. As described in the report published by the National Statistical Institute [20], a new spatial concept of 'city' is assigned to the territories in the Republic of Bulgaria, therefore, these 17 provinces are accepted as representing the country. Thus the population data from that report can be used as input in this paper. Table 1 indicates a number of examples of provinces and the population in thousands (i.e. demand) related.

Table 1. Population data of the provinces [20].

Examples of provinces	Population in thousands (demand)
Area with centre city Blagoevgrad	96.873
Area with centre city Burgas	279.478
Area with centre city Varna	411.302
Area with centre city Veliko Tarnovo	88.455
Area with centre city Vidin	66.818
Area with centre city Vratsa	73.767
Area with centre city Dobrich	114.053
Area with centre city Pazardzhik	117.293
Area with centre city Pleven	166.683
Area with centre city Plovdiv	550.289

The other input material, which is the spatial information (i.e. coordinate evidences), was taken from Google Maps. The spatial coordinates of the central cities of the provinces were provided. Then the location information is converted to a rectangular projection as discussed earlier, in order to input the data into WM.

3.2 Data of municipalities

According to the report [20], there is a total of 58 municipalities forming those 17 provinces in the country. In this paper, therefore, the data of these 58 municipalities is also used. Since each province is composed of particular municipalities and their particular areas, the total population of the municipalities must be equal to the population of a specific province. For instance, the total population of the province of Blagoevgrad (i.e. area with central city Blagoevgrad in Table 1) comprises the population of three municipalities named Blagoevgrad, Simitli, and Kocherinovo (Table 2). As can be seen in Table 2, the population data in thousands which are demands from a couple of municipalities are indicated. Additional to the demand data, the coordinate information of the municipalities were again taken from Google Maps.

Table 2. Population data examples of the municipalities [20].

Examples of municipalities	Population in thousands (demand)
Blagoevgrad	76.927
Simitli	14.486
Kocherinovo	5.460
Aytos	29.508
Burgas	209.224
Kameno	11.258
Sozopol	14.020
Sredets	15.468
Avren	8.835
Aksakovo	21.218

4 Results

In order to obtain the results the data were processed in an optimisation software called LINGO 17.0 x64 [17]. The objective function (5) was entered into the software to get the results using a computer with the properties of Intel® Core™ i7-2640M CPU @ 2.80GHz.

As mentioned earlier, the optimisation processes are carried out for both the data of the provinces (Table 1) and the municipalities (Table 2). When the process is completed by the inputs of the 'provinces', the coordinate of (X_{loc}, Y_{loc}) (i.e. the optimum location) is found as (0.7426740, 0.4432869). This spatial location (Location-A) corresponds to (42°33'07.5"N, 24°36'19.2"E) in Mercator projection (Table 3). In Location-A, there is not a facility for delivering emergency support in the case of a large scale disaster in the country (Fig. 1).

Table 3. Summary of Location-A and Location-B results.

Data	Location	Location coordinates	
		Latitude	Longitude
Provinces	A	42°33'07.5"N	24°36'19.2"E
Municipalities	B	42°33'15.8"N	24°35'39.2"E

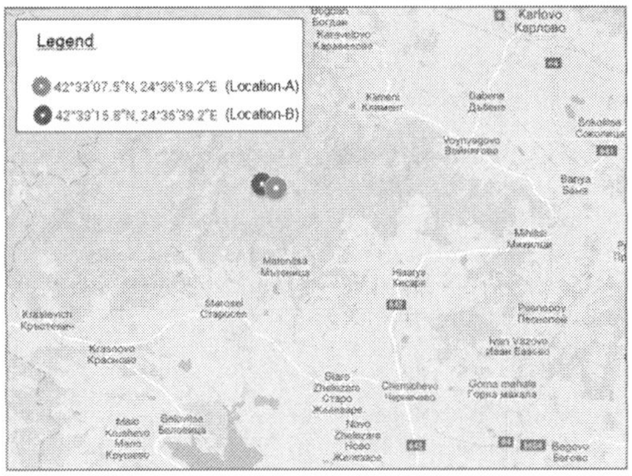

Fig. 1. Ideal locations for siting a support facility according to the data of both provinces and municipalities.

Using the data of the municipalities in Table 2, the location (X_{loc}, Y_{loc}) which makes the transportation distances minimum is again calculated. The result shows that the ideal location (Location-B) is (0.7427141, 0.4430738) which corresponds to (42°33'15.8"N,

24°35'39.2"E) in Mercator projection (Table 3). One can see that, currently, a support facility is not located at the coordinate of (42°33'15.8"N, 24°35'39.2"E) as well (Fig. 1).

In addition to the results of optimum locations of Location-A and Location-B, the distance between the two results can be also identified. The Euclidean distance between the points is measured as almost 1 kilometre.

Moreover, the verification for both results of Location-A and Location-B, which are the optimal solutions for the datasets of provinces and municipalities respectively, has been done and reported by the LINGO software. For both datasets, the software outputs such as the digital reports of numeric information clearly indicate that the local optimal solutions are found with the number of 0.00 infeasibilities. Although the report (either for Location-A or Location-B) notes that this is a local minima, this local minima is the global minima because our problem is convex. Thus, the results in this research are verified as they are the optimal solutions.

On the basis of these results, the next section presents observations and future work related to the research.

5 Concluding remarks and further research

The paper provides an insight to the positioning of a support facility for transporting supplies in emergency cases during or after a large scale catastrophe. By this study, the most suitable transportation location for an intervention against a large scale disaster in the Republic of Bulgaria was determined. The objective to minimise the cost of transporting emergency supplies and reach the area of event sooner from the facility was satisfied. The problem was simulated by a common location problem of WP. Accordingly, WM was applied to find the most optimum solution of the facility location. The main reason of using Euclidean distances between the locations was that the emergency support could be delivered quicker by air transportation (e.g. the transportation mean of helicopter). Last but not least, this paper may contribute to the state planning which is important for the National Disaster Protection Program and also the European Union Civil Protection Mechanism.

From the results of the research, it can be concluded that two locations (i.e. Location-A and Location-B) proposed for siting the facility are relatively close to one another in terms of their spatial correspondence. In other words, the distance (i.e. nearly 1 kilometre) between them is relatively short compared to the size of the country. Because of this topological reason, the site is proposed to erect in between them according to the results of this study. Building the facility which is fully equipped for delivering immediate supply and equipment is highly recommended to help decrease the number of casualties.

For further research, the population of towns and cities included in the census can be surely taken into account in order to rerun WM by using more powerful computer software. Thus, the location difference between the recent result and the future result can be revealed, and therefore further discussion on the facility location can be made.

References

1. D. Saliev, P. Asenov, Investigation and analysis for work-travels in direct from Pernic to Sofia, *BulTrans-2009 Proceedings*, Sozopol, Bulgaria, 199-202 (2009)
2. J. Chmielewski, T. Szczuraszek, Planning the development of spatial management and transport system of a city from the point of view of environmental impact, *MATEC Web of conferences – ECCE 2018*, **174**, 04009 (2018)
3. D. Clark, B. Cushing, Rural and urban traffic fatalities, vehicle miles, and population density, Accident analysis and prevention, **36**, 967-972 (2004)
4. S. Hapsari, T. Sjafrizal, R. Anugraha, Designing train passenger seat by kansei engineering in Indonesia, *MATEC Web of conferences – ECCE 2018*, **135**, 00017 (2017)
5. W. Qin, A. Lin, J. Fang, L. Wang, M. Li, Spatial and temporal evolution of community resilience to natural hazards in the coastal areas of China, Natural hazards, **89**(1), 331-349 (2017)
6. G. Bakioğlu, H. Karaman, Accessibility of medical services following an earthquake: a case study of traffic and economic aspects affecting the Istanbul roadway, *International journal of disaster risk reduction*, **31**, 403-418 (2018)
7. National Disaster Protection Program 2014-2018, Council of Ministers of the Republic of Bulgaria, Sofia, Bulgaria (2013)
8. A. Weber, *Über den standort des industrien*, *"Alfred Weber's Theory of the Location of Industries"*, C. J. Friedrich (Interpreter), (University of Chicago Press, Chicago, IL, USA, 1929)
9. E. Weiszfeld, Sur le point pour lequal la somme des distances de n pointes donnes est minimum, The Tohoku mathematical journal, **43**, 335-386 (1937)
10. R. Church, A. Murray, *Business site selection, Location Analysis, and GIS*, (John Wiley & Sons, Inc., 2009)
11. L. Tellier, The Weber problem: solution and interpolation, Geographical analysis, **4**(3), 215-233 (2017)
12. L. Cooper, I. Katz, The Weber problem revisited, Computers and mathematics with applications, **7**(3), 225-234 (1981)
13. J. Brimberg, The Fermat-Weber location problem revisited, *Mathematical programming*, **71**(1), 71-76 (1995)
14. H. Venceslau, M. Venceslau, A. Xavier, N. Maculan, A geometric perspective of the Weiszfeld algorithm for solving the Fermat-Weber problem, *RAIRO - Operations research*, **50**(1), 157-173 (2016)
15. U. Eckhardt, Weber's problem and Weiszfeld's algorithm in general spaces, Mathematical programming, **18**(1), 186-196 (1980)
16. R. Chen, Location problems with costs being sums of powers of Euclidean distances, *Computers and operations research*, **11**(3), 285-294 (1984)

17. http://www.lindo.com/index.php/products/lingo-and-optimization-modeling, LINDO Systems Inc. (06.07.2018)
18. K. Aftab, R. Hartley, J. Trumpf, Generalized Weiszfeld algorithms for Lq optimization. IEEE Transactions on Pattern Analysis and Machine Intelligence, **37**(4), 728-745 (2015)
19. H. Üster, R. Love, The convergence of the Weiszfeld algorithm, Computers and mathematics with applications, **40**(4), 443-451 (2000)
20. B. Bogdanov, Y. Kalchev, V. Angelova, I. Gavazki, *Cities and their urbanised areas in the Republic of Bulgaria*, National Statistical Institute, Sofia, Bulgaria (2016)
21. J. Snyder, *Map projections – a working manual*, U.S. Geological Survey Professional Paper 1395, Washington, USA (1987)

An approach for choosing the optimal route and type of transport for freight carriage using Compromise programming

Svetla Stoilova[1,*]

Technical University – Sofia, Faculty of Transport, 8 Kliment Ohridski Blvd., 1000 Sofia, Bulgaria

Abstract. This study elaborates a methodology for choosing a transport strategy based on the selection of the optimal route and type of transport. The research examines carriage by containers and includes two ways of transportation: by container block trains and by trucks. The methodology consists of three steps. The first step defines the determination of alternatives routes and criteria for assessment. Eight criteria have been studied: environmental protection; economic criteria; additional transportation fare; the duration of transportation; transhipment operations; security; reliability; stability. In the second step three approaches have been used and compared to determine the weights of the criteria: the Shannon entropy method, which is based on information from data, expert assessments by means of the Analytic hierarchy process (AHP) method and the Decision Making Trial and Evaluation Laboratory (DEMATEL) method. The third step determines the optimal solution based on the Compromise programming method. The methodology is approbated for routes from the railway and road network of Bulgaria. The results show that the criteria with the greatest impact are: the economic criteria; the duration of transportation and environmental protection. It was found out that the technologies that include rail transport are the best variant of transportation. The presented methodology could be used for choosing transportation for other destinations in the transport network.

1 Introduction

Different factors influence the choice of route and type of transport. For transport companies it is important that carriage is carried out with minimal transport costs. For transport users it is important to carry out the transport operations with the minimum delivery time, and minimum reloading operations. Considering the policy of the European Union, transport should be carried out with minimal environmental pollution. It can be summed up that, apart from the operating costs which are of major importance for selecting a route and the type of transportation, additional criteria such as environmental, technological and other ones also have impact.

Different authors have studied the factors of transportation. As key indicators in transport have been established the following factors: transport cost, transport time, flexibility, reliability, quality and sustainability, [1, 2]. In [3] the environmental aspects of transport have been considered. In [4] has been studied the problems for establishment of freight intermodal terminals. In [5] a model has been developed to optimize the collection and delivery of parcels on routes in an urban environment. The methods of multi-criteria analysis have been applied to assess the weights of the criteria and to choose the best alternatives. Some of methods use priority (AHP method), outranking techniques (PROMETHEE method), distance function and mathematical programming (Compromise programming), or mixed solutions. The multi-criteria decision making approach Analytic Hierarchy Process (AHP) is applied in many studies. In [6] it has been applied for best routes prioritization and selection. The authors examine criteria costs, time, risk of freight damaged, risk of infrastructure and equipment, risk of other factors for multimodal routes selection and they elaborate a decision support system using AHP Method and Zero-One Goal Programming. In [7] the AHP method is used to choose the best alternative route. The criteria: risk of explosion or spill, risk of road accidents, consequences of an incident, travel time and travel cost have been considered. By using the AHP method in [8, 9] are examined the factors: distance, terrain, road condition, convenience, safety and accessibility in the route selection. In [10] four groups of main criteria – environmental, economic, technological and social have been determined to assess the transportation with intermodal transport. The Decision Making Trial and Evaluation Laboratory (DEMATEL) multi-criteria method was used two determine the weights of the criteria and also to study the mutual influences of the criteria. It has been used in transport to evaluate criteria for tourism marketing [11]; in supply chain management [12]. In [13] a three-phase model for selection of the route of a road train has been presented. The criteria that have been studied are criteria of the road, criteria related to the throughput of the road, criteria related to the convenience of travel. The PROMETHEE method has been used to evaluate the alternatives of route selection in [14].

* Corresponding author: stoilova@tu-sofia.bg

Study [15] elaborates a methodology for determining optimal routes for dangerous goods transportation by means of the compromise programming approach. The case study is carried out for the transportation on the road network of Hong Kong. The studied criteria are travel time, probability of an incident, road users at risk, off-road population at risk, people with special needs at risk, negative impact on the economy: industrial, commercial and transport facilities at risk. The Compromise programming method has been applied also for research in air transport management [16], in water transportation [17] for economic research [18], to assess ecosystems [19]. This method is suitable for application when the criteria have a different type of optimization.

It can be said that the methods of multi-criteria analysis are appropriate for choosing the best alternative of transportation.

This paper aims to propose an approach for selecting a route and type of transportation for carriage of containers between two points by taking into account multiple factors relevant for transport and by using the Compromise programming method. This proposal is motivated by the fact that Compromise programming is a mathematical decision support model to obtain scores of alternatives by minimizing the weighted distances to an ideal point.

2 Methodology

The methodology for selection of a route and the type of transportation includes the following steps, Fig.1.

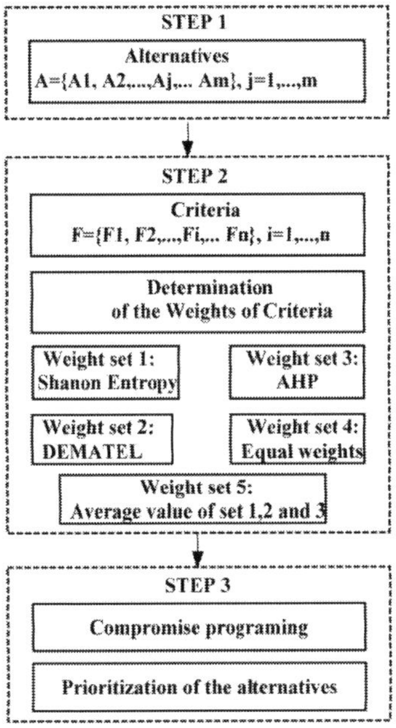

Fig. 1. Scheme of the methodology.

Step 1: Selection of alternatives variants of transportation. The carriage options including rail and road transport are examined in this research. The set of alternatives is: $A = \{A1, A2, \dots, Aj, \dots, Am\}$.

Step 2: Definition of quantitative and qualitative criteria for the assessment of alternatives routes for railway and road transport of containers. The set of criteria is: $F = \{F1, F2, \dots, Fi, \dots, An\}$.

The following criteria are studied in the research:

- F1: Carbon dioxide emissions, g;
- F2: Economic Criteria. This includes transport costs and infrastructure charges, BGN;
- F3: Presence of additional transportation fare, coeff.;
- F4: The duration of transportation, h;
- F5: Presence of transhipment operations, coeff.;
- F6: Security, coeff.;
- F7: Reliability, coeff.;
- F8: Stability, coeff.

Criteria F1, F2 and F4 are quantitative and values need to be set for them. Criteria F3, F5, F6 and F7 are qualitative. They have to be set with "1" and "0". The value is "1" if the answer is "yes", and "0", otherwise.

The study proposes five sets to define the criteria weights:

- *Set 1*: Shannon Entropy method

$$W_E = \{w_{E1}, w_{E2}, \dots, w_{Ei}, \dots, w_{En}\}, \quad (1)$$

where w_{Ei} are the weights determined by using Shannon entropy method;

- *Set 2*: DEMATEL Method

$$W_D = \{w_{D1}, w_{D2}, \dots, w_{Di}, \dots, w_{Dn}\}, \quad (2)$$

where w_{Di} are the weights determined by using DEMATEL method;

- *Set 3*: AHP Method

$$W_{AHP} = \{w_{AHP1}, w_{AHP2}, \dots, w_{AHPi}, \dots, w_{AHPn}\}, \quad (3)$$

where w_{AHPi} are the weights determined by using AHP method;

- *Set 4*: Equal weights:

$$W_{EQ} = \{w_{EQ1}, w_{EQ2}, \dots, w_{EQi}, \dots, w_{EQn}\}. \quad (4)$$

The values of equal weights are determined as:

$$w_{EQ1} = w_{EQ2} = \dots = w_{EQi} = \dots = w_{EQn} = \frac{1}{n}. \quad (5)$$

- *Set 5*: Average values by set 1, 2 and 3:

$$W_{AV} = \{w_{AV1}, w_{AV2}, \dots, w_{AVi}, \dots, w_{AVn}\}. \quad (6)$$

The average value of each weight is determined as follow:

$$w_{AV,i} = \frac{w_{E,i} + w_{D,i} + w_{AHP,i}}{3}. \quad (7)$$

The Shannon Entropy method allows us to determine the weights of criteria based of the information of the data, [20-22]. In this case the received results are not subjective. Both multi-criteria methods AHP and DEMATEL use expert's assessment to evaluate the criteria. AHP method applies Saaty's scale for pairwise comparison of criteria, the DEMATEL method uses another scale to assess the influence between criteria. The determined weights are dependent on the opinion of the

https://doi.org/10.1051/matecconf/201823406002

experts. The case with equal weights permits us to investigate the changes in rating the alternatives. This approach is identical with the case where the weights of criteria are not taken into account. The average value of the weights given by the Shannon entropy method, DEMTEL method and AHP method permits to reduce subjectivism when making a decision.

The application of these sets of weights assessment makes it possible to do a sensitivity analysis of the results.

Step 3: Ranking the alternative variants of the transportation by using Compromise programming. Choice of optimal variant of transportation.

Compromise programming is a distance method of multi-criteria analysis based on mathematical programming. This method is appropriate when it is necessary to decide on contradictory criteria (minimum and maximum optimization).

2.1 Determining the weights of the criteria using Shannon entropy

The information entropy for each criterion F_j is determined as follows, [20-22]:

$$E_j = -k \sum_{i=1}^{n} p_{ij} \, ln p_{ij}; \tag{8}$$

$$k = 1/ln n; \tag{9}$$

$$0 \le E_j \le 1, \tag{10}$$

where: k is a constant; p_{ij} – the normalized values of decision matrix $(x_{ij})_{mxn}$, $i = 1, \ldots, n$ – number of criteria, $j = 1, \ldots m$ – the number of alternatives.

The decision matrix $(x_{ij})_{mxn}$ is formed according to the data of the criteria for the studied alternatives.

The normalized values are determined as follows:

$$p_{ij} = \frac{x_{ij}}{\sum_{i=1}^{n} x_{ij}}. \tag{11}$$

To determine the weights of the criteria it is necessary to calculate parameter d_j for each alternative as follows:

$$d_j = 1 - E_j. \tag{12}$$

The weights of the criteria are determined as follows:

$$w_j = \frac{d_j}{\sum_{j=1}^{m} d_j}; \tag{13}$$

$$0 \le w_j \le 1; \tag{14}$$

$$\sum_{j=1}^{m} w_j = 1. \tag{15}$$

2.2 Determining the weights of the criteria through the AHP Method

The AHP method is a multi-criteria decision approach. It is based on a pairwise comparison between criteria using a standardized comparison scale of nine levels named Saaty's scale. The comparison scale is shown in Table 1, [23].

Table 1. Saaty's scale for AHP score.

Importance	Score
Equal	1
Moderate	3
Strong or essential	5
Very strong	7
Extreme	9
Values for intermediate comparison	2, 4, 6, 8

By using the AHP method is formed the A (n, n) evaluation square matrix based on the pairwise comparison on n criteria using Saaty's scale. For the matrix elements are used the following dependencies:

$$a_{ii} = 1;\ a_{ij} \neq 0;\ a_{ji} \neq 0; a_{ji} = 1/a_{ij}. \tag{16}$$

To determine the weights of the criteria it is necessary to normalize the evaluation matrix.

The weights of the criteria are determined according to the following equation:

$$A\,W = \lambda_{max}\,W, \tag{17}$$

where $W = \{w_1, \ldots, w_n\}^T$ is the normalized right eigenvector; λ_{max} – the largest eigenvalue of the evaluation square matrix A:

$$\lambda_{max} = \sum_{i=1}^{n} \left[\left(\sum_{j=1}^{n} a_{ij} \right) . w_i \right]. \tag{18}$$

With the AHP method the results are mathematically validated using the consistency ratio CR.

$$CR = CI/RI \le 0{,}1\,, \tag{19}$$

where: CI is the consistency index; RI – a random index. The random matrix is given by [20].

The consistency index is:

$$CI = \frac{\lambda_{max} - n}{n-1}. \tag{20}$$

If the condition (19) is met, the results are considered to be satisfactory. If the condition (19) is not met, it is necessary to make some revisions of scores.

2.3 Determining the weights of criteria by using DEMATEL Method

The DEMATEL method is a multi-criteria decision approach which determines the weights of the criteria and also the mutual influences of the criteria.

The procedure of the DEMATEL method includes the following, [24]: determination of the perception matrix; calculation of the average normalized perception matrix and the total relation matrix; determination of the normalized degree of influence of each criterion; calculation of the threshold value to obtain a cause and effect relationship diagram.

The scale of evaluation of criteria consists of the following scores: 0 – no influence; 1 – low influence; 2 – medium influence; 3 – high influence; 4 – very high influence. This scale is used by each of the experts who evaluate the direct influence between any two factors and thus a perception matrix is created. The diagonal elements of the perception matrix for $i = j$, are set to zero. When the scores of each expert are obtained, the average perception matrix A is formed as the average values of experts' assessments.

The average normalized perception matrix D_{nxn} is calculated as:

$$D = A/S, \quad (21)$$

where A is the average perception matrix; S – the major value of the sum of each column j of the matrix A and the sum of each row i of the matrix A.

The values of each element in matrix D are between zero and one.

The total relation matrix T_{nxn} is determined with this method as follows:

$$T = D(I - D)^{-1}, \quad (22)$$

where I is an nxn identity matrix.

For each criterion the normalized degree of influence is determined as follows:

$$e_i = \frac{r_i + c_i}{\sum_{i=1}^{n}(r_i + c_i)} 100, \%, \quad (23)$$

where r_i, c_i are the elements of vector R and vector C.

The degree of influence corresponds to the weights of the criteria.

Vector C represents the sum of the columns of the T matrix:

$$R = [r_i]_{nx1} = \left[\sum_{j=1}^{n} t_{ij}\right]_{nx1}; \quad (24)$$

$$C = [c_j]'_{1xn} = \left[\sum_{i=1}^{n} t_{ij}\right]_{1xn}, \quad (25)$$

where r_i is the sum of the i-th row in matrix T; c_j – the sum of the j-th column in matrix T ; $'$ – symbol meaning the transposed matrix.

The elements of vector R present both the direct and indirect effects by the i-th criterion on the other criteria. The elements of vector C show both direct and indirect effects by criterion j from the other.

With the DEMATEL method we determine a threshold value that serves to determine the relationships between the criteria in the considered system. The threshold value v is determined as an average value of the elements of matrix T, [25]:

$$v = \frac{\sum_{i=1}^{n}\sum_{j=1}^{n}[t_{ij}]}{N}, \quad (26)$$

where N is the total number of elements in matrix T.

After determining the values of the threshold value all elements of matrix T are compared with its value. The elements that are smaller or are equal to the threshold value v, are set to zero. The elements that are larger than the threshold value v, retain their value.

2.4 Ranking the alternative variants of trans-porttation by using Compromise programming

In this research prioritization of the alternatives is made using the Compromise programming (CP) method. This method defines the best/suitable solution as the one set of efficient solutions whose point is at the smallest distance from an ideal point. The objective is to obtain a solution that is as close as possible to some ideal solution. The criterion of optimization is, [19]:

$$minL_p = \left[\sum_{j=1}^{m} w_j^p \cdot \left|\frac{x_j^* - x_j}{M_j - m_j}\right|^p\right]^{1/p}; \quad (27)$$

$$\sum_{j=1}^{m} w_j = 1; \quad (28)$$

$$w_j \in \{0,1\}; \quad (29)$$

$$1 \leq p < \infty, \quad (30)$$

where L_p is a metric for alternative; x_j – value of criterion j for alternative A; M_j – the maximum value of criterion j in set N; m_j – the minimum value of criterion j in set N; x_j^* – ideal value of criterion j; w_j is the weight of the criterion j; p – parameter. The parameter p name also balancing factor and use to present the attitude of the decision-maker to the optimal solution.

The values of parameter p give a different measure of the distance from an efficient point to the reference point. The calculation with different values of parameter p permits us to have a perfect compensation among the objectives. When $p = 1$ is the so-called street-block distance, or Manhattan distance. The Manhattan distance between two points presents a grid found on a strictly horizontal and/or vertical path. When $p = 2$, it represents the Euclidean distance. The Euclidean distance is a diagonal of the horizontal and vertical components. When $p = \infty$, it is Tchebyshev distance, and corresponds to min-max problem.

The most popular value of solving Compromise programming is $p = 2$, as it is the shortest distance.

The optimal alternative is the one with lowest value for the L_p metric. By this method this will be the best compromise solution because it is the solution nearest to the ideal point.

4 Results end discussion

4.1 Definition of the alternatives and criteria

The methodology is approved for container transport by rail and road for direction Sofia – Varna in Bulgaria. Figures 2 and 3 show maps of the alternatives in railway and road network.

This research studied the alternatives given in [14]:

- A1: Freight block train: Iliyantsi-Gorna Oryahovitsa-Varna;
- A2: Freight block train: Iliantsi-Karlovo-Karnobat-Varna;
- A3: Road train: Route Sofia-Veliko Tarnovo-Varna;
- A4: Road train: Route Sofia-Plovdiv-Burgas-Varna;
- A5: Road train: Route Sofia-Plovdiv-Karnobat-Shumen-Varna.

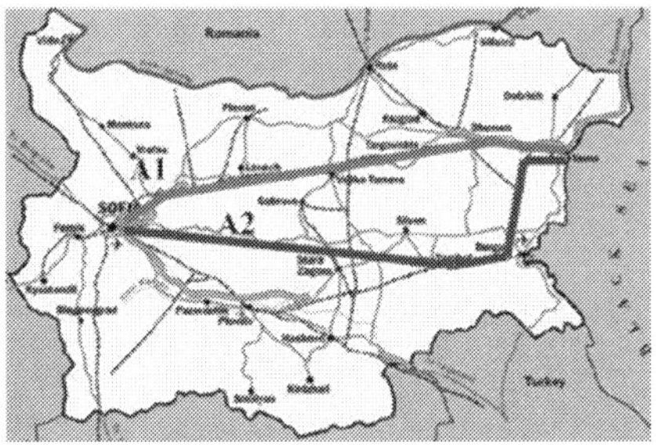

Fig. 2. Scheme of the alternatives - railway transport.

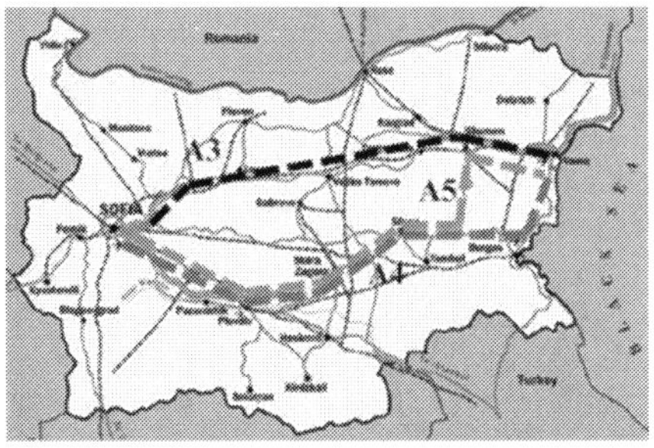

Fig. 3. Scheme of the alternatives - road transport.

The values of the quantitative criteria have to be determined for each alternative. This applies to the criteria of carbon dioxide emissions (F1), economical criteria (F2), travel time (F3). The qualitative criteria have to be set with "1" and "0". The value is "1" if the answer is "yes", and "0", otherwise.

Table 2 gives the values of the criteria. In order to compare the results, the criteria are set for one road train. The last row of Table 2 gives the type of optimization (minimum or maximum). The values of the carbon dioxide emissions, economic criteria and the duration of transportation are as in [14].

Table 2. Criteria values of Decision matrix.

F_i	F1	F2	F3	F4	F5	F6	F7	F8
A_j	Co2, g/RT	BGN/RT	-	h	-	-	-	-
A1	238257	431	1	9,05	1	1	1	1
A2	247250	413	1	8,57	1	1	1	1
A3	442887	469	0	7,14	0	0	0	0
A4	519180	544	0	8,12	0	0	0	0
A5	545931	468	0	8,62	0	0	0	0
type	min	min	min	min	min	max	max	max

4.2 Determining the weights of the criteria

Table 3 and Table 4 show the normalized decision matrix and results of the Shannon Entropy method.

Table 3. Normalized values p_{ij} of decision matrix.

A_j	F1	F2	F3	F4	F5	F6	F7	F8
A1	0,12	0,25	0,50	0,22	0,50	0,50	0,50	0,50
A2	0,12	0,24	0,50	0,21	0,50	0,50	0,50	0,50
A3	0,22	0,24	0,00	0,17	0,00	0,00	0,00	0,00
A4	0,26	0,28	0,00	0,20	0,00	0,00	0,00	0,00
A5	0,27	0,24	0,00	0,21	0,00	0,00	0,00	0,00

Table 4. Parameters of Shannon entropy method.

Fi	F1	F2	F3	F4	F5	F6	F7	F8
E_j	0,32	0,43	0,69	0,30	0,69	0,69	0,69	0,69
d_j	0,68	0,57	0,31	0,70	0,31	0,31	0,31	0,31
w_j	0,20	0,16	0,09	0,20	0,09	0,09	0,09	0,09

The scores of the AHP method are presented in Table 5. In the study the assessments have been made by five experts who gave a general assessment. The last column of the table shows the weights of the criteria using the AHP method. In the last row of the table is the value of the consistency ratio. It can be seen that according to formula (19) the experts' assessments are satisfactory, (CR = 0,06; CR < 0,1).

Table 5. Prioritization Matrix resulting from using the AHP method.

F_i	F1	F2	F3	F4	F5	F6	F7	F8	w_j
F1	1	1/3	3	1	1	1	1	1	0,10
F2	3	1	5	5	5	2	2	2	0,32
F3	1/3	1/5	1	1	4	1	1	1	0,11
F4	1	1/5	1	1	5	1	2	2	0,14
F5	1	1/5	1/4	1/5	1	1	1	1	0,06
F6	1	1/2	1	1	1	1	2	1	0,11
F7	1	1/2	1	1/2	1	1/2	1	1	0,08
F8	1	1/2	1	1/2	1	1	1	1	0,08
CR=0,06									

Table 6 presents the average matrix of the experts' scores resulting from using the DEMATEL method. In the study the assessments have been given by five experts. They are given independently of each other. Table 7 and Table 8 show the average normalized matrix and total relation matrix.

Table 6. Average matrix resulting from using the DEMATEL method.

Fi	F1	F2	F3	F4	F5	F6	F7	F8
F1	0,00	1,96	0,00	0,09	0,00	0,43	0,00	0,00
F2	3,96	0,00	3,96	3,96	0,04	1,96	0,04	0,43
F3	0,04	2,87	0,00	0,43	0,04	0,04	0,04	0,04
F4	0,61	1,96	1,04	0,00	0,04	0,43	1,48	0,43
F5	0,00	0,61	0,87	2,04	0,00	0,00	0,43	0,00
F6	0,00	1,87	0,13	0,74	0,61	0,00	0,43	0,43
F7	0,04	1,48	0,04	0,87	0,74	0,04	0,00	0,26
F8	0,26	1,04	0,61	0,00	0,00	1,13	0,26	0,00

Table 7. The average normalized matrix.

F_i	F1	F2	F3	F4	F5	F6	F7	F8
F1	0,000	0,136	0,000	0,006	0,000	0,030	0,000	0,000
F2	0,276	0,000	0,276	0,276	0,003	0,136	0,003	0,030
F3	0,003	0,200	0,000	0,030	0,003	0,003	0,003	0,003
F4	0,042	0,136	0,073	0,000	0,003	0,030	0,103	0,030
F5	0,000	0,042	0,061	0,142	0,000	0,000	0,030	0,000
F6	0,000	0,130	0,009	0,052	0,042	0,000	0,030	0,030
F7	0,003	0,103	0,003	0,061	0,052	0,003	0,000	0,018
F8	0,018	0,073	0,042	0,000	0,000	0,079	0,018	0,000

Table 8 presents the relation matrix. The number of elements of relation matrix is 64. The value of threshold value is determined by formula (26).

https://doi.org/10.1051/matecconf/201823406002

Table 8. Total relation matrix.

F_i	F1	F2	F3	F4	F5	F6	F7	F8
F1	0,050	0,171	0,053	0,059	0,004	0,058	0,009	0,009
F2	**0,350**	**0,204**	**0,364**	**0,361**	0,016	**0,191**	0,049	0,055
F3	0,077	**0,250**	0,078	**0,106**	0,007	0,044	0,017	0,016
F4	**0,106**	**0,214**	**0,141**	0,076	0,013	0,069	**0,115**	0,044
F5	0,036	**0,101**	**0,103**	**0,178**	0,005	0,022	0,050	0,010
F6	0,056	**0,183**	0,073	**0,116**	0,047	0,034	0,046	0,041
F7	0,049	**0,146**	0,056	**0,113**	0,055	0,030	0,015	0,027
F8	0,053	**0,118**	0,080	0,043	0,006	0,099	0,026	0,009

The values that are equal or greater than the threshold value (v=0,084) are given in bold. This parameter serves to determine the relationships in the considered system.

Table 9 shows the direct and indirect influence of criteria. The last column presents the weights of the criteria.

Table 9. Direct and indirect impact of the criteria.

Criterion	P	R	P+R	P-R	e%
F1	0,412	0,777	1,190	-0,365	11%
F2	1,590	1,387	2,977	0,202	28%
F3	0,594	0,947	1,541	-0,353	14%
F4	0,779	1,053	1,831	-0,274	17%
F5	0,505	0,153	0,658	0,352	6%
F6	0,596	0,546	1,142	0,049	10%
F7	0,492	0,328	0,820	0,164	8%
F8	0,434	0,210	0,644	0,224	6%

The prioritization is: F2>F4>F3>F1>F6>F7>F5>F8. The biggest impact in selecting the route and type of transportation for carriage of containers is that of the economic criteria (F2=28%), the duration of transportation (F4=17%), additional transportation fare (F3=14%) and carbon dioxide emissions (F1=11%).

Figures 4-7 present the cause and effect diagram for the criteria according to the results given in Table 8.

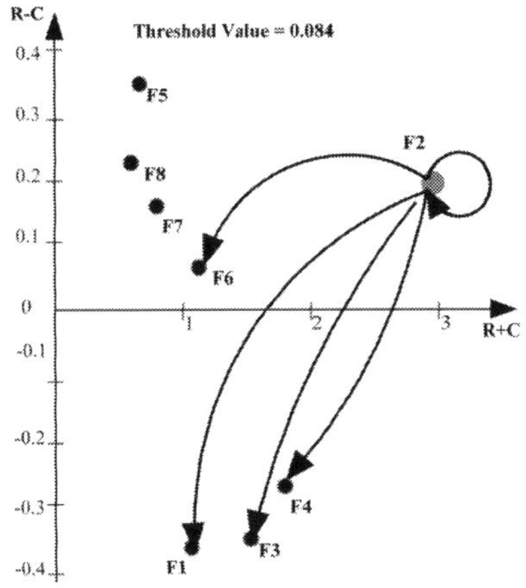

Fig. 4. Cause and effect diagram for criterion F2.

The relative importance of each criterion is presented by the values of (P+R). This impact is the same as the values of the criteria. Criteria F1, F3 and F4 are in the effect group. The values of (P-R) for these criteria are negative. Criteria F2, F5, F6, F7 and F8 are in the cause group. The values of (P-R) for these criteria are positive.

According to the results criterion F2 influenced criteria F1, F2, F3, F4 and F6; criterion F3 influenced criteria F2 and F4; criterion F4 influenced F1, F2, F3 and F7; criterion F5 influenced F2, F3 and F4; criterion F6 influenced criteria F2 and F4; criterion F7 influenced F2 and F4; criterion F8 influenced criterion F2.

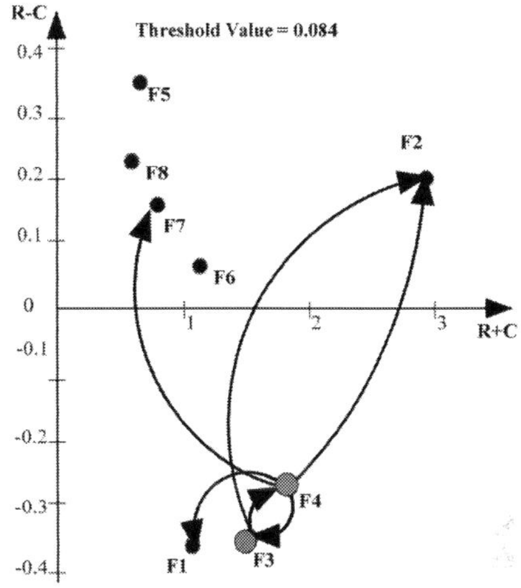

Fig. 5. The cause and effect diagram for criterion F3 and F4.

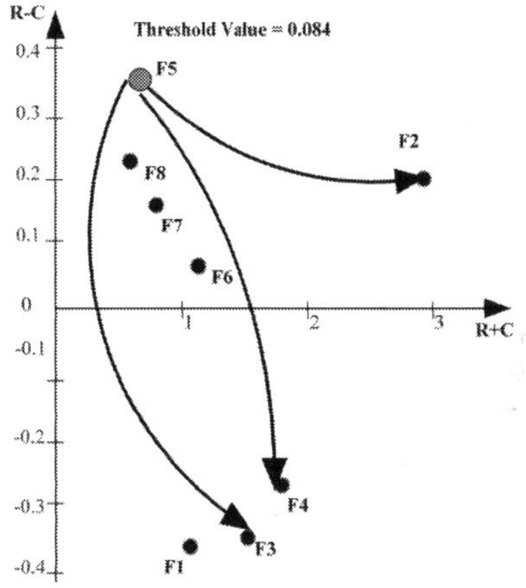

Fig. 6. Cause and effect diagram for criterion F5.

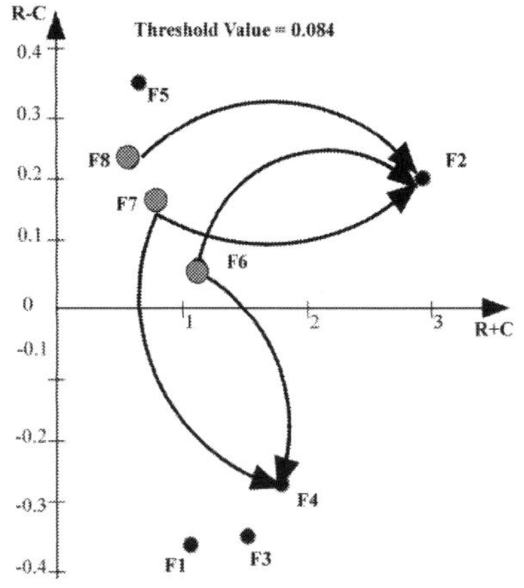

Fig.7. Cause and effect diagram for criterion F6, F7 and F8.

Table 10 and Figure 8 present a comparison of criteria weights given by using the different methods and also the average value according to formula (7). The results show that the average absolute error between the values obtained using the Shannon Entropy method and the average values is 3%; the average absolute error between the values obtained using the DEMATEL method and the average values is 1%, and the average absolute error between the values obtained using the AHP method and the average values is 2%.

Table 10. Comparison of the weights of the criteria.

Criteria	F1	F2	F3	F4	F5	F6	F7	F8
SHANNON ENTROPY	0,20	0,16	0,09	0,20	0,09	0,09	0,09	0,09
DEMATEL	0,11	0,28	0,14	0,17	0,06	0,10	0,08	0,06
AHP	0,10	0,32	0,11	0,14	0,06	0,11	0,08	0,08
AVERAGE	0,14	0,25	0,11	0,17	0,07	0,10	0,08	0,08

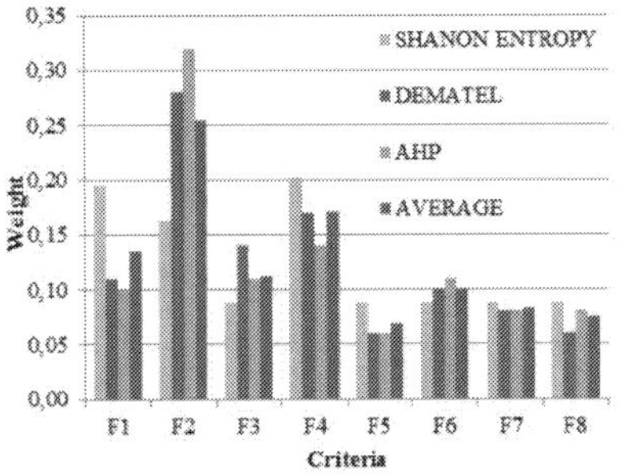

Fig. 8. Weights of the criteria.

4.3 Ranking the alternatives by using Compromise programming

The study examines two values of parameter p ($p = 1$, and $p = 2$). For both types of transport the charges for the use of transport infrastructure are different. The charges for using road infrastructure in Bulgaria are lower than those in the European Union, and also they are lower as compared to the charges for railway infrastructure. In many European countries, tolls have also been introduced [26]. The government policy provides for heavy goods vehicles to be covered by a toll system. The price will be determined by several components – vehicle category, toll distance, road characteristic. Two variants have been studied to evaluate the impact of infrastructure charges in choosing transport technology. Variant 1 includes values of infrastructure charges in the current situation. Usage of weekly vignette 2 times a week it has been studied. Variant 2 is with toll charges. In the research have been used the value of toll charges 0,2 BGN/km, as at the time of the study is not known tariff tolls. For comparison with Poland tolls charges for vehicles with maximum permissible gross weight over 12 tons and Euro 5 emission classes is 0,27 euro/km, [26].

Tables 11 and 12 present the results of optimization by using the Compromise programming method with parameter $p = 1$. Tables 13 and 14 show the results of optimization by Compromise programming method with parameter $p = 2$.

The results show that for parameter value $p = 1$, for four weights set (AHP, Shannon entropy, Equal weights and Average weights) the best technology is carriage by container block trains Sofia-Karlovo-Varna (alternative A2). According to the DEAMATEL method the best technology is carriage is by road trains via – Sofia-Veliko Tarnovo-Varna (alternative A3) route. The results for the variant with toll charges shows that the best technology is carriage of containers with block trains Sofia-Karlovo-Varna (alternative A2).

Table 11. Compromise programming (p=1), Vignette charges.

Set	1		2		3		4		5	
Aj	*Lj*	*r*	*Lj*	*r*	*Lj*	*r*	*Lj*	*r*	*Lj*	*r*
A1	0,42	3	0,44	3	0,39	3	0,4	2	0,41	3
A2	0,34	1	0,34	2	0,28	1	0,35	1	0,31	1
A3	0,4	2	0,32	1	0,33	2	0,46	3	0,35	2
A4	0,71	5	0,71	5	0,75	5	0,68	5	0,72	5
A5	0,62	4	0,48	4	0,48	4	0,6	4	0,53	4
1 – Shannon Entropy; 2 – DEMATEL; 3 – AHP; 4 – Equal weights; 5 – Average weights; *r* – rang of alternative										

Table 12. Compromise programming (p = 1), Toll charges.

Set	1		2		3		4		5	
Aj	*Lj*	*r*	*Lj*	*r*	*Lj*	*r*	*Lj*	*r*	*Lj*	*r*
A1	0,39	2	0,4	2	0,34	2	0,39	2	0,38	2
A2	0,33	1	0,33	1	0,27	1	0,35	1	0,31	1
A3	0,47	3	0,48	3	0,49	3	0,52	3	0,48	3
A4	0,7	5	0,71	5	0,75	5	0,68	5	0,72	5
A5	0,69	4	0,7	4	0,64	4	0,66	4	0,66	4
1 – Shannon Entropy; 2 – DEMATEL; 3 – AHP; 4 – Equal weights; 5 – Average weights'; *r* – rang of alternative										

Table 13. Compromise programming (p= 2), Vignette charges.

Set	1		2		3		4		5	
Aj	*Lj*	*r*	*Lj*	*r*	*Lj*	*r*	*Lj*	*r*	*Lj*	*r*
A1	0,24	3	0,22	3	0,2	3	0,22	2	0,22	3
A2	0,2	2	0,18	2	0,16	2	0,2	1	0,18	2
A3	0,2	1	0,16	1	0,16	1	0,23	3	0,17	1
A4	0,3	5	0,34	5	0,37	5	0,28	5	0,33	5
A5	0,29	4	0,22	4	0,21	4	0,27	4	0,24	4
1 – Shannon Entropy; 2 – DEMATEL; 3 – AHP; 4 – Equal weights; 5- Average weights; *r* – rang of alternative										

Table 14. Compromise programming (p=2), Toll charges.

Set	1		2		3		4		5	
Aj	*Lj*	*r*	*Lj*	*r*	*Lj*	*r*	*Lj*	*r*	*Lj*	*r*
A1	0,24	3	0,23	3	0,19	3	0,19	2	0,21	2
A2	0,2	1	0,2	1	0,16	1	0,16	1	0,18	1
A3	0,22	2	0,21	2	0,23	2	0,23	3	0,22	3
A4	0,3	5	0,34	5	0,37	5	0,37	5	0,33	5
A5	0,3	4	0,26	4	0,27	4	0,27	4	0,27	4
1 – Shannon Entropy; 2 – DEMATEL; 3 – AHP; 4 – Equal weights; 5- Average weights; *r* – rang of alternative										

The results for the value of parameter $p = 2$ show that for four weights set (AHP, Shannon entropy, Equal weights and Average weights) the best technology is carriage by road trains on the Sofia-Veliko Tarnovo-

Varna (alternative A3) route. In the case when the toll charges are introduced the best technology is transportation with container block trains Sofia-Karlovo-Varna (alternative A2) for all weights sets.

Using the equal weights gives the same results for the optimal solution.

Figures 9 and 10 present the results of ranking the alternatives according to the type of road infrastructure charges. It can be seen that the prioritization of the alternatives according to the weights sets is similar. This is due to the small percentage difference in the results obtained for weights by using the various applied methods.

It can be concluded that when the value of the parameter $p = 1$ is used, no difference is reported in the optimal solution for both variants of the type of road infrastructure charges. When value $p=2$, which represents the Euclidean distance, is applied there is a difference in the optimal solution for both studied variants.

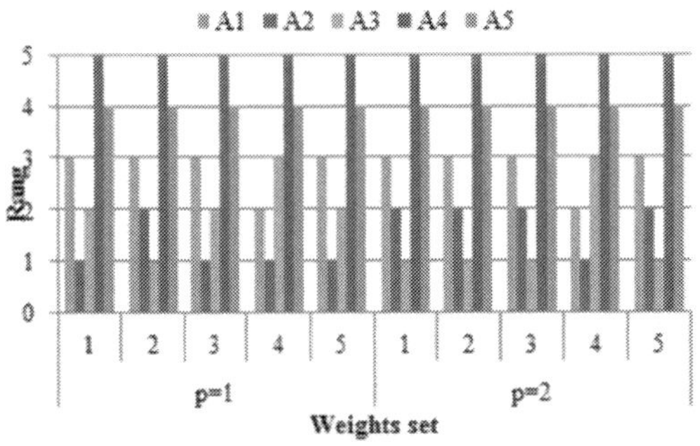

Fig. 9. Comparison of ranking of the alternatives, for the variant with vignette.

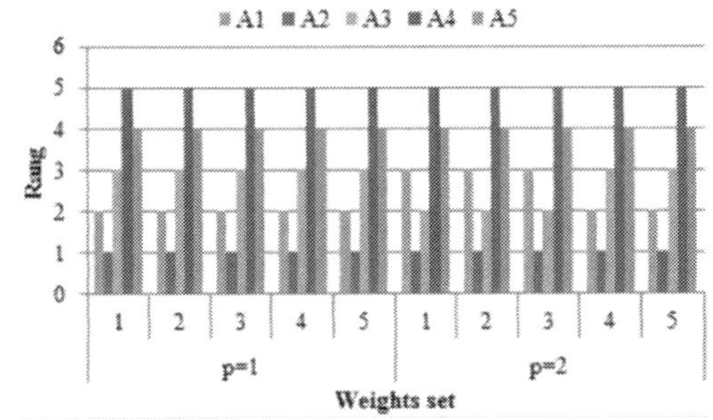

Fig. 10. Comparison of ranking of the alternatives, for the variant with toll charges.

5 Conclusion

- A methodology for selecting a route and a mode of transport between the initial and final points based on the Compromise programming method has been developed. The advantages of this method are the application of the type of optimization for each of the criteria, and usage of the minimum distance to the ideal solution as an optimization criterion.
- Information and expert approaches for decision-making are used to determine the criteria weights. An approach is proposed to determine the weightings of the criteria by determining the average weight for each factor from the application of two experts and one information method.
- The determination of average weights by combining expert and informative method allows us to reduce subjectivism when making a decision. This approach is recommended for use in determining the weights of the criteria.
- The effect of the Compromise programming parameter p on the choice of an optimal alternative has been examined. It was found out that at the Manhattan distance ($p = 1$) the introduction of the toll charges does not affect the optimal solution. The results show that carriage with container block trains is the optimal transport technology. This is due to the fact that this distance is not the shortest to the ideal solution. At Euclidean distance ($p = 2$), the introduction of this charge affects the optimal solution. It is appropriate to use the Euclidean distance to make a decision. The results obtained at this distance are identical to those obtained in [14].
- The influence of the road infrastructure charges (vignette and toll) on the choice of route and mode of transport is studied. It has been established that the introduction of toll charges influences the choice of optimal transport technology.

References

1. D. Islam, T. Zunder, R. Jorna, Performance evaluation of an online benchmarking tool for European freight transport chains, Benchmarking: An International Journal, **20**(2), 233-250 (2013)
2. B. Abramovic., I. Lovric, V. Supalo, Analysis of intermodal terminals service quality in the republic of Croatia, Promet – Traffic&Transportation, **24**(3), 253-260 (2012)
3. T. Skrucany, J. Ponicky, M. Kendra, J. Gnap, Comparison of railway and road passenger transport in energy consumption and GHG production, *12th International Scientific Conference Of Young Scientists On Sustainable, Modern and Safe Transport, TRANSCOM 2017*, Procedia Engineering, **192**, 806-811 (2017)
4. S. Martinov, Evaluation model of railway infrastructure potential for establishment of freight intermodal terminals, *Proceedings of the 5th International Conference on Road and Rail Infrastructures – CETRA 2018*, 911-916 (2018)
5. V. Pencheva, A. Asenov, D. Topchu, I. Beloev, B. Evstatiev, Organisation of the work on collecting routes in postal activity through an automated system for collection of information, Transport problems, **12**(3), 147-157 (2017)
6. A. Kengpol, W. Meethom, M. Tuominen. The development of a decision support system in multimodal transportation routing within Greater Mekong sub-region countries, International Journal of Production Economics **140**(2), 691-701 (2012)

7. W. Sattayaprasert, S. Hanaoka, P. Taneerananon, R. Pradhananga, Creating a risk-based network for hazmat logistics by route prioritization with AHP: Case Study: Gasoline Logistics in Rayong, IATSS Research, **32**(1), 74-87 (2008)
8. V. Podvezko, H. Sivilevicius, The use of AHP and rank correlation methods for determining the significance of the interaction between the elements of a transport system having a strong influence on traffic safety, Transport, **28**(4), 389 (2013)
9. H. Sivilevicius, Modelling the interaction of transport system elements, Transport, **26**(1), 20 (2011)
10. S. Stoilova, L. Kunchev, Study of criteria for evaluation of transportation with intermodal transport, *Proceedings of 16th International Scientific Conference Engineering for Rural Development,* Jelgava, Latvia, 349-357 (May 24-26, 2017)
11. C. Chen, Developing Taiwan into the tourist transport center of East Asia, Tourism Economics, **18**(6), 1401-1411 (2012)
12. S. Gandhia, S. Mangla, P. Kumar, D. Kumar, Evaluating factors in implementation of successful green supply chain, International strategic management review, **3**, 96-109 (2015)
13. S. Stoilova, L. Kunchev, Methodology for optimal transportation using a three-phase model, C. R. Acad. Bulg. Sci., **69**(7), 903-908 (2016)
14. S. Stoilova, Evaluation efficiency of intermodal transport Using multi-criteria analysis, *Engineering for rural development,* Jelgava, 2030-2039 (May 23-25, 2018)
15. R. Li, Y. Leung, Multi-objective route planning for dangerous goods using compromise programming. J. Geogr. Syst., **13**, 249-271 (2011)
16. Q. Ma, H. Song, W. Zhu. Low-carbon airline fleet assignment: A compromise approach, Journal of Air Transport Management, **68**, 86-102 (2018)
17. J. Shiau, F. Wu, Compromise programming methodology for determining instream flow under multiobjective water allocation criteria, Journal of the American water resources association JAWRA, Paper No. 05013, April 2007, 1179-1191 (2007)
18. F. Andre, M. Cardenete, C. Romero, Using compromise programming for macroeconomic policymaking in a general equilibrium framework: theory and application to the Spanish economy, Journal of the Operational Research Society, **59**, 875-883 (2008)
19. B. Poff, D. Tecle, G. Neary, B. Geils, Compromiise programming in forest management, Journal of the Arizona-Nevada academy of science, **42**(1), 44-60 (2010)
20. R. Srinivasa, D. Kumar, *Multicriterion analisis in engineering and management,* 266 (PHL Learning Private Limited, Delhi, 2014)
21. A. Haddadha, A. Namazian, S. Yakhchali, Project selection problem by combination of Shannon Entropy and MCDM techniques, *International Conference on Literature, History, Humanities and Social Sciences (LHHSS-17),* Dubai, 32-35 (Jan. 1-2, 2017)
22. R. Mavi, M. Goh, N. Mavi, Supplier selection with Shannon Entropy and Fuzzy TOPSIS in the context of supply chain risk management, Procedia-Social and Behavioral Sciences, **235**, 216-225 (2016)
23. T. Saaty, *Fundamentals of Decision Making and Priority Theory with the Analytic Hierarchy Process* (RWS Publications, 2000)
24. D. Vujanovic, V. Momcilovic, N. Bojovic, V. Papic, Evaluation of vehicle fleet maintenance management indicators by application of DEMATEL and ANP, Expert Systems with Applications, **39**, 10552-10563 (2012)
25. S. Khanam, J. Siddiqui, F. Talib. A DEMATEL approach for prioritizing the TQM enablers and IT resources in the Indian ICT industry, International Journal of Applied Management Sciences and Engineering, **3**(1), 11-26 (2016)
26. P. Lewandowski, User charges for road infrastructure in certain European Union member states, Scientific Journals of the Maritime University of Szczecin, **48**(120), 138-145 (2016)

Investigation of the operation mode at Ruse-Danube Bridge border checkpoint considered to be a mass service system with incoming flow of automobiles at a non-stationary mode of operation

Dimitar Grozev[1*], *Velizara* Pencheva[1], *Ivan* Georgiev[2], and *Ivan* Beloev[1]

[1]Ruse University "Angel Kanchev", Transport Department, Bulgaria
[2]Ruse University "Angel Kanchev", Natural sciences and education Department, Bulgaria

Abstract. The present work reviews the Ruse-Danube Bridge border checkpoint and the work carried out for a year, by examining the average daily incoming flows of cargo vehicles. On the basis of the information collected, the daily irregularity of the work at the Ruse-Danube Bridge border checkpoint has been assessed and the system has been modelled as a mass service system to assess capacities in existing infrastructure and organization of work. The type of incoming flow has been specified as a non-stationary Poisson flow. Under non-stationary conditions, the basic values of the system parameters have been calculated and a MatLab application has been created. A system of differential-algebraic equations has been selected for a model according to Kolmogorov model for stochastic processes.

1 Introduction

Transport is an important element of the international trade activities of a country. The integration of the country into the European space has led to a significant increase in the traffic on the main road sections. The efficient sustainable integration of the national infrastructure into the European one aims to improve the cohesion and the connections between the Republic of Bulgaria and the remaining EU member-states. Besides, it poses new challenges in terms of the completion, maintenance and optimization of our roads. [1-3].

The relatively high share of road transport in the volume of transport work on import and export as well as the indicated increase in the volume of transport work was the prerequisite for the envisaged increase in the number of road border checkpoints during the preparation period for accession of the country to the EU [4].

Border checkpoints are detached territories with a special escape and protection regime, through which only the crossing of the state border is allowed, unless otherwise provided in an international treaty.

By 2018 there are 37 border checkpoints in Bulgaria. The total number of points for heavy goods vehicles is 21 [5].

There are usually large queues of trucks at the border checkpoints. The reasons are different: poor interaction between the internal services; poor international interaction with the respective neighboring state or with a third country; insufficient throughput of border checkpoints; poor organization of work; in some cases, not using capacity capabilities in the border checkpoints; technical reasons, e.g. problems in information systems, etc.

Assessing capacity capabilities at border checkpoints may provide preliminary information on the possible number of cars processed and waiting times for available infrastructure and work experience. The lack of methodology for such an assessment does not allow for pre-planning. In this publication, on the basis of collected information, the seasonal inconsistency of work at the border checkpoints is assessed and their mass service system work is modelled. The time for vehicles to stay at different incoming flows, as well as the limit value of the number of cars that can be handled is determined.

The main purpose of this publication is to research the incoming flow of a four channel mass service system and to establish its effectiveness. The following tasks have been set:

- input modelling;
- defining the type of mass service system;
- solving the system of differential-algebraic equation [6-8].

2 Exposition

In order to describe the operating regime of Rousse border checkpoints, considered as a mass service system, it is necessary to know the characteristics of the incoming flow of trucks considered as stochastic process, the intensity of service and the number of service units [9, 10].

For the incoming flow of trucks, we can set the following prerequisites:

* Corresponding author: dgrozev@uni-ruse.bg

- *ordinary flow* – the probability of two or more trucks entering the system for an elementary time interval is infinitesimal, compared to the probability of only one truck doing it; i.e. the property of ordinariness means that automobiles enter one at a time rather than two or three, etc.
- *flow without consequences* – the number of trucks arriving in the system for time interval Δt does not depend on how many vehicles have already been received, that is, does not depend on the history of the studied phenomenon (the flow without action afterwards (Poisson flow);
- *stationarity / non-stationarity of the flow* [11-12] – for sufficiently long periods of time – 1 month, 6 months, 1 year, etc. it is possible to assume a steady-state of the incoming stream, that is to say, with certain conventions. The probability of occurrence of a certain number of cars in a given, sufficiently long interval depends only on the length of that interval. Generally, at random times, flow λ is non-stationary $\lambda = \lambda(t)$. This non-stationarity is well marked for 1 business day (24 hours).

The incoming and outgoing stream is considered and modelled. Depending on the measurements and observations made, an approximation function of the input flow density is determined. The same applies to service speed (outflow).

Table 1 shows the number of trucks waiting in the queue, as well as the number of arrivals each hour.

Table 1. Number of trucks waiting in the queue for two randomly selected days.

Time interval, h	Number of trucks in the queue on 5.7.2018	Number of trucks in the queue on 17.7.2018
0 - 2	74	69
2 - 4	22	28
4 - 6	11	15
6 - 8	6	9
8 - 10	3	4
10 - 12	7	5
12 - 14	14	13
14 - 16	26	25
16 - 18	53	49
18 - 20	88	90
20 - 22	107	102
22 - 24	93	89

Concerning the intensity of servicing trucks, according to the Head of the Border Checkpoint, the service time of one truck is a constant value and is 7 minutes (scheduled service time), μ = const.

Regarding the number of serving channels (servers), it can, if necessary, reach 5 ($1 \le s \le 5$). In order to provide 24 hours of continuous work per channel, three teams are required (one every 8 hours). For system performance testing, it is necessary to determine the probability that the system will have k automobiles at a moment t with s running servers [13, 14]:

$$P_k(t) = ?, k = \overline{0, \infty}, t \in [0, T], 1 \le s \le 5, \quad (1)$$

where for a single period T, a full working day T = 1 of 24 hours was taken. The beginning of the working day t = 0 coincides with the astronomical start of the day (0 hours and 0 minutes), the end of the working day $t = T$, with the end of the astronomical day (24 hours and 0 minutes).

For the mass service system model, the following can be summarized as follows: a non-stationary stream of requests with density $\lambda(t)$ enters a mass service with s service channels. Request service time is a random variable with an indicative distribution and parameter μ = const. A car that arrives at a busy channel time stands in the waiting line and "patiently" waits to be served, that is, a system with unlimited waiting and an unlimited number of trucks in the queue.

Given the characteristics of the incoming and outgoing flow, as well as the behavior of the system itself (multi-channel system without failures, endless queuing time, behavior in the "first arrived" queue, etc.), a specific type of differential equation system (of Kolmogorov) is applied. The probabilities of having a certain number of automobiles in the system at any given moment represent solutions to the differential equation system. From these probabilities, predictions are made for the waiting time of a vehicle arriving at a specific time of the day, the number of cars in the system at that time, and other characteristics. The theoretical results are compared with the measurements. If necessary, the functions that describe the incoming and outgoing stream are corrected.

From what has been said so far, we can say that the system is of type $(M/M/s)$ in non-stationary mode. To describe a system of this type, the following system of differential equations of Kolmogorov (Erlang-Kolmogorov) is in effect [15-16]:

$$\begin{gathered} \frac{dP_0(t)}{dt} = -\lambda(t)P_0(t) + \mu P_1(t) \\ \dots\dots \\ \frac{dP_i(t)}{dt} = \lambda(t)P_{i-1}(t) - (\lambda(t) + i\mu)P_i(t) + (i+1)\mu P_{i+1}(t) \\ \dots\dots \\ \frac{dP_{s+n}(t)}{dt} = \lambda(t)P_{s+n-1}(t) - (\lambda(t) + s\mu)P_{s+n}(t) + s\mu P_{s+n+1}(t) \\ 1 \le i \le s-1, n = \overline{0, \infty}. \end{gathered} \quad (2)$$

n indicates the number of cars in the queue when all servers are busy. The system of differential equations is infinite and for the numerical solution it is necessary to take a finite number of equations. The number of equations is such that from a given number N, the sum $\sum_{i=N}^{\infty} P_i(t)$ must not exceed a given number e. In this paper, a system of N = 505 differential equations are solved, with a system "slicing" error less than $\varepsilon = 10^{-8}$. After selecting a finite number of equations, the number of probabilities $P_k(t)$ is one less than the number of equations, so it is necessary to introduce the additional algebraic equation for normality $\sum_{i=0}^{N-1} P_i(t) = 1$. The computational features associated with a system (2) after "cutting" and introducing an algebraic equation are as follows:

- the system is of great size;
- the system is differential-algebraic;
- the system is of the "rigid system" type.

This can be summarized as: a system of solid, differential-algebraic equations with a large dimension. Numerical methods have been developed to overcome these difficulties. A MatLab program was developed to

 https://doi.org/10.1051/matecconf/201823406003

solve a system (2), using the built-in "solver" *ode15s* [15, 20], implementing the Gir method. When entering $\lambda(t), \mu, s, N$, the application returns a numerical solution to $P_k(t)$. Accuracy is artificially made greater than that implied for the solvent – from 10^{-6} to 10^{-9} for absolute error and 10^{-3} to 10^{-5} for a relative error, respectively.

The input stream $\lambda(t)$, generally, is different every day, but some tendency and periodicity is noted. Measurements made in several non-attendance days in July 2017 show that roughly 50% (k = 0.5) of the daily number of cars is between approximately 13:00 (t_0 = 13/24) and 21:00 (t_2 = 21/24) is about 50% (k = 0.5) of the daily number of cars, with the peak being around 17:00 (t_1 = 17/24), and the smallest between 4:00 and 6:00. The average daily number of cars for these few days is between 700 and 750 ($A \in [700; 750]$).

For modelling $\lambda(t)$, it is advisable to choose a relatively elementary function having a periodicity to reach the largest (lowest value) at peak times, and the integral sum at the busiest period is about 50% of the full-day flow. For a model, a Fourier line has been selected with three members [17, 18]:

$$\lambda(t) = a_0 + a_1 \cos(2\pi t) + b_1 \sin(2\pi t). \quad (3)$$

Coefficients a_0, a_1, b_1 are determined from the above comments:

1) $\int_0^1 \lambda(t)dt = A$ – the sum of all cars of the day is A;

2) $\frac{d\lambda(t)}{dt} = 0, at\ t = t_1$ – a necessary condition for the extreme in t_1;

3) $\int_{t_0}^{t_2} \lambda(t)dt = kA$ – in the peak interval $[t_0, t_2]$ a part of all A cars equal to k passes.

1), 2) and 3) result in the following system of linear algebraic equations with the unknown coefficients a_0, a_1, b_1:

$$\begin{cases} a_0 = A \\ -a_1 \sin(2\pi t_1) + b_1 \cos(2\pi t_1) = 0 \\ a_0(t_2 - t_0) + a_1 \frac{\sin(2\pi t_2) - \sin(2\pi t_0)}{2\pi} + b_1 \frac{\cos(2\pi t_0) - \cos(2\pi t_2)}{2\pi} = kA \end{cases} \quad (4)$$

Figures 1 and 2 show the graphs of $\lambda(t)$ with specific meanings of the other parameters.

For the intensity of service from one channel – μ, taking into account the service time of one car from one channel (7 min), we receive for one day:

$\mu = 24\frac{60}{7} = 205.714285$ automobiles per day per channel.

The integration of the system needs to be done not for a period of time but for a sufficient number of periods. In this way the probabilistic functions $P_k(t)$ begin to cling to their regular values. After multiple integration with different end times it was found out that only after 5-6 periods the functions $P_k(t)$ entered into a regular mode (for two adjacent periods, remain the same). Accuracy is also increased here, with integration being done for 20 periods, with the difference of all $P_k(t)$ in the last and penultimate periods being less than 10^{-8} for each t [19].

The following graphs reflect the results of the system decision (2) at the following values:

$$A = 730, k = 0.5, t_0 = \frac{13}{24}, t_1 = \frac{17}{24}, t_2 = \frac{21}{24}, \quad (5)$$

$$\lambda(t) = 730 - 114.2318 \cos(2\pi t) - 426.3189 \sin(2\pi t)\ \mu = 205.714285$$

$s = 4$ (4 running servers).

Fig. 3 illustrates the probability of having exactly $k \in [0; 3]$ cars in the system. It is noteworthy that the most likely values for a small or zero number of cars are between 4:00 and 10:00.

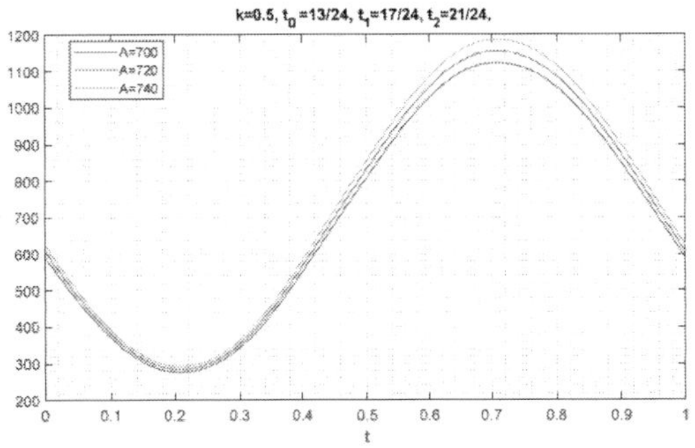

Fig. 1. Changing the coefficient λ at different coefficient values A.

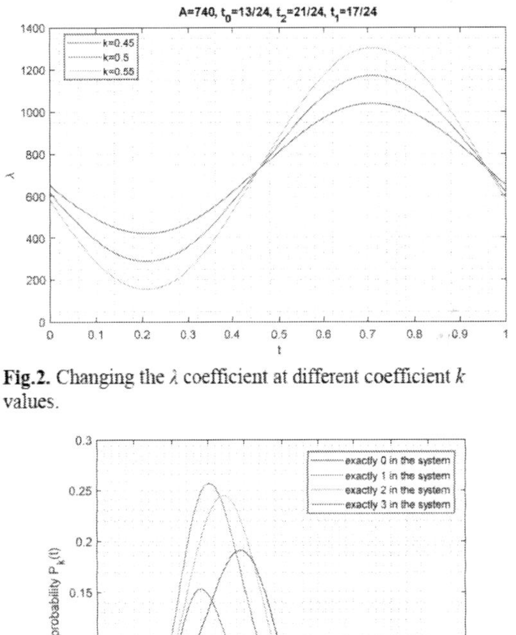

Fig.2. Changing the λ coefficient at different coefficient k values.

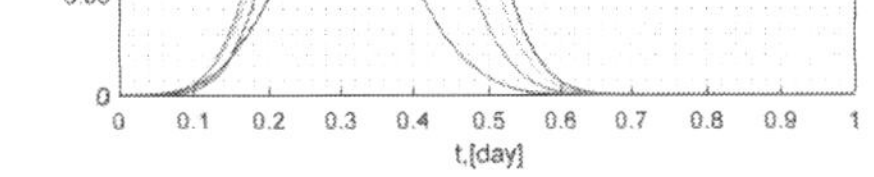

Fig. 3. Graph of the probability of having exactly $k \in [0; 3]$ automobiles in the system.

The probability of having exactly $k \in \{50,100,150\}$ automobiles in the system is presented in Fig. 4. Although the probabilities are small, the maximum number of automobiles is observed between 4 p.m. and 7 p.m.

https://doi.org/10.1051/matecconf/201823406003

Fig. 5 shows the times when there are between 0 and 5, between 6 and 15, as well as between 16 and 30 automobiles. It is most likely to have between 0 and 5 automobiles at about 7:30 a.m.

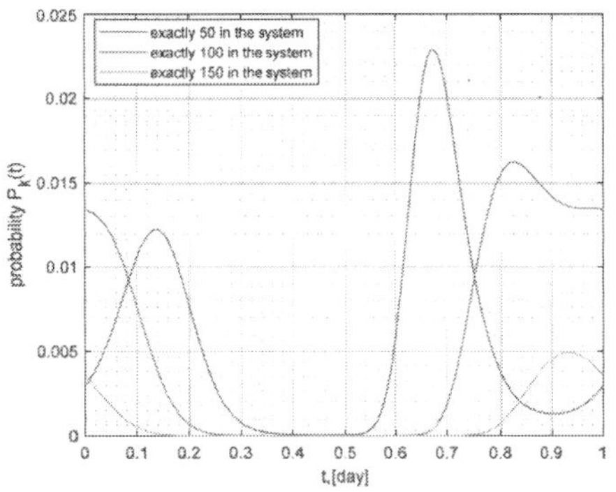

Fig.4. The probability of having exactly $k \in \{50,100,150\}$ automobiles in the system.

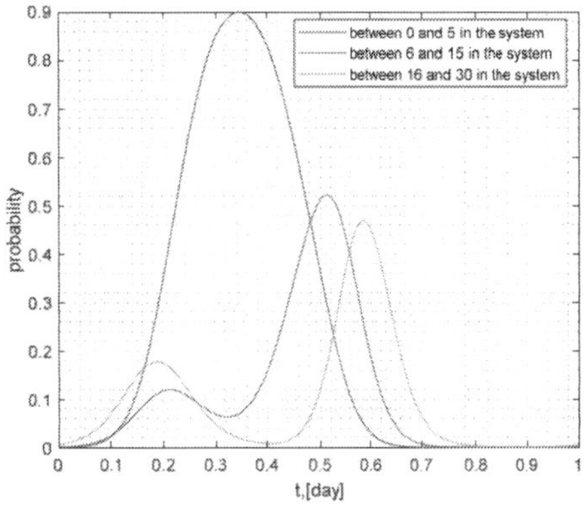

Fig.5. The probability of having between 0 and 5, between 6 and 15, as well as between 16 and 30 automobiles.

Fig. 6 shows the likelihood to have between 91 and 120, between 121 and 170 and over 170 vehicles It is most likely to have the greatest number of cars in the system between 5 p.m. and 1 a.m.

The average number of queueing cars is given (Fig. 7). The highest number (over 100) is observed for about 21:30 minutes, and the smallest (less than 5) about 8:30.

Average vehicle downtime (on the abscissa are the hours) is shown in Fig. 8. The automobiles arriving at around 10 p.m. have to wait the longest time (over 160 min.), while those arriving at about 9 a.m. have the shortest time of waiting (under 5 min.).

Research shows that when 4 pay desks are open constantly (12 teams), on a regular day (a day when the temperature is below 32 degrees), there is a marked imbalance in the main indicators of the system. The ineffective mode of operation of the system leads to the need of taking adequate measures to optimize it.

The observations made in July 2018, on regular days, as well as the data for the previous year, confirm that there is an increase in peak arrivals of cars between 4;30 p.m. and 6:30 p.m.

The smallest peak is observed between 4:00 and 6:00. With a permanent mode of operation of 4 cases (each with 3 teams on an 8-hour working day) there is a strong unevenness in the number of queuing cars as well as waiting times.

Between 20:00 and 0:00 there are more than 100 cars in the tail and a vehicle arriving during this interval has to wait over 160 minutes. And in the interval between 8:00 and 12:00 there are less than 5 cars in the tail, and a vehicle arriving during this interval, waits for 6-7 min.

Appropriate measures must be taken to reduce waiting times so that cars which arrived at a certain point have to wait for over 160 minutes, while others which arrived at another moment wait just 6-7 minutes, and regardless of the time of arrival, everyone would wait approximately the same time; this time should be as little as possible.

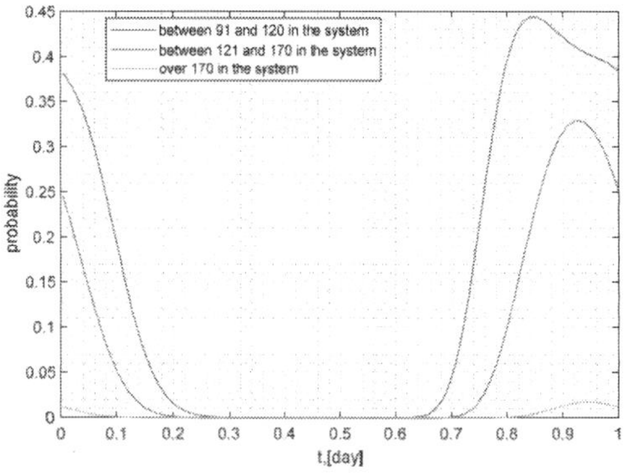

Fig. 6. The probability of having between 91 and 120, between 121 and 170 and over 170 vehicles.

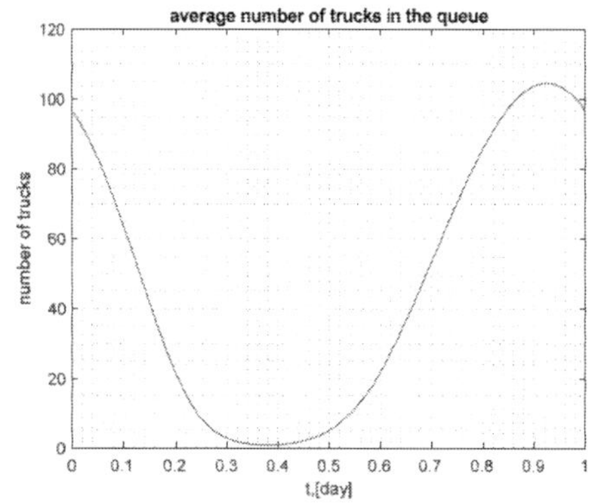

Fig. 7. Average number of cars queuing.

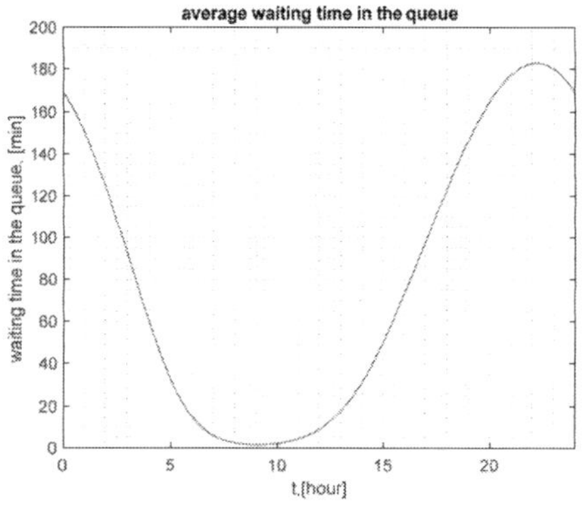

Fig.8. Average vehicle downtime per hour.

3 Conclusions

1. The theoretical results obtained are intended to clarify the behavior of the system at any time of the day for periods of approximately equal non-stationary intensity (accuracy to constant) of arrival and service.

2. A way is proposed to approximate the input flow density by the most general average statistics.

3. A methodology is proposed for modelling and examination of the characteristics of a checkpoint/ checkpoints with similar non-stationary behavior, as well as those of checkpoints having any number of service units.

4. When system characteristics are unsatisfactory, the theoretical results will determine what adjustments in service intensity and / or number of channels are to be made for better performance. This would save time and problems of the "trial and error" type.

5. For the Ruse-Danube Bridge checkpoint, given the behavior of the average daily intensity, the "average queue duration" characteristic is synchronized sufficiently with the theoretical predictions (Table 1 and Fig. 8). With sufficient statistics and measurements (a sufficiently accurate approximated input and output of the intensity) for other checkpoints, it will provide theoretical characteristics that are close enough to the real ones.

6. Improvement can be achieved without incurring additional costs (e.g. for opening new positions that will make it possible for a fifth pay desk to work continuously). This can be done by organizing the working hours of the 12 teams in such a way that a fifth pay desk will work at specific times only, while only two or three pay checks will work in the remaining time. This organization of working hours will have the necessary effect for minimizing and regulating the waiting time. It would lead to a much more even mode of work for the system itself. The waiting time will be approximately the same regardless of the vehicles' arrival. The total average waiting time will also be reduced and this will take the burden off the transport companies themselves. They would plan a more optimal schedule, saving time and unnecessary costs. This would result in a release of funds that could be reinvested.

The study was supported by a contract of University of Ruse "Angel Kanchev", № BG05M2OP001-2.009-0011-C01, "Support for the development of human resources for research and innovation at the University of Ruse "Angel Kanchev". The project is funded with support from the Operational Program "Science and Education for Smart Growth 2014-2020" financed by the European Social Fund of the European Union.

References

1. R. Angelova, P. Stoyanov, N. Bunchyev, D. Grozev, Analysis of the traffic intensity in the border towns of Ruse and Vidin, *Youth forums "Science, Technology, Innovation, Business"* autumn 2017, Plovdiv, (2017)
2. Strategy for integrated border management of the Republic of Bulgaria 2014-2017 http://www.strategy.bg/PublicConsultations/View.aspx?lang=bg-BG&Id=1281
3. Strategy for development of the transport system of the republic of Bulgaria by 2020 March 2010 Sofia, (2010)
4. EUROSTAT (2015), *Sustainable Development in the European Union: 2015 monitoring report of the EU sustainable development strategy*, (EUROSTAT Statistical Books, 2015)
5. Bulgarian Helsinki Committee, Annual Report on Border Surveillance in 2017
6. B. Li, W. Li, Modeling and simulation of container terminal logistics systems using Harvard architecture and agent-based computing, *Proceedings of the 2010 Winter Simulation Conference* (2010)
7. C. Hsu, H. Shih, W. Wang, Applying RFID to reduce delay in import cargo customs clearance process, Computers & Industrial Engineering, **57** (2009)
8. Exploring the possibilities for reducing the use of the TEN-T network in the cross-border region Romania-Bulgaria by optimizing the freight and passenger transport and development of a joint mechanism for intermodal connections support, http://www.adodunav.org/uploads/PDF/Brochure_BG_15.1.1.010-ilovepdf-compressed.pdf
9. J. Sztrik, *Basic Queueing Theory*, GlobeEdit, OmniScriptum GmbH, KG, Saarbrucken, Germany (2016)
10. M. Zukerman, *Introduction to queueing theory and stochastic teletraffic models*, EE Department, City University of Hong Kong (2016)
11. S. Ross, *Introduction to probability models*, (Elsevier, 2007)
12. L. Shampine, M. Reichelt, The MATLAB ODE Suite, SIAM Journal on Scientific Computing, **18**, 1-22 (1997)
13. L. Norov, D. Akbarov, Customs – time for a change, Problems of Economic Transition, **52**(2), 47-60 (2009)
14. A. Mitrofanova, *Continuous times Markov chains. Poisson process. Birth and death process*, (New York University, 2007)
15. N. Gautam, *Analysis of queues: Methods and applications* (CRC Press, 2012)
16. R. Markov, *Transport corridors South-North in Southeastern Europe. Sustainable development of Lower Danube regions. Variant "Son – 2" of Rousse – Veliko Tarnovo Motorway*, Sofia (2017)
17. M. Fitzmaurice, Optimization plan for freight movements at key commercial border crossings, Report compiled by Transport Logistics Consultants, August (2009)
18. H. Taha, *Operations research, an introduction*, New York (J. Wiley and Sons, 1993)
19. V. Todorov, I Georgiev, *Optimization methods with MATLAB*, (Avangard print, 2016)

Increasing the accuracy of evaluation in selecting a location for establishing an intermodal terminal

Svetoslav Martinov[1,*]

[1]Technical University – Sofia, Department of Railway Engineering, Bulgaria

Abstract. The present study is aimed at increasing the accuracy of multi-criteria evaluation of alternatives in selecting a location for establishing an intermodal terminal in the North-Central planning development region of Bulgaria. A model has been used in the paper that allows us to increase the accuracy of multi-criteria evaluation of all studied alternatives. This has been achieved by converting the values of various units into identical dimensionless units within a defined interval. It makes it possible to do an evaluation of the alternatives by using the real values of the criteria. The model has been used for multi-criteria evaluation of the alternatives in selecting a location for establishing an intermodal terminal in the city of Ruse. The results have been compared with the results of a pre-feasibility study of establishing an intermodal terminal in the city of Ruse in the North-Central planning development region of Bulgaria.

1 Introduction

Some of the guidelines of the Integrated Transport Strategy for the period of up to 2030 are related to the development of the transport network in Bulgaria as part of the Trans-European Transport Network (TEN-T) and to increasing the efficiency and competitiveness of the transport sector [1]. The investment strategy of the Operational program "Transport and Transport Infrastructure" 2014-2020, under priority axe 3 referring to intermodal transport, envisages the development of a network of intermodal terminals for improving intermodality [2] of freight transport. Part of the investment strategy is Project "Construction of an intermodal terminal in the North-Central planning development region in Bulgaria – Ruse". The preparatory activities for the establishment of the terminal are carried out under Project "Technical assistance for construction of intermodal terminal in North-Central planning development region in Bulgaria – Ruse" [3], including the selection of a terminal location. Five alternative locations for establishing the intermodal terminal have been discussed [3]. The Weighted Sum Method (WSM) has been applied in the project [3] to evaluate different alternatives and to select a leading alternative.

The Weighted Sum Method is one of the most commonly used multi-criteria decision making methods [4, 5], especially in single dimensional type problems [6]. The choice of a decision among many alternatives often depends on different parameters. This requires the use of multi-criteria analysis methods [6-9]. The weight multi-criteria analysis methods are used in different fields of transport, including for selection of a location of terminal [9, 10]. The evaluation of the alternatives in [3] has been carried out through different criteria that are estimated using a three-tier rating scale [3] and weight coefficients. The three-tier rating scale allows for alternatives with different values of the criteria to be put in the same group. That could lead to an inaccurate final rating of the alternatives. For alternatives evaluated through a lot of criteria [11, 12], the accumulated inaccuracies could lead to a significant error and, consequently, to an incorrect final ranking.

2 Methodology of research

2.1 Model for evaluation and selection of an alternative ("As-Is" scenario)

Five possible alternatives for the construction of the terminal have been studied in project [3]. In the paper we will call this scenario "As-Is" scenario. In the project, the alternative places for the establishment of the terminal are represented by designations A, B, C, D and E [3].

The Weighted Sum Method has been used in selecting a location for establishing an intermodal terminal – Ruse [3]. The method is based on a comparison between different alternatives by determining their weighted sums. The weighted sum for each alternative is calculated through the value of the alternative and its individual importance (weight coefficient) [5]. The Weighted Sum Method allows to compare variants with the same dimensions of the criteria [13].

The project [3] describes all the criteria and their relative weight. Criteria (C) are divided into five main groups that contain relevant sub-criteria (SC) [3]: Criterion 1 – Impact on existing infrastructure and the environment; Sub-criterion 1.1 – Strategic location of the site; Sub-criterion 1.2 – Purpose of the land; Sub-criterion

* Corresponding author: s.martinov@tu-sofia.bg

1.3 – Land ownership; Sub-criterion 1.4 – Environment and European protected areas – Natura 2000; Sub-criterion 1.5 – Effect on communal infrastructure; Criterion 2 – Opportunities for development of a place for intermodal terminal; Sub-criterion 2.1 – Size of the site; Sub-criterion 2.2 – Length of the longest arrival-departure railway track; Sub-criterion 2.3 – Average useful length of railway tracks used for loading/unloading; Sub-criterion 2.4 – Opportunity for further development; Criterion 3 – Transport access to the intermodal terminal; Sub-criterion 3.1 – Railway infrastructure; Sub-criterion 3.2 – Road infrastructure; Sub-criterion 3.3 – Port infrastructure; Criterion 4 – Costs and operating parameters of the intermodal terminal; Sub-criterion 4.1 – Investment costs; Sub-criterion 4.2 – Annual operating and maintenance costs; Sub-criterion 4.3 – Opportunities for development of additional services; Sub-criterion 4.4 – Quantity of manipulated Twenty-foot equivalent units (TEU); Criterion 5 – Economical parameters of the intermodal terminal; Sub-criterion 5.1 – Net present value (NPV) of the total costs; Sub-criterion 5.2 – Sensitivity of NPV to operating and maintenance costs; Sub-criterion 5.3 – NPV ratio of the total costs per TEU.

The estimates of the sub-criteria of the different alternatives and their weight coefficients are shown in Table 1. The type of criteria and sub-criteria and their weights are proposed in the documentation of [3]. The sub-criteria in the project are evaluated by a three-tier scale in which the preferred level is rated as 1, the average level is rated as 2 and the critical level is rated as 3 [3].

Table 1. Estimates and weight coefficients.

Criterion (C)/ Sub-criterion (SC)	Weight coefficient	Alternatives				
		A	B	C	D	E
C 1	0,15	-	-	-	-	-
SC 1.1	0,2	3	3	1	3	3
SC 1.2	0,3	1	2	1	1	3
SC 1.3	0,2	2	2	1	2	2
SC 1.4	0,2	1	1	1	1	2
SC 1.5	0,1	2	2	1	2	1
C 2	0,2	-	-	-	-	-
SC 2.1	0,25	2	2	1	2	2
SC 2.2	0,25	2	2	1	3	2
SC 2.3	0,25	2	1	1	2	1
SC 2.4	0,25	3	1	1	3	2
C 3	0,2	-	-	-	-	-
SC 3.1	0,4	1	2	1	2	2
SC 3.2	0,35	3	1	1	2	3
SC 3.3	0,25	2	2	2	1	1
C 4	0,2	-	-	-	-	-
SC 4.1	0,25	1	2	1	2	3
SC 4.2	0,25	1	2	2	2	3
SC 4.3	0,25	1	1	1	1	1
SC 4.4	0,25	2	2	1	3	2
C 5	0,25	-	-	-	-	-
SC 5.1	0,3	1	2	2	2	3
SC 5.2	0,3	1	2	1	3	3
SC 5.3	0,4	1	2	1	3	3

The project [3] does not contain information or data about the matrix of pairwise comparisons [7, 14] of the criteria and the sub-criteria. The weights of the criteria and the sub-criteria are described in project documentation and their values [3] are used in the paper. For that reason a sensitivity analysis of weights coefficients and values of the consistency ration [7, 14] is not presented in the research.

Evaluation of the criteria through the sub-criteria and of the alternatives through the criteria in the project [3] is carried out by:

$$a_{nr} = \sum_{s=1}^{S} a_{sr}\, w_s, n = 1, 2, \dots, N;\ r = 1, 2, \dots, R \quad (1)$$

and

$$A_r = \sum_{n=1}^{N} a_{nr}\, w_n, r = 1, 2, \dots, R, \quad (2)$$

where a_{nr} is the estimate of the n-th decision criterion for the r-th alternative; N – the number of the same type criteria; R – the number of the compared alternatives ($R \geq 2$); a_{sr} – the estimate of the s-th decision sub-criterion for the r-th alternative; w_s – weight coefficient of the s-th sub-criterion ($\sum_{s=1}^{S} w_s = 1$); S – the number of the same type sub-criteria; A_r – the value of the weighted sum for the r-th alternative; w_n – weight coefficient of the n-th criterion ($\sum_{n=1}^{N} w_n = 1$).

The selection of a leading alternative is made by [3]:

$$A_r \rightarrow min, r = 1, 2, \dots, R. \quad (3)$$

2.2 Model for increasing the accuracy of estimates ("To-Be" scenario)

For the purpose of increasing the accuracy of multi-criteria evaluation of the alternatives in selecting a location for construction of an intermodal terminal in the city of Ruse the values of the parameters are converted into dimensionless units. The parameters of the criteria that have a numerical value and a corresponding dimension are converted into dimensionless units within a defined interval [15]. In the paper we will call this scenario "To-Be" scenario. The conversion for $r = 1, 2, \dots, R$ is done through equation (4) for the direct ratio criteria and through equation (5) for the inverse ratio criteria [16]:

$$M_{sr} = L_{sr,max} \cdot \frac{(V_{sr} - V_{s,min})}{(V_{s,max} - V_{s,min})} + L_{sr,min} \quad (4)$$

$$M_{sr} = L_{sr,max} \cdot \frac{(V_{s,max} - V_{sr})}{(V_{s,max} - V_{s,min})} + L_{sr,min}, \quad (5)$$

where M_{sr} is the dimensionless estimate of the s-th decision sub-criterion ($s = 1, 2, \dots, S$) for the r-th alternative; $L_{sr,min}$ and $L_{sr,max}$ – respectively the minimum and the maximum values determining the length of the interval of conversion of the estimate of the s-th sub-criterion of the r-th alternative; V_{sr} – the current value of the s-th sub-criterion of the r-th alternative; $V_{s,min}$ – the minimum possible value of the s-th sub-criterion; $V_{s,max}$ – the maximum possible value of the s-th sub-criterion.

The values of $L_{sr,min}$ and $L_{sr,max}$ are:

$$L_{sr,min} \geq 0; \quad (6)$$

$$L_{sr,max} > 0. \quad (7)$$

The interval of limit [16] of estimate M_{sr} is determined by the values of $L_{sr,min}$ and $L_{sr,max}$:

$$M_{sr} \in \left[L_{sr,min}, L_{sr,min} + L_{sr,max}\right]. \quad (8)$$

When converted into dimensionless estimates the values of $V_{s,min}$ and $V_{s,max}$ for $s = 1, 2, \dots, S$ could be determined [16] using:

$$V_{s,min} = \min V_{sr}, r = 1, 2, \dots, R; \quad (9)$$

$$V_{s,max} = \max V_{sr}, r = 1, 2, \dots, R. \quad (10)$$

The limits [16] for the value of $V_{s,min}$ and $V_{s,max}$ are:

$$0 \le V_{s,min} < V_{s,max}. \quad (11)$$

When it is not necessary for some parameters to be converted into dimensionless units, the value of $M_{sr} = V_{sr}$ for $s = 1, 2, \dots, S$ and $r = 1, 2, \dots, R$ or it could be calculated with equations (4) and (5) for $V_{s,min} = L_{sr,min}$ and for $V_{s,max} = L_{sr,min} + L_{sr,max}$.

The evaluation of the criteria through the sub-criteria and of the alternatives through the criteria for the "To-Be" scenario is carried out for $r = 1, 2, \dots, R$ through:

$$a_{nr}^{c} = \sum_{s=1}^{S} M_{sr}\, w_s, n = 1, 2, \dots, N \quad (12)$$

and

$$A_r^c = \sum_{n=1}^{N} a_{nr}^{c}\, w_n \quad (13)$$

where a_{nr}^{c} is the estimate with a higher accuracy of the n-th decision criterion for the r-th alternative; A_r^c – the value of the weighted sum for the r-th alternative after applying the model for increasing the accuracy [16].

The choice of a leading alternative is carried out with:

$$A_r^c \rightarrow min, r = 1, 2, \dots, R. \quad (14)$$

3 Results and discussion

3.1 "As-Is" scenario

The estimates of the criteria and alternatives for the "As-Is" scenario are shown in Table 2. A graphical comparison between the estimates of the alternatives is shown in Fig. 1.

Table 2. Estimates of the criteria and alternatives

Criterion	Alternatives				
	A	B	C	D	E
Criterion 1	1,70	2,00	1,00	1,70	2,40
Criterion 2	2,25	1,50	1,00	2,50	1,75
Criterion 3	1,95	1,65	1,25	1,75	2,10
Criterion 4	1,25	1,75	1,25	2,00	2,25
Criterion 5	1,00	2,00	1,30	2,70	3,00
Estimate of alternative	1,60	1,78	1,18	2,18	2,33

As we could see from the results, the leading alternative in the project [3] is alternative C, followed by alternatives A, B, D and E.

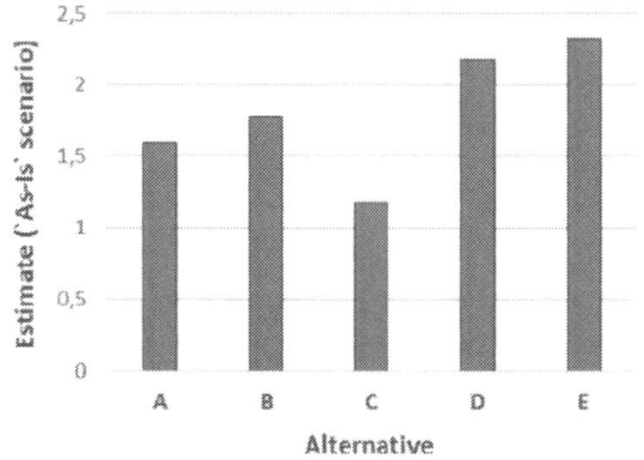

Fig. 1. Comparison of the alternatives.

3.2 "To-Be" scenario

The model for conversion of the values of the parameters into dimensionless estimates [16] is applied to increase the accuracy of the multi-criteria evaluation of the alternatives in choosing a location for the construction of an intermodal terminal in the North-Central planning development region in Bulgaria – Ruse.

The selection of a location of the terminal has been made by applying nineteen sub-criteria that are shown in Table 1. In the project documentation [3], all sub-criteria have been evaluated by using the three-tier scale with estimates of 1, 2 or 3. Nine of the sub-criteria (SC 2.1, SC 2.2, SC 2.3, SC 3.2, SC 3.3, SC 4.1, SC 4.2, SC 4.4 and SC 5.1) are parameters with the respective numerical value and dimension [3] shown in Table 3.

Table 3. Values of the sub-criteria.

Sub-criteria, dimension	Alternatives				
	A	B	C	D	E
SC 2.1, 1000 m^2	109,3	109,8	123,4	96,0	112,1
SC 2.2, m	760	750	774	636	750
SC 2.3, m	470	530	530	470	530
SC 3.2, m	1050	220	100	580	850
SC 3.3, m	5800	3500	5400	500	400
SC 4.1, million EUR	21,7	22,2	21,8	24,8	27,9
SC 4.2, 1000 EUR	399,0	419,4	423,7	468,2	554,2
SC 4.4, 1000 TEU	30,6	30,7	34,6	26,9	31,4
SC 5.1, million EUR	29,1	29,8	29,6	33,1	37,3

The choice of an optimal value, minimum or maximum for the target function, is being made depending on the type of target function – costs or benefits and the type of the evaluated criterion – proportionate (in direct proportion) or disproportionate (in inverse proportion) [16]. Table 4 shows the type (proportionate or disproportionate) of the used sub-criteria with a numerical value and dimension SC 2.1, SC 2.2, SC 2.3, SC 3.2, SC 3.3, SC 4.1, SC 4.2, SC 4.4 and SC 5.1 for which the accuracy of evaluation is increased.

The choice of a correlation to determine the value of the M_{sr} estimate is made according to the type of target function – costs or benefits and the optimal value for choosing an option – minimum or maximum. Table 5 shows a matrix with combinations for choosing a correlation (proportionate or disproportionate) to determine the value of estimate M_{sr} depending on the target function (costs or benefits) and the optimal value (minimum or maximum) of the target function. As can be seen in Table 5, with a minimum (min) value for choosing an option and a target function of the costs or benefit type we have to use proportionate (in direct ratio) and disproportionate (in inverse ratio) correlation for estimating the value of M_{sr}. On the contrary, with a maximum (max) value for choosing an option and a target of the costs or benefits type, a disproportionate (inverse ratio) or proportionate (direct ratio) correlation is used.

Table 4. Type of sub-criteria.

Sub-criterion	Type of sub-criterion
SC 2.1	inverse ratio
SC 2.2	inverse ratio
SC 2.3	inverse ratio
SC 3.2	direct ratio
SC 3.3	direct ratio
SC 4.1	direct ratio
SC 4.2	direct ratio
SC 4.4	inverse ratio
SC 5.1	direct ratio

Table 5. Matrix of combinations.

Target function	Choice of an option	Type of criteria
Cost	min	direct ratio
	max	inverse ratio
Benefit	max	direct ratio
	min	inverse ratio

Table 6 shows the values of sub-criteria SC 2.1, SC 2.2, SC 2.3, SC 3.2, SC 3.3, SC 4.1, SC 4.2, SC 4.4 and SC 5.1 by alternatives after their transformation into dimensionless estimate M_{sr}. The transformation is made by using the model described in point 2.2. The change of the estimates for the "To-Be" scenario is within a range with a lower limit 1 and upper limit 3. The range of the change of the interval coincides with the minimum and maximum estimate of the criteria in the project [3].

Table 6. Values of M_{sr}.

Sub-criterion	M_{sr}				
	Alternatives				
	A	B	C	D	E
SC 2.1	2,03	1,99	1,00	3,00	1,82
SC 2.2	1,20	1,35	1,00	3,00	1,35
SC 2.3	3,00	1,00	1,00	3,00	1,00
SC 3.2	3,00	1,25	1,00	2,01	2,58
SC 3.3	3,00	2,15	2,85	1,04	1,00
SC 4.1	1,00	1,16	1,03	2,01	3,00
SC 4.2	1,00	1,26	1,32	1,89	3,00
SC 4.4	2,03	1,99	1,00	3,00	1,82
SC 5.1	1,00	1,16	1,11	1,98	3,00

The estimates of the criteria and the results of the multi-criteria evaluation of the alternatives after increasing the accuracy of the calculations are shown in Table 7.

A comparison between the estimates of the alternatives after increasing their accuracy is shown in Fig. 2.

Table 7. Estimates of the criteria and alternatives.

Criterion	Alternatives				
	A	B	C	D	E
Criterion 1	1,70	2,00	1,00	1,70	2,40
Criterion 2	2,31	1,34	1,00	3,00	1,54
Criterion 3	2,20	1,78	1,46	1,76	1,95
Criterion 4	1,26	1,35	1,09	1,98	2,21
Criterion 5	1,00	1,75	1,03	2,69	3,00
Estimate of alternative	1,66	1,63	1,12	2,28	2,25

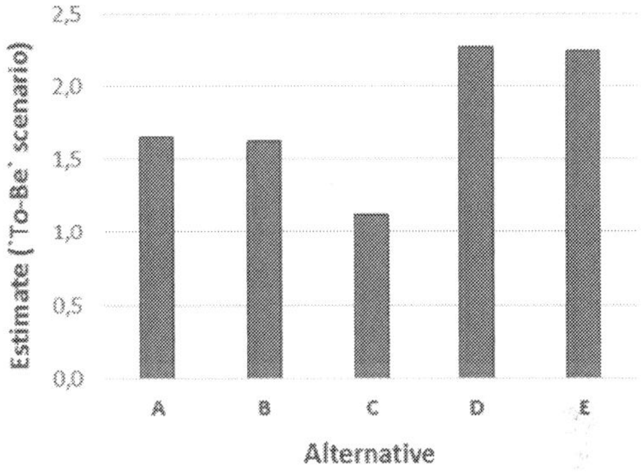

Fig. 2. Comparison of the alternatives.

The estimates in "To-Be" scenario are determined by using the same criteria, sub-criteria and weight coefficients as in "As-Is" scenario. The weights of the criteria are determined in [3]. The evaluation of the alternatives is made by using the same weight coefficients in both "As-Is" and "To-Be" scenarios.

A comparison of the rating of the alternatives before and after increasing the accuracy of the estimates is shown in Table 8.

Table 8. Comparison matrix.

Scenario		Rating of the alternatives				
		1st	2nd	3rd	4th	5th
"As-Is"	Alternative	C	A	B	D	E
	Estimate	1,18	1,60	1,78	2,18	2,33
"To-Be"	Alternative	C	B	A	E	D
	Estimate	1,12	1,63	1,66	2,25	2,28

The comparison of the results (Table 8) from the evaluation of the alternatives for both "As-Is" and "To-Be" scenarios shows that after applying the model for increasing the accuracy described in point 2.2, the ranking of four of the five studied alternatives in "To-Be" scenario (A, B, D and E) changes. In the process of selecting a site for establishing an intermodal terminal in the city of Ruse, the leading alternative in both scenarios is alternative C (Table 8). The rating of the remaining four compared alternatives (A, B, D, and E) changes with improving the accuracy of the estimates (Table 8).

The results of the study show that the accuracy of the estimate influences the choice of a leading alternative and can change the ranking of the variants. This is especially true for alternatives with close values of the estimates. As we can see from the results of the research (Table 8), the alternatives that exchange their places (alternatives A and

B and alternatives D and E), after improving the accuracy of the estimates, are with close values to those of the estimates for the study conditions. With the "As-Is" scenario the difference in the estimates of the alternatives (Table 2) is significant as compared to the difference in the estimates of the alternatives in the "To-Be" scenario (Table 7). The results of study show that the accumulated inaccuracy through the application of the three-tier scale [3] for evaluating the sub-criteria in selecting a place for construction of an intermodal terminal in the city of Ruse leads to a change in the ranking of the studied alternatives. It is recommended, if possible, to apply more accurate methods for evaluating the alternatives with close values of the multi-criteria estimates.

4 Conclusion

The proposed in point 2.2 model for increasing the accuracy of estimates allows us to broaden the scope for applying to the Weighted Sum Model through conversion of the estimates. By converting of dimension values of the criteria into dimensionless grades through the model described in point 2.2, the multi-criterion Weighted Sum Method becomes applicable to a wide range of dimensions. The model could be used for increasing the accuracy of evaluation in selecting a location for establishing an intermodal or other type terminal.

The results of the study show that the model enables comparability of the results obtained for the different options and the application of the model increases the accuracy of the estimates. This can change the final rating of the variants. The model described in the paper is a new approach for application of the Weighted Sum Method. The model could be used for multi-criteria evaluation and comparative analysis among many different alternatives through use the real values of the criteria. The model allows to be used in different types of studies based on the Weighted Sum Method.

References

1. Integrated Transport Strategy for the period until 2030. Ministry of Transport, Information Technology and Communications, Republic of Bulgaria (2017)
2. Terminology on combined transport. EU, ECMT, UN/ECE (2001)
3. Project "Technical assistance for construction of intermodal terminal in North-Central planning development region in Bulgaria – Ruse" – BG161PO004-3.0.01-0004, National Railway Infrastructure Company, Bulgaria (2012)
4. F. Helff, Le Gruenwald, L. d'Orazio, Weighted sum model for multi-objective query optimization for mobile-cloud database environments, *CEUR-WS, EDBT/ICDT 2016 Joint Conference*, **1558** (2016)
5. E. Mulliner, N. Malys, V. Maliene, Comparative analysis of MCDM methods for the assessment of sustainable housing affordability, Elsevier, Omega, **59**(B), March 2016, 146-156, DOI: https://doi.org/10.1016/j.omega.2015.05.013 (2016)
6. E. Triantaphyllou, B. Shu, S. Sanchez, T. Ray, *Multi-criteria decision making: An operations research approach*, Encyclopedia of Electrical and Electronics Engineering, J.G. Webster, Ed., **15**, 175-186 (John Wiley & Sons, New York, 1998)
7. S. Stoilova, An integrated approach for selection of intercity transport schemes on railway networks, Promet – Traffic & Transportation, **30**(4), 367-377, https://doi.org/10.7307/ptt.v30i4.2673 (2018)
8. S. Stoilova, Evaluation efficiency of intermodal transport using multi-criteria analysis. Proceedings of 17th International Scientific Conference Engineering for Rural Development, Jelgava, Latvia, 2030-2039, https://doi.org/10.22616/ERDev2018.17.N246 (2018)
9. A. Deluka-Tibljaš, B. Karleuša, N. Dragičević, Review of multicriteria-analysis methods application in decision making about transport infrastructure, Građevinar **65**(7), 619-631 (2013)
10. P. Site, F. Filippi, Weighting methods in multi-attribute assessment of transport projects, European Transport Research Review, **1**(4), 199–206 (2009)
11. S. Ananiev, S. Martinov, Opportunities of the railway infrastructure in the crossborder region Bulgaria-Romania for differentiation of freight intermodal centers, Almanac of „Todor Kableshkov" University of Transport, **9**, Sofia (2018)
12. P. Jenček, E. Twrdy, Development of regional transport logistics terminal – transport logistics approach, Promet – Traffic & Transportation, **20**(4), 239-249 (2008)
13. G. Odu, O. Charles-Owaba, Review of multi-criteria optimization methods – Theory and applications, IOSR Journal of Engineering (IOSRJEN), **3**(10), 01-14 (2013)
14. R. Saaty, The analytic hierarchy process – what it is and how it is used, Mathematical Modelling, **9**(3-5), 161-176 (1987)
15. B. Miljković, M. Žižović, A. Petojević, N. Damljanović, New weighted sum model, Filomat, Faculty of Sciences and Mathematics, University of Niš, Serbia, **31**(10) 2991-2998 (2017)
16. S. Martinov, Evaluation model of railway infrastructure potential for establishment of freight intermodal terminals, *Proceedings of the 5th International Conference on Road and Rail Infrastructures – CETRA 2018*, 911-916, DOI: https://doi.org/10.5592/CO/CETRA.2018.907 (2018)

AUTHOR INDEX

AUTHOR INDEX